AF581883

New Frontiers in Cementitious and Lime-Based Materials and Composites

New Frontiers in Cementitious and Lime-Based Materials and Composites

Editors

Cesare Signorini
Antonella Sola
Sumit Chakraborty
Valentina Volpini

MDPI • Basel • Beijing • Wuhan • Barcelona • Belgrade • Manchester • Tokyo • Cluj • Tianjin

Editors

Cesare Signorini
Technical University of Dresden
Germany

Valentina Volpini
University of Modena and
Reggio Emilia
Italy

Antonella Sola
Commonwealth Scientific
and Industrial Research
Organisation (CSIRO)
Australia

Sumit Chakraborty
JIS Institute of Advanced
Studies and Research
(JISIASR)
India

Editorial Office
MDPI
St. Alban-Anlage 66 4052
Basel, Switzerland

This is a reprint of articles from the Special Issue published online in the open access journal *Crystals* (ISSN 2073-4352) (available at: https://www.mdpi.com/journal/crystals/special_issues/Lime-Based_Materials).

For citation purposes, cite each article independently as indicated on the article page online and as indicated below:

LastName, A.A.; LastName, B.B.; LastName, C.C. Article Title. *Journal Name* **Year**, *Volume Number*, Page Range.

ISBN 978-3-0365-3543-2 (Hbk)
ISBN 978-3-0365-3544-9 (PDF)

Contents

About the Editors

Cesare Signorini (Ph.D.) is currently a postdoctoral researcher at the Institute of Construction Materials at the Technical University of Dresden, Germany, in the framework of the Research Training Group GRK2250 "Mineral-bonded composites for enhanced structural impact safety", supported by the German Research Society (DFG). Dr. Signorini achieved his Ph.D. in Industrial Innovation Engineering at the University of Modena and Reggio Emilia, in Italy. His main research interests include fibre-reinforced concrete (FRC) and Textile Reinforced Mortar/Concrete (TRM/TRC) for structural retrofitting of masonry and concrete structures. Within these areas, his main focus was on the improvement of the interphase bond between fibers and inorganic matrices to optimize the mechanical behavior of the composites under quasi-static and dynamic regimes. In the framework of a European Regional Development Fund (ERDF) Project, Dr. Signorini recently investigated the inclusion of recycled synthetic aggregates and fibers in concrete.

Antonella Sola (Ph.D.) is the Science Leader in Active Materials at the Commonwealth Scientific and Industrial Research Organization, CSIRO, Australia. With more than 100 scientific publications, she has been working on composites, advanced coatings, and biomaterials. Currently, she is investigating additive manufacturing and new processing technologies for the obtainment of composite systems with improved functionalities. She is also the head of CSIRO's research facility for the development of printable composites for fused filament fabrication, FFF.

Sumit Chakraborty (Ph.D) is currently an Assistant Professor at the JIS Institute of Advanced Studies and Research (JISIASR), Kolkata, India. He was formerly a Marie Curie Fellow at the University of Sheffield in the UK. His research vision is to develop sustainable mineral and textiles composite materials utilizing natural and industrial wastes and CO_2 sequestration, followed by exploring their underlying chemistry. He is currently involved in producing cement-less binder for concrete utilising industrial wastes and aqueous based carbonation of the mineral wastes for permanent storage of CO_2. He also worked on the development of surface-modified natural fibre reinforced spalling-proof geopolymer mortar/concrete.

Valentina Volpini (Ph.D.) is a postdoctoral research fellow at the Interdepartmental Research Center "En&Tech" of the University of Modena and Reggio Emilia. She achieved her Ph.D. in Civil and Environmental Engineering, International cooperation and Mathematics in 2020 at the University of Brescia, Italy, and also collaborated with the School of Engineering of Cardiff University, Wales. Her research activities are primarily concerned with modelling the multiphysical behaviour of smart materials and studying the propagation of waves in microstructured media by means of analytical (perturbative analyses and homogenisation procedures) and numerical (finite element analyses) methods. Recently, she also focused on the experimental investigation of Fibre Reinforced Cement Composites (FRCCs), with applications in construction engineering.

Editorial

New Frontiers in Cementitious and Lime-Based Materials and Composites

Cesare Signorini [1,*], Antonella Sola [2], Sumit Chakraborty [3] and Valentina Volpini [4]

1 Institute of Construction Materials, Technical University of Dresden, Georg-Schumann-Straße 7, 01187 Dresden, Germany
2 Manufacturing Business Unit, Commonwealth Scientific and Industrial Research Organization (CSIRO), Research Way, Clayton, VIC 3168, Australia; antonella.sola@csiro.au
3 Department of Civil and Structural Engineering, University of Sheffield, Sheffield S1 3JD, UK; sumit.chakraborty@sheffield.ac.uk
4 Interdepartmental Research Centre "En&Tech", University of Modena and Reggio Emilia, P.le Europa 1, 42124 Reggio Emilia, Italy; valentina.volpini@unimore.it
* Correspondence: cesare.signorini@tu-dresden.de; Tel.: +49-351-46336565

Abstract: Cement and lime currently are the most common binders in building materials. However, alternative materials and methods are needed to overcome the functional limitations and environmental footprint of conventional products. This Special Issue is entirely dedicated to "New frontiers in cementitious and lime-based materials and composites" and gathers selected reviews and experimental articles that showcase the most recent trends in this multidisciplinary field. Authoritative contributions from all around the world provide important insights into all areas of research related to cementitious and lime-based materials and composites, spanning from structural engineering to geotechnics, including materials science and processing technology. This topical cross-disciplinary collection is intended to foster innovation and help researchers and developers to identify new solutions for a more sustainable and functional built environment.

Keywords: cement; lime; sustainable materials; fibre-reinforced composite; recycled aggregates

Citation: Signorini, C.; Sola, A.; Chakraborty, S.; Volpini, V. New Frontiers in Cementitious and Lime-Based Materials and Composites. *Crystals* **2022**, *12*, 61. https://doi.org/10.3390/cryst12010061

Received: 9 December 2021
Accepted: 21 December 2021
Published: 4 January 2022

Cement and lime have been the predominant binders in the construction sector since ancient times, owing to their worldwide abundance, affordability, and well-established physical and mechanical performance. By tuning the relative amounts of binders with aggregates, water, and additives, a variety of conglomerates can be designed to address specific service requirements. Cement and lime-based conglomerates range from fine-grained mortars for plasters to structural concrete with or without fibre reinforcement for buildings, bridges, tunnels, pavements, girders, precast members, walls, screeds, etc.

Prompted by the necessity of developing smart and reliable infrastructures and more energy-efficient urbanised areas, new frontiers are opening to identify viable materials and methods with the inclusion of industrial by-products, alternative aggregates, and natural reinforcements [1]. The partial or total replacement of conventional cement and virgin mineral aggregates is pivotal to reducing the environmental issues and carbon footprint of customary conglomerates, since the conventional manufacturing chain is known to require a substantial amount of energy and resources. Further, intensive investigation is under-way to formulate novel inorganic composite materials for lightweight precast elements and for strengthening laminates, which are expected to transform present-day approaches to the preservation and restoration of historical buildings and architectural heritage [2–4]. As new materials with embedded functionalities are being proposed every day, scientists gain a deeper understanding of the relationship existing between composition, manufacturing, and in-service behaviour of cementitious and lime-based materials and composites.

This Special Issue aims to shed light on the latest research outcomes in this multidisciplinary field, spanning through the vast areas of structural, geotechnical, and environmental engineering, mineralogy and materials science, nanotechnology, fibre and textile technology, and design criteria. The high-quality peer-reviewed contributions gathered in this Special

Issue include both experimental papers, which discuss lab-based activities and/or theoretical modelling, as well as review articles that extensively analyse the state-of-the-art within specific fields of interest. The topics covered in this Special Issue include fibre-reinforced composites (e.g., textile-reinforced mortar/concrete (TRM/TRC) [5,6] and fibre-reinforced concrete (FRC) [7,8]), sustainable cementitious conglomerates where the ordinary Portland cement (OPC) binder is replaced by recycled by-products [9] (silica fume, kiln ashes, biochar, slag, biomass ashes, geopolymers, alkali-activated concrete, etc.), with the possible implementation of self-healing properties. In addition, studies regarding the incorporation of recycled aggregates providing concrete with increased eco-compatibility and additional functional properties, such as thermal and acoustic insulation, are also presented [10,11]. Moreover, durability is a key issue, especially for emerging materials whose behaviour in the long run is still unknown. Long-term stability is still the subject of debate for some of the recycled constituents, such as plastic fibres coming from low-grade sources, and also for composite systems in which the unoptimised interphase between the reinforcement and the surrounding inorganic matrix may represent the weakest link [12,13]. Reliability becomes critical in harsh environmental conditions, e.g., in coastal areas or underground applications [14]. Therefore, updated guidelines are sought to regulate the safe usage of these structural materials.

In this Special Issue, readers will find a wealth of information regarding existing literature and emerging research in all areas of cementitious and lime-based materials and composites, as summarised in the following paragraphs.

- Glass-fibre reinforced cement (GRC) panels may be superimposed on existing precast concrete (PC) walls for decorative and structural purposes. However, the GRC layer is likely to crack and break off as a consequence of the different shrinkage that GRC and PC experience due to changes in temperature and humidity, especially in outdoor settings. As proven by Chen et al. [15], adjusting the formulation of the GRC is key to reducing the shrinkage mismatch, with the addition of rubber powder, expanding agent, and metakaolin being advantageous to minimise the internal porosity and improve the compactness of the mortar structure. Additionally, a smooth GRC-PC interface is is useful for increasing the crack resistance as compared to a rough interface;
- In principle, the addition of end-of-life tyre (ELT) rubber can improve the mechanical properties of concrete, thus offering a viable approach to valorising this type of waste material. Nonetheless, the actual reinforcing effect is often undermined by the weak cementitious matrix–rubber interaction. The contribution by García et al. [16] compares different surface treatments aimed at improving the performance of concrete–ELT rubber composites. The experimental outcomes identify hydration as the most favourable treatment, as it makes it possible to add up to 5% ELT rubber (with respect to the aggregate weight) without compromising the design strength. The adoption of a pozzolanic Portland cement, with local (Chilean) fly ash waste, also contributes to these promising results;
- The paper by Li et al. [17] investigates the structural behaviour of concrete joints, where an on-site cast concrete element is jointed with a precast wall. An extensive experimental activity accounts for the role played by (*i*) the strength grade of the concrete mixtures, (*ii*) the interface between precast and on-site cast concrete elements, and (*iii*) the storage time for the precast element. Remarkably, it is observed that the mechanical behaviour of the joints benefits from the higher strength of the on-site cast concrete. Moreover, interface roughness is beneficial to the mechanical behaviour of the whole joint. The paper is mainly centred on structural issues and provides some interesting design inputs that can be successfully implemented in the practice;
- The contribution by Yao et al. [18] presents an experimental research on hybrid fibre-reinforced concrete specifically designed to manufacture vertical shafts, which are frequently found in coal mining to create the load-bearing structure of tunnels for the movement of operators, ventilation, and drainage. As expected, these structural ele-

ments are exposed to extremely harsh environmental conditions, which may represent a challenge for the adoption of newly designed composite materials. For this reason, the paper by Yao et al. [18] thoroughly investigates the long-term performance of a new cement-based composite system based on the synergistic interaction of polyvinyl alcohol and polypropylene-steel hybrid fibres. It is found that the combination of these two kinds of fibres enhances the impermeability of the wall, thus bringing in superior resistance to corrosion and freeze-thaw cycles;

- Al-Ameri et al. [19] discuss the residual strength to impact of concrete subjected to thermal ageing. This topic is of particular interest because the greatest part of the available literature elaborates on the resistance of concrete to fire and heat, whereas just a few papers address the resistance to high-rate loading, i.e., blast or impact. These loading conditions may be regarded as extraordinary, nonetheless the vulnerability of structures to dynamic loading is dramatically brought to light when disastrous events occur. Remarkably, according to the experimental work conducted by Al-Ameri et al. [19], thermal ageing has a deleterious effect on the capacity of concrete to withstand impact loading, especially for temperatures higher than 100 °C, as a consequence of the weight loss experienced by the conglomerate;
- The review by Almajed et al. [20] provides a detailed summary of the existing techniques for soil bio-stabilisation. As opposed to chemical and mechanical methods, bio-stabilisation techniques use either bacteria or enzymes to induce calcite precipitation and, hence, to bind and consolidate the soil grains. Bio-stabilisation is environmentally friendly, sustainable, and conducive to remediation of soil contamination. In addition to a survey of the published literature, the paper details practical guidelines for implementing different bio-stabilisation strategies in the field, according to specific site conditions;
- Pertaining to the geotechnical field, the review paper by Gudla et al. [21] explores the viability of granite dust as a soil stabiliser. Although granite dust comes from the processing of natural aggregates, disposing of this kind of industrial by-product is particularly challenging and poses severe issues for the environment. In addition, the pros and cons of using granite dust are thoroughly discussed, alongside some specific technological aspects, such as the interaction with weak soils and the choice of the optimal dosage in different applications;
- Song et al. [22] investigate the variability of the mechanical behaviour of coastal cemented soil for foundations as a result of the addition of iron tailings and nano-silica. Since nano-silica establishes a stable bond with clay, the addition of small amounts of nano-silica is sufficient to sharply increase the mechanical properties of cemented soils. On the other hand, the addition of iron tailings up to 20% improves the strength and stiffness of cemented soils. However, if the filler loading exceeds the 20% critical threshold, the material becomes extremely porous, with detrimental effects on the mechanical strength. This experimental study is particularly useful to foster understanding on how coastal cemented soils may attain satisfactory properties for real-scale applications;
- Zhang et al. [23] focus their attention on high-strength concrete and give an insight into its complex constitutive relations and failure behaviour under different stress states, both experimentally and numerically. Firstly, Zhang et al. [23] conduct conventional triaxial compressive tests on different high-strength concrete samples, and analyse the failure modes as a function of the applied confining pressure. Secondly, the experimental findings are discussed according to a statistical damage constitutive model based on the thermodynamic theory. The model accurately captures the stress-strain response of the tested samples and proves to be a useful tool to describe the performance of high-strength concrete;
- The substitution of natural coarse aggregates (NCAs) with recycled concrete aggregates (RCAs) in building materials represents a potential solution to improving the sustainability of the construction industry. Makul et al. [24] examine the state of the art on this subject and thoroughly analyse the performance of the composite material

as a function of source, type, or chemical and physical characteristics of the RCAs, as well as composition of the mixture, water content, curing conditions, and several other parameters. Makul et al. [24] also outline the open issues and future research that must be carried out to optimize the design of "green" composites. In conclusion, RCAs are a valuable resource that can be safely employed in traditional concrete, but their uptake to produce high-performance structural concrete, which must meet stringent mechanical standards, still requires additional efforts;

- As demonstrated by Vasovic et al. [25], the addition of cement to lime mortar makes it possible to effectively protect the cultural heritage, speeds up the restoration process, improves the materials' compatibility, and enhances the mechanical properties of mortar. Interestingly, Vasovic et al. [25] report that the presence of 20 wt% white Portland cement to replace lime and the reduction in the water-to-binder ratio, in combination with the incorporation of an air-entraining agent in air lime mortar, increase the strength of mortar without influencing the permeability. However, further research is sought to optimize the exact replacement of lime, evaluate the microstructural changes, and identify the compatibility of the blended cement-lime mortar in the masonry unit;
- Considering the climate emergence and and the increasing need to develop new cementitious materials, Landa-Ruiz et al. [26] put forward an interesting technical solution that uses sugarcane bagasse ash (SCBA) in combination with silica fume (SF) for the partial replacement of cement. Since several factors are relevant for developing new binder materials as an effective alternative to Portland cement, Landa-Ruiz et al. [26] consider three different variables, namely, (*i*) physical, mechanical, and durability properties of the concretes; (*ii*) CO_2 emissions; and (*iii*) recycling of waste materials. The use of eco-friendly ternary blend concretes prepared with 50% SCBA and SF leads to a resistance of 20.09 MPa at 180 days, significantly reduces the CO_2 emissions caused by Portland cement, and promotes a culture of waste recycling;
- Concrete cracking is an inevitable phenomenon and incorporating autogenous healing capability is a new branch of science that offers endless research scope. Luo et al. [27] consider the self-healing capacity of concrete prepared using several supplementary cementitious materials, such as fly ash (FA) and blast furnace slag (BFS), and cured in different conditions, reporting that the highest self-healing efficiency is found for water-incubated specimens. Additionally, BFS-based concrete shows better healing efficiency than FA-based concrete. The microstructural analysis demonstrates that the prime healing product, with and without supplementary cementitious materials, is micron-sized calcite crystals with a typical rhombohedral morphology. However, in-depth analysis is still needed to clarify the complete self-healing process.

Author Contributions: Conceptualization, C.S., A.S., S.C. and V.V.; resources, C.S.; writing—original draft preparation, C.S.; writing—review and editing, A.S., S.C. and V.V. All authors have read and agreed to the published version of the manuscript.

Funding: This research received no external funding.

Institutional Review Board Statement: Not applicable.

Informed Consent Statement: Not applicable.

Data Availability Statement: Not applicable.

Acknowledgments: A.S. is supported by the Commonwealth Scientific and Industrial Research Organisation (CSIRO) Research Office through the "Science Leader in Active Materials" grant.

Conflicts of Interest: The authors declare no conflict of interest.

References

1. Kundu, S.P.; Chakraborty, S.; Chakraborty, S. Effectiveness of the surface modified jute fibre as fibre reinforcement in controlling the physical and mechanical properties of concrete paver blocks. *Constr. Build. Mater.* **2018**, *191*, 554–563. [CrossRef]
2. Signorini, C.; Nobili, A.; Siligardi, C. Sustainable mineral coating of alkali-resistant glass fibres in textile-reinforced mortar composites for structural purposes. *J. Compos. Mater.* **2019**, *53*, 4203–4213. [CrossRef]

3. Ban, M.; Aliotta, L.; Gigante, V.; Mascha, E.; Sola, A.; Lazzeri, A. Distribution depth of stone consolidants applied on-site: Analytical modelling with field and lab cross-validation. *Constr. Build. Mater.* **2020**, *259*, 120394. [CrossRef]
4. Donnini, J.; Maracchini, G.; Lenci, S.; Corinaldesi, V.; Quagliarini, E. TRM reinforced tuff and fired clay brick masonry: Experimental and analytical investigation on their in-plane and out-of-plane behavior. *Constr. Build. Mater.* **2021**, *272*, 121643. [CrossRef]
5. Signorini, C.; Sola, A.; Nobili, A.; Siligardi, C. Lime-cement textile reinforced mortar (TRM) with modified interphase. *J. Appl. Biomater. Funct. Mater.* **2019**, *17* , 2280800019827823. [CrossRef]
6. Wang, L.; Rehman, N.U.; Curosu, I.; Zhu, Z.; Beigh, M.A.B.; Liebscher, M.; Chen, L.; Tsang, D.C.; Hempel, S.; Mechtcherine, V. On the use of limestone calcined clay cement (LC3) in high-strength strain-hardening cement-based composites (HS-SHCC). *Cem. Concr. Res.* **2021**, *144*, 106421. [CrossRef]
7. Fraternali, F.; Ciancia, V.; Chechile, R.; Rizzano, G.; Feo, L.; Incarnato, L. Experimental study of the thermo-mechanical properties of recycled PET fiber-reinforced concrete. *Compos. Struct.* **2011**, *93*, 2368–2374. [CrossRef]
8. Signorini, C.; Volpini, V. Mechanical Performance of Fiber Reinforced Cement Composites Including Fully-Recycled Plastic Fibers. *Fibers* **2021**, *9*, 16. [CrossRef]
9. Mandal, R.; Chakraborty, S.; Chakraborty, P.; Chakraborty, S. Development of the electrolyzed water based set accelerated greener cement paste. *Mater. Lett.* **2019**, *243*, 46–49. [CrossRef]
10. Wang, J.; Du, B. Experimental studies of thermal and acoustic properties of recycled aggregate crumb rubber concrete. *J. Build. Eng.* **2020**, *32*, 101836. [CrossRef]
11. Signorini, C.; Nobili, A. Durability of fibre-reinforced cementitious composites (FRCC) including recycled synthetic fibres and rubber aggregates. *Appl. Eng. Sci.* **2022**, *9*, 100077. [CrossRef]
12. Donnini, J. Durability of glass FRCM systems: Effects of different environments on mechanical properties. *Compos. Part B Eng.* **2019**, *174*, 107047. [CrossRef]
13. Signorini, C.; Nobili, A. Comparing durability of steel reinforced grout (SRG) and textile reinforced mortar (TRM) for structural retrofitting. *Mater. Struct.* **2021**, *54*, 1–15. [CrossRef]
14. Signorini, C. Durable and Highly Dissipative Fibrous Composites for Strengthening Coastal Military Constructions. Key Engineering Materials. *Trans. Tech. Publ.* **2021**, *893*, 75–83.
15. Chen, D.; Li, P.; Cheng, B.; Chen, H.; Wang, Q.; Zhao, B. Crack resistance of insulated GRC-PC integrated composite wall panels under different environments: An experimental study. *Crystals* **2021**, *11*, 775. [CrossRef]
16. García, E.; Villa, B.; Pradena, M.; Urbano, B.; Campos-Requena, V.H.; Medina, C.; Flores, P. Experimental Evaluation of Cement Mortars with End-of-Life Tyres Exposed to Different Surface Treatments. *Crystals* **2021**, *11*, 552. [CrossRef]
17. Li, C.; Yang, Y.; Su, J.; Meng, H.; Pan, L.; Zhao, S. Experimental Research on Interfacial Bonding Strength between Vertical Cast-In-Situ Joint and Precast Concrete Walls. *Crystals* **2021**, *11*, 494. [CrossRef]
18. Yao, Z.; Fang, Y.; Zhang, P.; Huang, X. Experimental Study on Durability of Hybrid Fiber-Reinforced Concrete in Deep Alluvium Frozen Shaft Lining. *Crystals* **2021**, *11*, 725. [CrossRef]
19. Al-Ameri, R.A.; Abid, S.R.; Murali, G.; Ali, S.H.; Özakça, M. Residual Repeated Impact Strength of Concrete Exposed to Elevated Temperatures. *Crystals* **2021**, *11*, 941. [CrossRef]
20. Almajed, A.; Lateef, M.A.; Moghal, A.A.B.; Lemboye, K. State-of-the-Art Review of the Applicability and Challenges of Microbial-Induced Calcite Precipitation (MICP) and Enzyme-Induced Calcite Precipitation (EICP) Techniques for Geotechnical and Geoenvironmental Applications. *Crystals* **2021**, *11*, 370. [CrossRef]
21. Gudla, A.; Baig Moghal, A.A.; Almajed, A. A State-of-the-Art Review on Suitability of Granite Dust as a Sustainable Additive for Geotechnical Applications. *Crystals* **2021**, *11*, 1526.
22. Song, X.; Xu, H.; Zhou, D.; Yao, K.; Tao, F.; Jiang, P.; Wang, W. Mechanical performance and microscopic mechanism of coastal cemented soil modifed by iron tailings and nano silica. *Crystals* **2021**, *11*, 1331. [CrossRef]
23. Zhang, L.; Cheng, H.; Wang, X.; Liu, J.; Guo, L. Statistical damage constitutive model for high-strength concrete based on dissipation energy density. *Crystals* **2021**, *11*, 800. [CrossRef]
24. Makul, N.; Fediuk, R.; Amran, M.; Zeyad, A.M.; Murali, G.; Vatin, N.; Klyuev, S.; Ozbakkaloglu, T.; Vasilev, Y. Use of Recycled Concrete Aggregates in Production of Green Cement-Based Concrete Composites: A Review. *Crystals* **2021**, *11*, 232. [CrossRef]
25. Vasovic, D.; Terzovic, J.; Kontic, A.; Okrajnov-Bajic, R.; Sekularac, N. The Influence of Water/Binder Ratio on the Mechanical Properties of Lime-Based Mortars with White Portland Cement. *Crystals* **2021**, *11*, 958. [CrossRef]
26. Landa-Ruiz, L.; Landa-Gómez, A.; Mendoza-Rangel, J.M.; Landa-Sánchez, A.; Ariza-Figueroa, H.; Méndez-Ramírez, C.T.; Santiago-Hurtado, G.; Moreno-Landeros, V.M.; Croche, R.; Baltazar-Zamora, M.A. Physical, Mechanical and Durability Properties of Ecofriendly Ternary Concrete Made with Sugar Cane Bagasse Ash and Silica Fume. *Crystals* **2021**, *11*, 1012. [CrossRef]
27. Luo, M.; Jing, K.; Bai, J.; Ding, Z.; Yang, D.; Huang, H.; Gong, Y. Effects of Curing Conditions and Supplementary Cementitious Materials on Autogenous Self-Healing of Early Age Cracks in Cement Mortar. *Crystals* **2021**, *11*, 752. [CrossRef]

MDPI

Review

Use of Recycled Concrete Aggregates in Production of Green Cement-Based Concrete Composites: A Review

Natt Makul [1], Roman Fediuk [2], Mugahed Amran [3,4,*], Abdullah M. Zeyad [5], Gunasekaran Murali [6], Nikolai Vatin [7], Sergey Klyuev [8], Togay Ozbakkaloglu [9] and Yuriy Vasilev [7]

1 Department of Civil Engineering Technology, Faculty of Industrial Technology, Phranakhon Rajabhat University, Bangkok 10220, Thailand; natt@pnru.ac.th
2 School of Engineering, Far Eastern Federal University, 8, Sukhanova Str., 690950 Vladivostok, Russia; roman44@yandex.ru
3 Department of Civil Engineering, College of Engineering, Prince Sattam Bin Abdulaziz University, Alkharj 11942, Saudi Arabia
4 Department of Civil Engineering, Faculty of Engineering and IT, Amran University, Amran 9677, Yemen
5 Department of Civil Engineering, Faculty of Engineering, Jazan University, Jazan 45142, Saudi Arabia; azmohsen@jazanu.edu.sa
6 School of Civil Engineering, SASTRA Deemed to Be University, Thanjavur, Tamil Nadu 613401, India; murali@civil.sastra.ac.in
7 Moscow Automobile and Road Construction University, 125319 Moscow, Russia; vatin@mail.ru (N.V.); yu.vasilev@madi.ru (Y.V.)
8 Department of Theoretical Mechanics and Strength of Materials, Belgorod State Technological University Named after V.G. Shukhov, 308012 Belgorod, Russia; klyuyev@yandex.ru
9 Ingram School of Engineering, Texas State University, San Marcos, TX 78666, USA; togay.oz@txstate.edu
* Correspondence: m.amran@psau.edu.sa or mugahed_amran@hotmail.com

Citation: Makul, N.; Fediuk, R.; Amran, M.; Zeyad, A.M.; Murali, G.; Vatin, N.; Klyuev, S.; Ozbakkaloglu, T.; Vasilev, Y. Use of Recycled Concrete Aggregates in Production of Green Cement-Based Concrete Composites: A Review. *Crystals* **2021**, *11*, 232. https://doi.org/10.3390/cryst11030232

Academic Editor: Cesare Signorini

Received: 9 February 2021
Accepted: 24 February 2021
Published: 26 February 2021

Abstract: Recycled concrete aggregates (RCA) are used in existing green building composites to promote the environmental preservation of natural coarse aggregates (NCA). Besides, the use of RCA leads to potential solutions to the social and economic problems caused by concrete waste. It is found that insufficient information on the longevity and sustainability of RCA production is a serious issue that requires close attention due to its impact on changing aspects of the sector. However, more attention has been paid to explaining the effect of RCA on concrete durability, as well as the properties of fresh and hardened concrete. Therefore, this study aims to provide a critical review on the RCAs for the production of high-performances concrete structures. It begins by reviewing the source, originality, types, prediction of service life, features and properties of RCA, as well as the effect of RCA on concrete performance. In addition, this literature review summarizes the research findings to produce complete insights into the potential applications of RCA as raw, renewable, and sustainable building materials for producing greener concrete composite towards industrializing ecofriendly buildings today. Further, it has also highlighted the differences in the current state of knowledge between RCAs and NCAs, and offers several future research suggestions. Through this critical and analytical study, it can be said that RCA has the possible use in the production of high-performance structural concrete depending on the source and type of recycled aggregate while the RCA can be used widely and safely to produce traditional green concrete.

Keywords: durability; hardened properties; green composite; fresh properties; recycled concrete aggregates; natural coarse aggregates

1. Introduction

One of the most Sustainable development is now a major development challenge for the entire world and has become the guiding standard for the construction sector [1]. Technological progress in the production of concrete and reinforced concrete requires additional sources of raw materials, in particular, the use of high-quality aggregates [2].

The implementation of this problem is impeded by the constantly growing shortage of mineral and energy resources, as well as environmental requirements for environmental protection [3]. Therefore, at present, the task is to comprehensively use the deposits of low-quality raw materials and waste from related industries [4].

Reusing and recycling concrete waste can be a successful strategy for achieving sustainability along the way [5]. Waste concrete is collected and crushed and then used in structural concrete, in which it replaces natural coarse aggregates (NCA) [6]. Many administrations worldwide have introduced various control measures to minimize the use of virgin aggregate and improve the recycling of concrete waste for reuse as materials when environmentally, technically, and economically acceptable [7]. Environmental problems are known to be exacerbated by rising landfill fees and land scarcity. The use of concrete waste in sustainable development can alleviate such problems [8]. However, it has been observed that most concrete plants were reluctant to produce recycled concrete aggregates (RCAs) and make full use of them [9]. Manufacturing plants have not yet mastered RCA's use, not only due to its unclear characteristics for concrete but also due to unexplored manufacturing processes, which, however, have yet to be determined [10]. It has been observed that most concrete batching plants are hesitant to manufacture and use RCA at their optimum [11]. However, it has become necessary to study this problem.

As a result of human-made and natural anomalies occurring on the Earth, there are many destroyed cities, settlements, and houses (see Figure 1 that shows the process of destroyed buildings wastes production as RCAs). The question is how to rebuild these cities and how to use parts of the destroyed buildings and structures [12]. One of the ways is to take everything out to relatively low forms of the earth's surface, store it, cover it with soil, plant a forest on top, produce new building materials and from them rebuild cities and settlements [13].

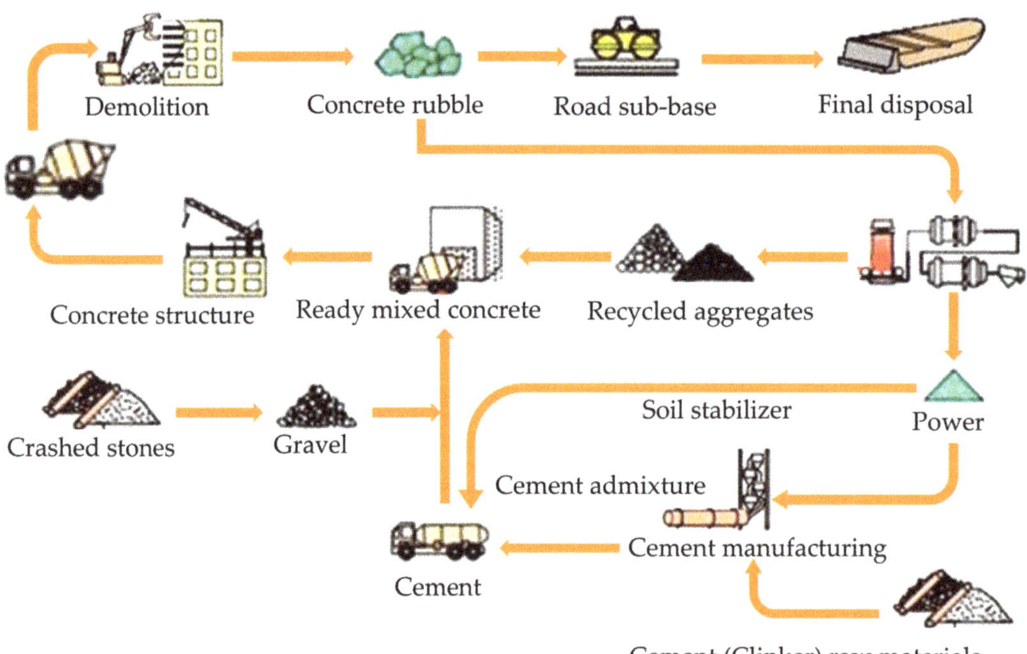

Figure 1. Process of destroyed buildings wastes production as RCAs.

But this is a very expensive undertaking [5]. The second way out of this situation is the use of fragments of destroyed buildings and structures to create building materials, using which to repair and build new buildings and structures in place of the destroyed ones [14]. Demolition of old buildings and construction of new ones is common practice due to natural disasters, expansion of traffic routes, urban redevelopment, structural destruction and change of purpose [15]. In the European Union, about 850 tons of construction waste is generated annually, which is about 30% of the total waste [16]. Debris from demolition alone is about 123 tons annually in the United States [17].

Huge concrete waste is generated during the demolition of old buildings; then, it is most often disposed of in landfills, which poses significant health risks and damages the environment [18]. More than forty years ago, research began on the characteristics of RCA [19]. In the past, most of the research carried out was mainly limited to the production of unstructured concrete due to the deleterious physical properties of RCA, for example, high water absorption, which increases the need for water for a certain workability (Table 1) [20–22].

Insufficient information on the longevity and sustainability of RCA production is a serious issue that requires close attention due to its impact on changing aspects of the sector [23]. It is unknown whether the manufacturing, quality control, and production costs in standard RCA manufacturing industries outweigh the benefits from RCAs purchased as concrete components at a lower cost than natural concrete aggregates [24]. One explanation for the slow acceptability of production for RCA production is the already existing plants for the production of bulk ready-mixed concrete [25].

However, the current models, in which RCA increasingly replaces NCA in various structural designs, have gradually gained in importance for specific reasons [26]. For example, RCA manufacturing provides sustainability for concrete waste and encourages recycling rather than landfill [27]. In addition, it focuses on the lack of natural aggregates, minimizes the need for them, and ultimately allows the preservation of the NCA natural aggregates mined in the open pit [28]. Despite the cost, these and many other benefits have led to increased interest in manufacturing and using RCA in design [29,30]. It is found that insufficient information on the longevity and sustainability of RCA production is a serious issue that requires close attention due to its impact on changing aspects of the sector. Therefore, this study aims to provide a critical overview on the RCAs for the production of high-performances concrete structures. This study reviews the source, originality, types, service life prediction, features and properties of RCA. However, more attention has been paid to explaining the effect of RCA on concrete durability, as well as the properties of fresh and hardened concretes. In addition, this literature review summarizes the research findings to produce complete insights into the potential applications of RCA as raw, renewable and sustainable building materials for producing greener concrete composite towards industrializing ecofriendly buildings today. The paper evaluates high performance concrete structures that favor RCA's manufacturing activities rather than natural materials [31,32]. Therefore, it has also been highlighted the differences in the current state of knowledge between RCAs and NCAs, and offers some suggestions for future research.

Table 1. Approval criteria about RCAs.

Country	Standard	Criterion of Oven-Dry Density (kg/m^3)	RCA Type	Absorption Ratio of the Criterion of Aggregate (%)	Refs.
Australia	AS1141.6.2	≥2100	Class 1A	≤6	[33]
		≥1800	Class 1B	≤8	
Germany	DIN 4226-100	≥1500	Type 1	No limit	[34]
		≥1800	Type 2	≤20	
		≥2000	Type 3	≤15	
		≥2000	Type 4	≤10	
Hong Kong	Works Bureau of Hong Kong	≥2000	HK, 2000	≤10	[34]
Japan	JIS A 5021, 5022 and 5023	≥2500	Fine-Class H	≤3.5	[35]
		≥2200	Fine-Class M	≤7	
		≥2300	Coarse-Class M	≤5	
		≥2500	Coarse-Class H	≤3	
		No limit	Fine-Class L	≤13	
		No limit	Coarse-Class L	≤7	
Korea	KS F 2573	≥2200	Fine	≤5	[36]
		≥2500	Coarse	≤3	
International	RILEM	≥2000	Type 2	≤10	[37]
		≥1500	Type 1	≤20	
Spain	EHE	≥2000	-	≤5	[38]
		≥2500	Type 3	≤3	

2. Source and Originality of RCA

RCA originate from the Portland cement concrete demolition. Aggregate pieces can be expected to vary considerably, given that the original concrete could be weathered or fresh, loose or dense, weak or strong [39]. In the process of crushing, agglomerates of concrete aggregate with adhered mortar are formed [40]. Typically, these agglomerations are less efficient than angles units [41]. Fines are also formed from crushed concrete [42].

The aggregates form a concrete frame. Aggregates often take up about seventy percent of the total volume of concrete [43]. A large percentage of the number of these aggregates is mainly a coarse aggregate [44]. However, in the construction sector, the demand for coarse aggregate is enormous [45]. The growing extraction of raw materials from natural resources is essential to meet these high demands [46]. The ever-increasing use of natural large aggregates creates an ecological imbalance [47]. Therefore, the use of alternative raw materials is essential in the construction industry [48]. The use of RCA derived from demolished concrete buildings is one approach to achieving this goal [49]. The use of reclaimed concrete materials in structures minimizes the need for natural coarse aggregate [50]. In turn, this minimizes the negative environmental impacts due to natural aggregate recovery [51]. Rising landfill costs and NCA shortages have also contributed to the use of RCA in concrete [52]. In addition, the increased distances between the construction site and the NCA quality source have forced contractors to consider replacing the NCA with RCA [53]

The use of RCA in high performance and high strength structural concrete is possible due to proper quality control and mixing, and the addition of pozzolanic additives. It has been noted that poverty imposes specific constraints that lead to delays and sometimes a lack of programs to carry out various engineering processes [54]. Cost issues are limited by the difficulty of implementing environmentally friendly and sufficient, safe solutions such as aggregate recycling. In particular, front-loading and top-hopper industries can be used to reduce additional RCA costs [55]. However, pricing and the supply and demand aspects of recycled materials pose various challenges that are rarely considered.

In addition, it is necessary to carefully study the characteristics of concrete with each type of RCA [56]. In particular, the strength of concrete is influenced by various elements such as the replacement rate, moisture content, RCA type, and water-to-cement ratio [57]. Unlike natural aggregates, RCA generally has a higher water absorption, which can significantly affect concrete properties, especially workability [58]. In addition, concrete with drier RCA had a higher slump and more rapid slump flow than water-saturated concrete [59]. It is clear that the strengths for completely replacing fine aggregate increase over time, and blends having a higher percentage of the combination get better composite performance [60]. This reflects the characteristics of the presence of reclaimed concrete, which from the available literature review show that the strength of the material increases with the life of the structure [61]. The only reliable way to minimize construction costs is to use available local resources and introduce innovative building materials [62]. The use of RCA as fine aggregates in concrete seems to be the best, especially in areas where demolition waste is freely available [63].

3. Service Life Prediction of RCA Concrete

The theoretical basis for the design of green composites using the specified raw materials is the transdisciplinary science of geomimetics [64–66], which uses the results of studies of natural processes to create high-strength concretes and building composites of a new generation. The results were tested on raw materials from destroyed buildings and structures in Iraq, which mainly consist of concrete, ceramic bricks, and limestone wall blocks. The performance of RCA concrete is influenced by various key aspects such as air entrainment, cement content, curing conditions, RCA humidity conditions, properties of the original RCA concretes, RCA physical characteristics, RCA size and type, RCA content, and water to cement ratio [67]. The aspects are explained below.

3.1. RCA Features and Percentage

The physical properties of RCA have a significant effect on the properties of fresh and hardened concrete. For example, fresh concrete with coarse and angular particles becomes tough and therefore difficult to cast [68]. In addition, high RCA absorption can affect the workability of concrete. In addition, large RCA pore volumes can affect the durability characteristics (permeability and water absorption), strength, and porosity of concrete.

The fresh and hardened properties of concrete are strongly influenced by the RCA percentage used as a complete or partial replacement for NCA. Using seven independent variables, [69] created a model of aggregate quantity and type to predict concrete performance when replacing 0.0–100% NCA with RCA. The researchers reported that concretes created using RCA had a lower modulus and compressive strength than NCA concretes. Higher RCA content also increases water absorption but decreases density, resulting in increased concrete porosity. The use of coarse RCA reduces concrete density by 50–100% and increases water absorption by about 2.11% to 3.50% and from 0.14% to 0.38%, respectively. In addition, it is noted that as the content of grounded RCA increases, the resistance to chloride ion penetration decreases, while the tensile splitting and compressive strength of concrete decreases [70]. In addition, the researchers noted that concrete drying shrinkage improved with increasing RCA content. It can be controlled by lowering the water to cement ratio.

3.2. RCA Sizes and Original Concrete Quality

It is used three different aggregate sizes to evaluate the effect of RCA size on concrete properties [71]. A more significant reduction in modulus of elasticity was obtained for concretes prepared with smaller RCA dimensions. On the other hand, they reported that the strength increases with the size of the RCA. In addition, they found that the water absorption of concrete decreases with increasing RCA dimensions. This is due to the relatively low content of weak solutions adhered to the coarse aggregate. It is investigated the effect of base concrete quality on RCA concrete performance [72]. Scientists reported that RCA water absorption increases with the strength of the base concrete. This is due to the fact that for concretes with higher strength, higher cement content is required in principle; therefore, the amount of mortar adhering to the aggregate increases. Thus, adjustments to mix water content are necessary for newer concretes, including RCA made from older, harder concretes, to obtain the preferred workability. Porous RCAs affect the strength of RCA concrete. The proportional loss of tensile or compressive strength of new concretes due to the use of RCA is more significant when it is obtained from weaker old concretes than from strong old concretes [45].

3.3. Influence of Cement Content and Water to Cement Ratio

RCA concretes with high cement content are reported to be highly resistant to carbonation. It is found that higher cement content in RCA concrete results in a preferred compressive strength [73]. It is also found that the RCA tensile strength of concrete increases with increasing cement content in concrete [74]. The degradation of RCA concrete is related to the water to cement ratio used in the mix design. It is reported that compared to the original RCA concrete, RCA concretes require a higher cement content and a lower water-to-cement ratio to achieve a certain compressive strength [75]. The resistance of the NCA to melting and freezing at water to cement ratio of 0.290 was exceptionally high. But for RCA concrete, the same water-to-cement ratio does not provide suitable freeze-thaw resistance.

3.4. Hydration of Cement in the Original Concrete

It has been established that the composition of the crushing concrete scrap contains about 30% of non-hydrated Portland cement, which makes it possible to use it as an active microfiller in the production of multicomponent highly active binders [34]. Aggregate from the concrete scrap has a partial or solid shell on the surface of its grains from the cement paste of crushed concrete, actively influences the process of forming both the structural characteristics of the cement paste and the dense interfacial transition zone between them [76]. The structure of concrete composites is characterized by lower water absorption (up to 3–7%) and the presence of rather small and uniform pores in size [77].

3.5. Sources or Types of RCAs

Several studies have been carried out to investigate the effect of sources or types of RCA on concrete performance [78–80]. With the exception of concretes made from recycled aggregates made from masonry ceramics, which reported an increase in compressive strength, concretes made with coarse RCA had lower compressive strengths [81]. The modulus of elasticity of concrete has been reduced for all RCA types. However, RCAs made from red ceramics had a greater effect on reducing the modulus of concrete due to the lower density.

3.6. Curing Conditions, RCA Moisture Conditions, and Air Entrainment

External curing in the environment is more detrimental to RCA concretes than NCA ones. It is shown that the differences in splitting tensile strength among NCA and RCA concretes are large when they are hardened in the external environment [79]. In addition, It is found that the carbonization depth of RCA concrete water-cured is almost twice that of RCA air-cured concrete [73]. The decrease in carbonization depth caused by curing in water may be partly due to the higher internal moisture content of the concrete. The moisture conditions of the aggregate affect the workability of the concrete. The initial slump of concrete (workability measurements) is highly dependent on the initial free water content of the concrete mix. It is shown that although air-dry and saturated surface-dry RCA exhibit typical initial slump and slump loss, kiln-dried RCA results in faster slump losses and higher initial slump [79]. Thanks to the appropriate air entrainment, durable concrete can be obtained from RCA [82]. Air entrainment for NCA concretes is as successful as for RCA concretes [83]. In addition, the use of entrained air is more effective than reducing the water-to-cement ratio in increasing freeze- resistance of RCA concrete [84].

4. Chemical Properties

The chemical characteristics of RCA affect the performance of the concrete. The durability of concrete should be affected by the presence of reactive chemicals such as chlorides, alkalis and sulphates in RCAs. The chemical properties of RCA are summarized below.

4.1. Soundness

Soundness is an indicator of the stability of aggregates to environmental influences, such as atmospheric influences [45,85–87]. The strength test for magnesium sulfates and sodium sulfates is mainly used to assess the integrity of the aggregate. RCAs usually pass magnesium sulfate strength tests, but do not pass sodium sulfate strength tests. For RCA and NCA, respectively, the strength loss of magnesium sulfate were 70% and 2.50% [88]. These figures are identical to the results obtained by several researchers worldwide [45,85–87]. It is also examined the reliability tests of sodium sulfates and found that the losses were 12.0% and 9.0%, respectively, for RCA and NCA [45,85–87]. Thus, the sodium sulfate leak test was in accordance with the standard. Moreover, the results are identical to those obtained by Lye et al. [87]. The failure of RCA is mainly associated with the distribution of pore sizes in aggregates.

4.2. Reactivity of Alkali-Aggregates

RCA concretes can undergo alkaline aggregate reactions (AAR) if the original concrete aggregate was prone to alkaline aggregate reactions. In addition, mortars adhered to RCA or the alkali content of the cement paste can have a significant impact on the AAR susceptibility of new concretes with RCA. The alkali content of RCA is related to the content of its solution. RCAs with more mortars lead to higher alkali content and are thus more susceptible to alkaline aggregate reactions.

4.3. Sulfate and Chloride Content

RCA can have high sulphate content due to sulphate substances present in adhered cement slurries. It is investigated the content of water-soluble sulfates for RCA and NCA [87]. The researchers set the sulfate content of 0.032 g/L and 0.0250 g/L for RCA and NCA, respectively. The sulfate content of RCA depends on the amount of adhered mortar/cement paste. Higher sulfate content in RCA indicates that higher amounts of mortar will attach to RCA.

High levels of chlorides have been found in RCA, formed from bases with prolonged exposure to chloride-oriented antifreeze agents. RCA with higher chloride content can affect the durability of new concrete due to corrosion of the steel reinforcement [45,85–87]. Since increased corrosion of steel can lead to premature failure of reinforced concrete structures, RCA made from old concrete with chloride content higher than 0.04 kg/m^3 should not be used in new concrete. Furthermore, polymer materials, rubbers, plastics, joint seals, textiles, wood, and paper can be present in RCAs [45,85–87]. In concrete, these materials become unbalanced when thawed, dried.

5. Physical Properties

The concrete properties and mix proportions are influenced by the physical properties of the RCA. Basic properties such as absorption, pore volume, bulk density, specific gravity, texture and shape of RCA are generally inferior to those of natural aggregate due to impurities and the presence of residual slurry/paste [87]. The magnitude of the impact will vary depending on the amount and nature of the recycled slurries or pastes present in RCA. From the available literature [89–91], a summary of the main physical characteristics of RCA is shown in Table 2.

Table 2. Fundamental physical properties of NCA and RCA [67–69].

Physical Properties	NCA	RCA
Pore Volumes (vol.%)	0.50–2	5–16.5
Absorptions (wt.%)	0.50–4	3–12
Compacted Bulk Densities (kg/m^3)	1450.00–1750	1200–1425
Specific Gravity	2.40–2.90	2.10–2.50
Shapes and Textures	Smooth and well rounded	Angular with rough surfaces

5.1. Particle Size Distribution

At present, during the development of fragments of destroyed buildings and structures, a large amount of unfractionated crushing concrete waste with a size of 0–5 mm is formed [67,92,93], the use of which in the production of concrete is difficult due to the presence of a significant amount of dusty fraction in their composition (Table 3). However, judging by the chemical composition, the fine fraction of concrete scrap can be used as an additional cementitious material [94]. The comminuted form of the particles, as well as the high content of silica and clinker minerals in them, will contribute to their high activity, which is confirmed by the studies of other authors [95–97]. RCA have very rough and angular due to the crumbling of the concrete and the presence of grout adhered to the original rough surfaces of the aggregate (Figure 2). Consequently, recycled concrete aggregate has better adhesion to the cement matrix than natural one. Depending on the size of the aggregates, RCA particles typically contain 30 to 60% of mortar. RCAs are identical in particle shape to crushed stones, but the types of grinding tools affect the gradation and other characteristics of fine concrete. Also, Figure 3 shows the fundamental components of demolition wastes [80].

Table 3. The mineral composition of various fractions of demolition wastes.

Particle Sizes, mm	Mineral Composition, %					
	SiO_2	$Ca(OH)_2$	$CaCO_3$	CSH	C_3S	C_2S
0.00–0.16	48.4	11.5	10.0	5.8	12.0	12.3
0.16–0.315	55.2	7.4	11.0	4.4	11.0	11.0
0.315–0.63	56.4	11.0	3.9	12.0	6.7	10.0
0.63–1.25	65.1	12.0	6.0	5.9	5.0	6.0
1.25–2.5	64.4	10.5	6.9	3.0	7.6	7.6
2.5–5	62.8	11.0	0	6.0	9.2	11.0
5–20	60.8	8.0	3.0	4.0	9.0	11.2
20–40	56.0	11.4	3.0	12.9	6.7	10.0
40–70	64.5	10.4	6.0	3.9	7.6	7.6

Figure 2. Original rough surfaces of the RCAs.

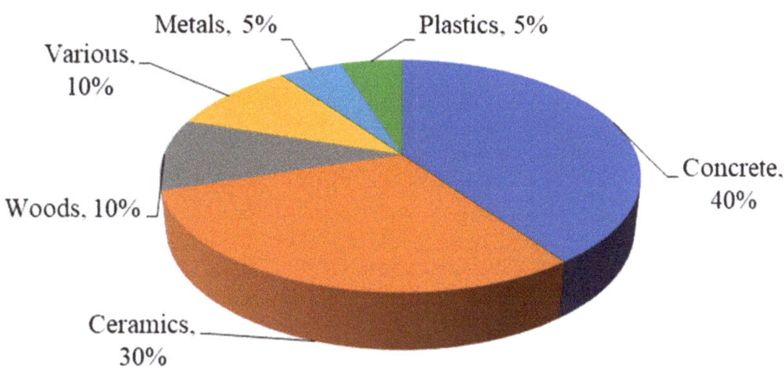

Figure 3. Fundamental components of demolition wastes [80]. Reprinted with permission from Elsevier [80].

5.2. Specific Gravity and Bulk Density

The specific gravity and bulk density of RCA are often lower, and the pore volume and absorption, respectively, are higher than those of natural aggregate. The lower specific gravity of RCA is due to the presence of aggregate particles in the old cement mortar/paste, which makes it less dense than NCA due to its higher porosity [98]. In saturated dry conditions, RCA distinctive gravity ranges from 2.10 to 2.50, which is 5–10% less than NCA as shown in Table 2. Experimental data from Aliabdo et al. [99] found that RCA bulk densities are 9.80% less than primary gravel aggregates. Compared to virgin concrete aggregate, the higher pore volume of RCA makes it less dense and fragile.

5.3. Aggregate Grounded, Abrasion and Effect Values

Aggregate ground values (AGV) are measures of the aggregate resistance to commination under progressively applied compressive loads. It is reported that the lower the value, the stronger the aggregates. It was found that ACV for NCA (14% to 22%) is significantly less than for ACV for RCA (20–30%) from the literature [54,100–102] and as indicated in Table 3. This is expected due to the relatively weak mortars and cement pastes attached to the RCA particles.

The aggregate abrasion values (AAV) are a measure of the aggregate wear resistance. When material loss due to wear becomes higher, a higher AAV is obtained. Generally, the cumulative abrasion of RCA is higher than NCA. Classic RCA values range from 20% to 45%, which is higher than the values for pure concrete aggregates, as shown in Table 3 [54,100–102]. Despite its origins, RCA's cumulative abrasion rates are nevertheless generally below the acceptable optimum limits for structural applications (50% by weight).

Furthermore, aggregate effect values (AEV) are the aggregate strength values exposed to impact. AEVs indicate the aggregate resistance to dynamic loads. As shown in Table 4, AEV of NCA (15% to 20%) are lower than RCA (20% to 25%) [54,100–102]. Attached cementitious and mortar pastes make RCA less durable and thus lead to higher toughness values for RCAs that were frozen or wetted. About 0.150% by weight RCA is the acceptable limit for organic matter.

Table 4. The main RCA and NCA mechanical properties [54,100–102].

Mechanical Properties (wt.%)	RCA	NCA
Aggregate Grounded Values	20–30	14–22
Aggregate Abrasion Values	20–25	15–30
Aggregate Effect Values	20–25	15–20

6. Fresh Properties

RCA can affect the performance of fresh concrete due to their higher porosity, absorption, surface roughness, and angularity [103]. The higher size and angularity of RCAs will reduce the workability of the concrete and make it difficult to lay it properly [104].

6.1. Workability

Workability decline rates increase with the improvement in the proportion of RCA in concrete mixes [105]. Thus, RCA concretes require more water to obtain similar workability to NCA concretes [106]. Concrete mixes that integrate RCA generally meet the initial settlement requirements [107]. Also, Table 5 summaries the RCA effects on the concrete hardened features. However, greater RCA uptake can lead to a rapid loss of workability, which limits the time required for paving and completing concreting [108]. Problems associated with rapid loss of workability should be addressed by changing and controlling the RCA water content before mixing or increasing the amount of superplasticizer, rather than adding more water on construction sites [109]. In general, the greater sharpness and surface coarseness of RCA particles reduce the concrete workability and lead it more problematic to finish appropriately.

Table 5. The RCA effects on the concrete hardened features [7,110,111].

Properties	Comparison with NCA Concrete
Thermal Expansions	10–30% more
Creeps	30–60% more
Drying Shrinkages	20–50% more
Penetrations of Chlorides	0–30% more
Water Absorptions	0–40% more
Permeability	0–50% more
Porosities	10–30% more
Elasticity Modulus	10–45% more
Strength of Bonds	9–19% less
Flexural Strengths	0–10% less
Splitting Tensile Strengths	0–10% less
Compressive Strengths	0–30% less
Dry Densities	5–15% less

6.2. Wet Density

Several studies have been conducted to investigate the effect of RCA on the wet density of concrete [112]. Typically, the wet density of RCA concrete is lower than that of virgin concrete aggregates, as noted in Table 4 [7,110,111]. Compared to NCA concrete, it has been observed that the wet density of RCA concrete is 5–15% less [111]. This RCA contains adhered old cement pastes or mortars that are less dense than NCA [113]. As a rule, the density of hardened RCA concrete is 5–15% less than that of concrete with natural aggregates [114]. This is due to the used solutions attached to the RCA [115]. Depending on the size of the aggregate, the amount of mortar attached to the re-concrete aggregates ranges from 30% to 60% by volume RCA [116]. The density of the secondary mortar is much lower than that of most natural concrete materials [117]. This results in a lower density of RCA concrete [118].

6.3. Stability

The stability of the concrete mix enhanced as a result of abridged bleeding and augmented cohesiveness. Therefore, the resistance to segregation of NCA concrete could be equivalent to that of RCA concrete. As a rule, the soaking of NCA concrete is greater than that of RCA concrete [6]. During mixing, some of the ancient cement pastes are wiped off the RCAs and form additional fines in the concrete mix. Thus, these fine particles minimize the soaking of the concrete after some water has been adsorbed in the mixture. With lower free water content, more fine particles also increase the adhesion of the concrete mix. In addition, the increased surface roughness and angularity at higher RCA levels contribute to better concrete adhesion. The stability of the concrete mix is enhanced by improved cohesiveness and minimal soaking [119].

6.4. Air Content

The air content in concrete is significantly affected by the volume of their mortars. RCA influences the air content of concretes as they have higher mortar content. Fresh concrete having RCA is usually 60% more than the air content of fresh NCA concrete [62]. This is due to air entrained in recycled RCA mortars. Thus, when determining the target air content in RCA concrete, the existing air content in the mortar must be taken into account.

7. Mechanical Properties

The mechanical properties of concrete depend on the properties of the aggregate. It was found that the mechanical characteristics of RCA are lower in comparison with the mechanical characteristics of primary concrete aggregates from the available literature [87,98,99]. The main mechanical characteristics of RCA and NCA are listed in Table 3 and then briefly explained [120–122].

7.1. Compressive Strength

Silva et al. [123] studied the effect of adding a small percentage of waste plastic fiber and waste building material on some of the mechanical properties of concrete. In his study, the volume fraction of waste was 0.1–0.2% by volume. The results obtained proved the improvement in compressive and flexural strength. The results also showed an increase in the density of the fiber-reinforced concrete specimens compared to the control mix. As shown in Figure 4, the compressive strength of RCA concrete is often 5–10% lower than that of virgin concrete [124].

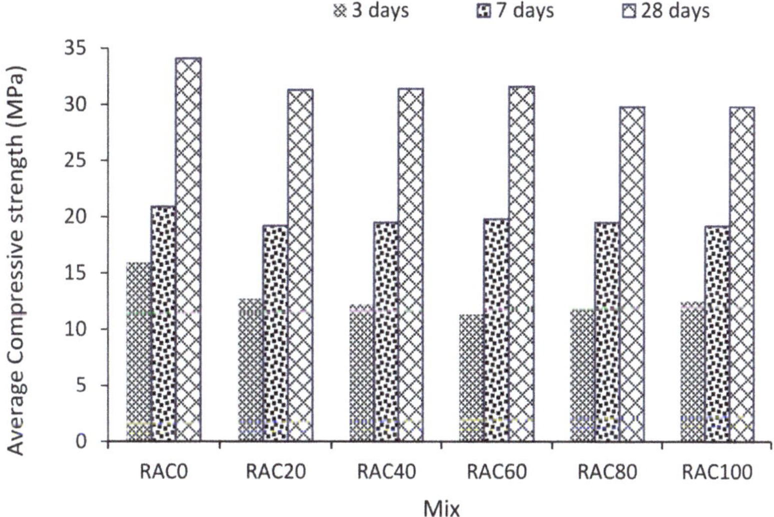

Figure 4. Compressive strength versus level of replacements of NCA by RCA [124]. Reprinted with permission from Elsevier [124].

Depending on the RCA quality, it can also be reduced by up to 25%. Typically, higher air content in concrete mixes with RCA can also lead to lower strength values [99]. RCA concretes, however, can have the same and sometimes higher compressive strength than NCA concretes if the reclaimed concrete is obtained from old sources of concrete that were originally produced with a lower water to cement ratio than new concretes.

Duan et al. [125] found that RCA does not have any effect on the compressive strength of concrete up to a degree of substitution of 30% by weight, after which it decreases. It is also showed that the compressive strength of concrete was much lower when RCA was used in dry state [87]. The reduction in compressive strength from 20% to 30% was realized due to the use of RCA in the case of high strength concretes. Similar results have been noted by other researchers [87,98,126]. Seethapathi et al. [127] studied the self-compacting properties of concretes made with RCA and compared them with those of NCA concretes. Nitesh et al. [128] found that at the same age, the variation in compressive strength was negligible.

In addition, fine RCAs can affect the compressive strength of concrete. The compressive strength of RCA concrete depends on the ratio of the coarse aggregate to the fine aggregate of the original RCA concrete according to [129]. A lower ratio of coarse aggregate to fine aggregate results in more mortar adhered to the coarse RCA particles, and therefore a decrease in the strength of RCA concrete. This reduction is even greater when using recovered fines. Thus, the use of fine RCA in concrete is generally not recommended. Nevertheless, according to Maria et al. [130], the reduction in compressive strength did not occur for concretes containing up to 20% fine RCA. Strength decreases with increasing RCA content above this level.

The strength of RCA concrete can be increased either by soaking up a portion of the reconstituted aggregates for mixing without or with pozzolanic fluids during mixing, or by soaking RCA in mixtures of pozzolanic fluids such as colloidal silica or water before mixing the concrete. It is expected that microcracks in RCA will be filled with absorbed pozzolanic fluids or water absorbed cement gels during pozzolanic reactions or cement hydration. Consequently, the strength of RCA concrete can be increased.

7.2. Splitting Tensile and Flexural Strengths

There is limited literature on the effect of RCA on the tensile strength of concrete [39,46,100]. As shown in Figure 5 [124], tensile strength at cracking of RCA concrete is lower than that of NCA concrete. Various researchers have reported that the splitting tensile strength of RCA concrete is 0–10% less than that of NCA concrete [39,46,100]. Over a period from 90 to 365 days, there was no statistically significant decrease in tensile strength. Guo et al. [46], in contrast, noted that RCA concretes have higher tensile strength than NCA concretes. Therefore, more research is needed to investigate the effect of RCA on concrete cracking toughness.

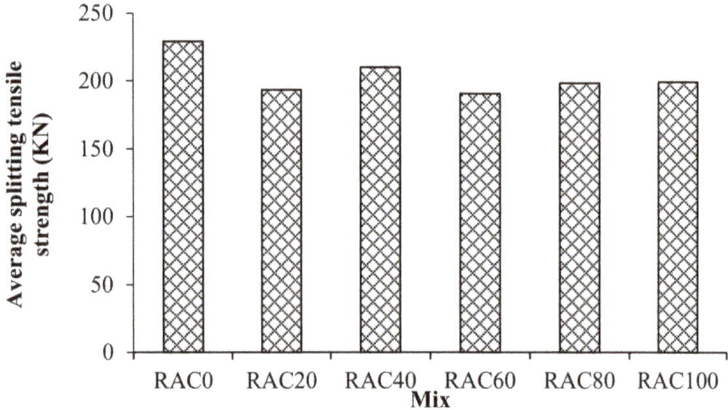

Figure 5. Splitting tensile strength versus level of replacement of NCA by RCA [124]. Reprinted with permission from Elsevier [124].

Typically, as shown in Figure 6, the flexural strength of RCA concrete is less than that of NCA concrete [124,131–133]. However, the flexural strength of 3 day RCA concrete was higher than that of NCA concrete, but at 28 days the strength was lower according to [131]. In their studies, NCA concrete gradually increased in strength and had greater flexural strength than RCA concrete at a later age. It is noted that RCA has never had a noticeable negative effect on the flexural strength of concrete [133]. However, given sufficient strength, RCA concretes can be produced for various purposes, sometimes even with 100% NCA replacements.

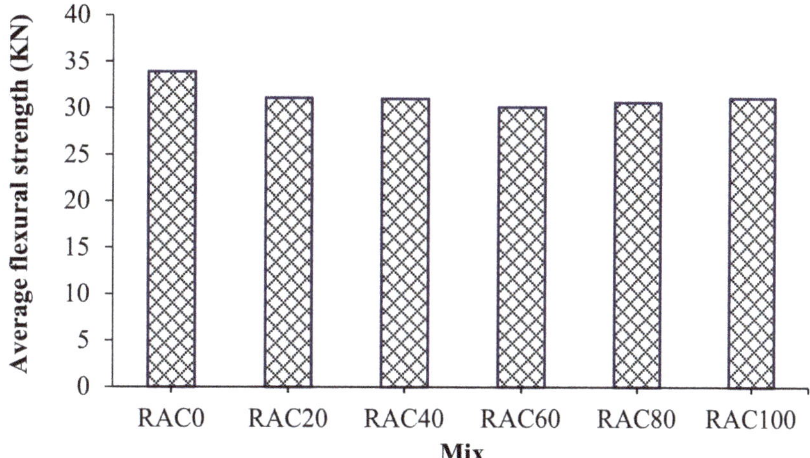

Figure 6. Flexural strength versus level of replacement of NCA by RCA [124]. Reprinted with permission from Elsevier [124].

7.3. Bond Strength and Impact Strengths

The bond strength of concrete is an indicator of the interrelated properties of pastes and aggregates. Rough RCA surfaces provide better adhesion than pure concrete aggregates. To test the bonds between reinforcement and concrete, which included 100%, 50% and 0% RCA, in [134], cylindrical specimens 150 mm × Ø100 with fixed soft and ribbed reinforcement were used (diameter = 12 mm and embedment length = 150 mm). The data obtained showed that RCA connections between reinforcement and concrete are not strongly influenced by RCA inclusions in concrete. It is, however, showed that the bond strength of NCA concrete was 9–19% higher than that of RCA concrete [135]. These conflicting results mean more research is needed to investigate the effect of RCA on concrete bond strength.

The impact of RCA on the hardened concrete performance can be significant or negligible depending on their physical characteristics, grades, content, types and sources [136]. The performance of RCA hardened concrete declines with the NCA substitution rate due to reclaimed concrete aggregates [3]. Without significant effect on the properties of hardened concrete, up to 30 wt. % natural filler can be replaced with RCA. As noted in the available literature, various changes in the performance of RCA hardened concrete are presented in Table 4.

7.4. Elasticity Modulus

The elasticity modulus of concrete is increased by an aggregate with a higher elasticity modulus. As shown in Figure 7, the elasticity modulus of concrete thus decreases with increasing RCA content in the concrete [129,136,137]. As a rule, the elasticity modulus of RCA concrete is 10–33% less than that of NCA concrete. It is demonstrated that the use of 30% RCA in concretes led to a decrease in the elasticity modulus by about 15% [129].

However, compared to NCA concrete, RCA concrete's elasticity modulus can be as low as 46%. The decrease in the modulus of elasticity of concrete is due to the fact that RCA generally have a lower elasticity modulus compared to NCA. In addition, the decrease in the elasticity modulus of concrete is associated with an improved total content of mortars (recycled and new), which have lower elastic moduli than most concretes based on natural aggregates.

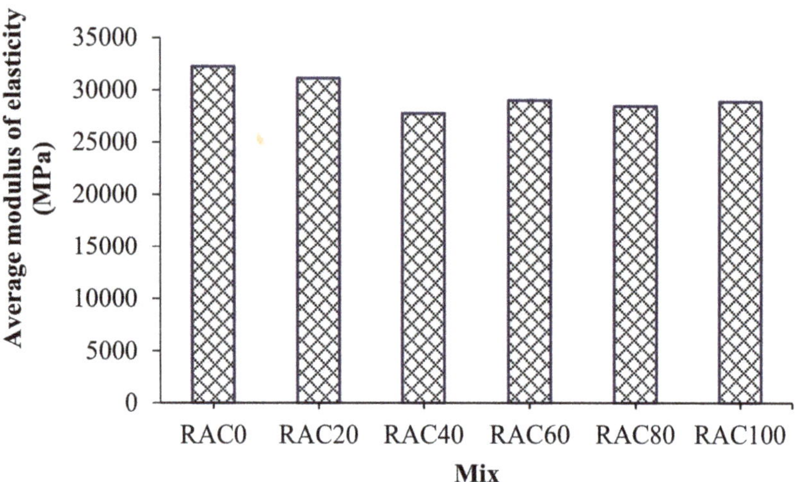

Figure 7. Modulus of elasticity versus level of replacement of NCA by RCA [124]. Reprinted with permission from Elsevier [124].

7.5. Creep and Thermal Expansion

Limited research has been done on the creep of RCA concrete. As a rule, the creep of RCA concrete is greater than the creep of NCA concrete [76,86,87]. This is because creep depends on the paste content, which can be 51% higher in RCA concretes.

Thermal expansion coefficients mainly depend on the content and types of aggregates. It is noted that the coefficients of thermal expansion of RCA concrete are usually 10–30% higher than that of NCA concrete [138]. However, further studies of the effect of RCA on the thermal expansion behavior of concrete are required to confirm these results.

7.6. Drying Shrinkage

Drying shrinkage commonly depends on the ratio of water to cement and paste content and is controlled by the aggregate particles. In some works, it was found that shrinkage is 20–50% higher than that of NCA concretes, because RCA concretes have a high paste content [45,86,87,139]. On the other hand, several studies have reported relatively lower RCA drying shrinkage values. It is reported low shrinkage strain at different curing ages of concrete when replacing 30% virgin concrete aggregate with recycled concrete aggregate, as shown in Figure 8 [140]. Conflicting results suggest that more research is needed to investigate the effects of RCA on concrete shrinkage when drying.

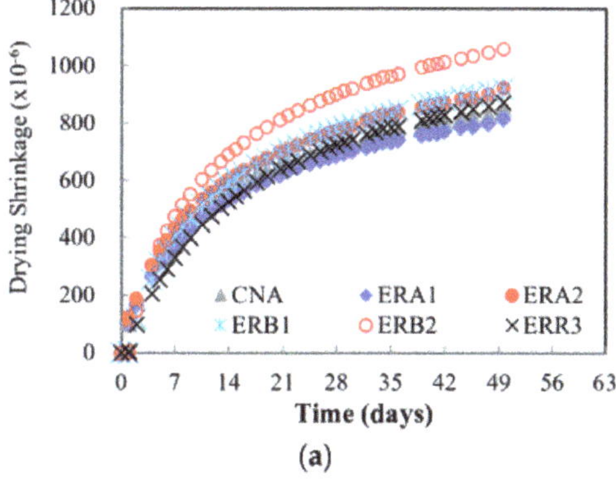

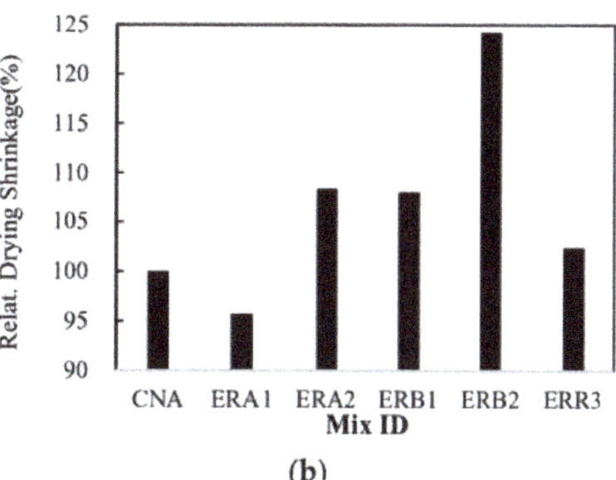

Figure 8. (**a**) Drying shrinkage versus time of exposure, and (**b**) drying shrinkage versus type of mix (ID) [140]. Reprinted with permission from MDPI [140].

8. Durability and Functional Properties

Concrete's ability to resist abrasion, chemical attack, weathering, and other adverse operating conditions is durability. Even when RCA is constructed from concretes with strength problems, RCA concretes can be significantly durable, provided the quality is maintained during construction and the mix ratio is correct [86,87,118,141]. The various strength characteristics of RCA concrete are described below.

8.1. Permeability

As shown in Figure 9, the porosity of the aggregate was less than that of the control concretes with aggregates from virgin concrete [55,142–144]. It is also reported that for all concretes the total porosity was increased at 50% RCA [143]. However, at 100% NCA substitution, class 20 concrete had a slightly lower porosity than traditional class 20 concrete. The porosity of RCA concrete can be 10–30% higher than that of NCA concrete, depending on the strength class according to [142]. These characteristics of RCA concrete are due to differences in the composition of concrete mixes such as RCA and cement content, aggregate amount and percentage of pozzolan.

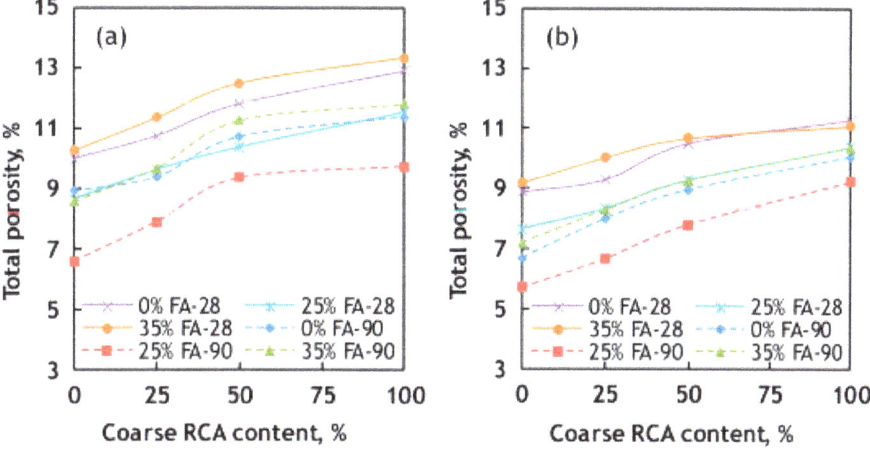

Figure 9. Total porosity versus level of replacement of NCA by RCA, with (**a**) Water/cement (w/c) ratio of 0.55; (**b**) w/c ratio of 0.45 [144]. Reprinted with permission from Elsevier [144].

The permeability of concrete depends both on the permeability of the concrete matrix (binder and cement paste) and on the absorptive capacity of the aggregate. In addition, the permeability of concrete is influenced by pore continuity, distribution, size, and porosity. The coefficients of intrinsic gas permeability of RAC are shown in Figure 10. The gas permeability of RAC decreased significantly with the carbonated RCAs; it is also showed an increase on the carbonation pressure from 0.1 Bar to 5.0 Bar still helped to reduce the gas permeability of RAC with new RCAs [72].

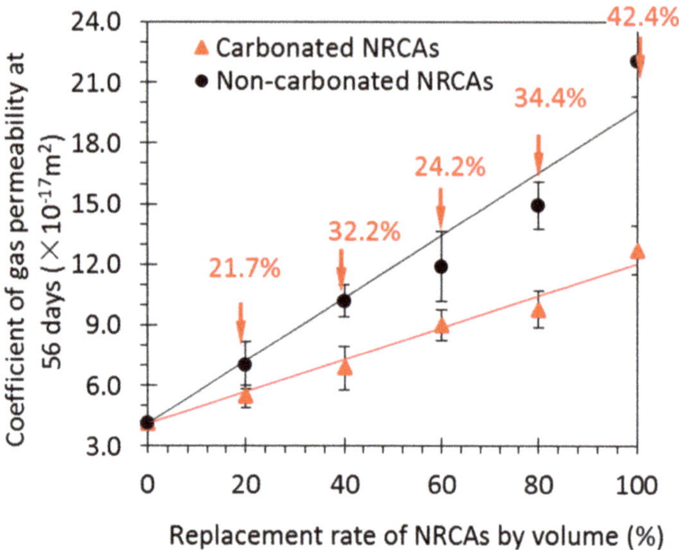

Figure 10. Gas permeability versus curing age [72]. Reprinted with permission from Elsevier [72].

8.2. Water Absorption

The water absorption of RCA concrete is expected to be greater than that of NCA one. This is due to the very high air permeability and water absorption of RCA. As shown in Figure 11, water absorption of concrete increased with increasing RCA content [145]. The water absorption of RCA concrete is 0–40% higher than that of NCA concrete, depending on the strength grades according to [146]. The author did not explain the reason for the lower water absorption at 20% RCA content, but suggested further research.

8.3. Sulfate and Chloride Resistance

Limited research has shown that RCA concrete's sulfate resistance is about the same or slightly lower than that of natural concrete aggregates [147–149]. In general, the sulfate resistance of RCA concrete can be improved by using silica fumes, fly ash and fine-grained blast-furnace slags and using the correct mixing ratio. Good quality control of the construction and correct dosing of mixes can reduce the corrosion rate of steel reinforcement in RCA concretes. It is applied polarization approaches to determine the corrosion behavior of steel anchored in RCA concretes [148]. The researchers reported that the corrosion rate never depended on aggregates for lower chloride levels (0.50%) and binder type. Compared to NCA concrete, the researchers also reported that RCA concrete with 65% fine blast furnace slag and 30% pulverized fuel ash was more successful in mitigating corrosion reactions at higher chloride levels.

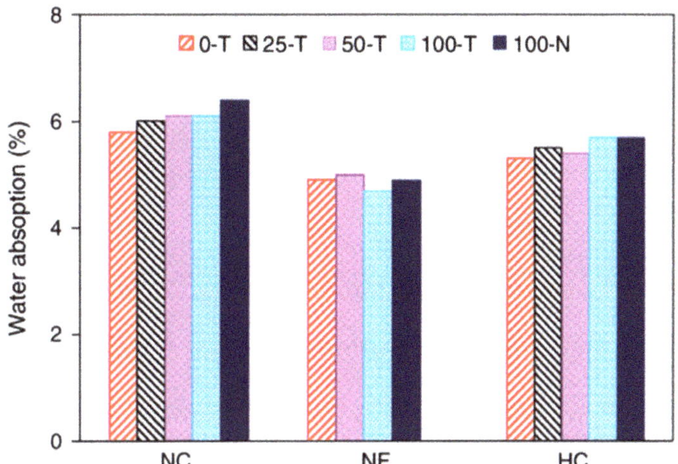

Figure 11. Water absorption versus level of replacement of NCA by RCA [145]. Reprinted with permission from Ozbakkaloglu et al. [145].

It is noted that the high chloride penetration rates in concrete made with and without RCA were very similar [139]. The use of 100% RCA in concretes reduces their ability to resist the penetration of chloride ions by about 30% compared to NCA concrete [143,150]. However, the chloride penetration resistance of RCA concrete can be significantly improved by increasing additional binders such as crushed granular blast furnace slag and fly ash. It is disclosed that, in comparison with fly ash, crushed granular blast furnace slag reduced the diffusion coefficient of chloride in concrete by about 80–200% [147,150,151] (See Figures 12 and 13).

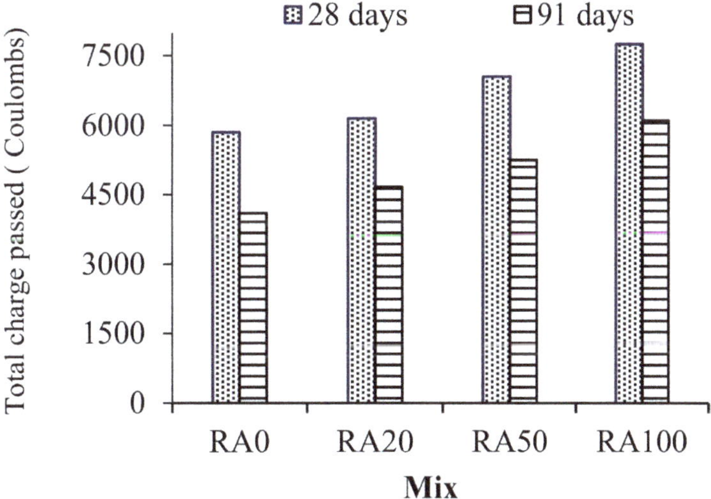

Figure 12. Chloride-ion penetration at 28 days and 90 days [151]. Reprinted with permission from Silva et al. [151].

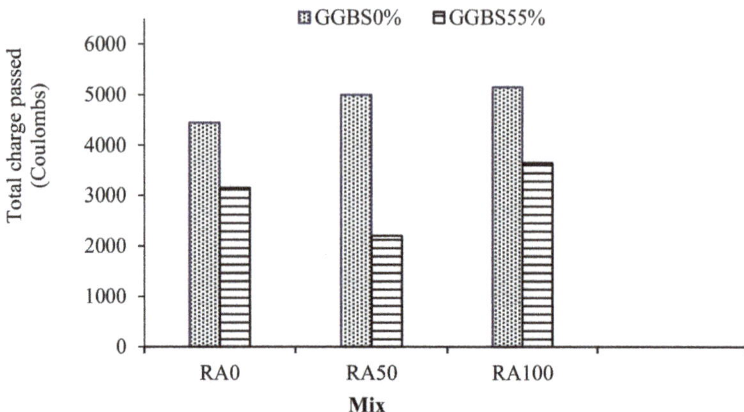

Figure 13. Chloride-ion penetration with 55% of GGBS at 28 days [151]. Reprinted with permission from Silva et al. [151].

8.4. Carbonation Resistance

With regard to concrete carbonation, existing research points to conflicting effects of RCA. The researchers [102] reported that carbonation depth decreases with increasing RCA content, resulting in better performance except for 100% replacement. RCA concretes require a higher cement content to achieve identical strength to NCA concrete at the same water to cement ratio [102]. The higher the cement content, the higher the alkali reserves, which affect the depth of concrete carbonization. In contrast, it is found that at the same water-to-cement ratio, RCA concrete has a greater carbonization depth than NCA concretes [152,153]. In addition, the carbonization rate of RCA concrete is four times that of NCA concrete. Increased carbonation can increase the risk of reinforcing steel corrosion in RCA concretes. Increased concrete pavement, appropriate additional cementitious materials, corrects curing, and a lower water to cement ratio can counterbalance such risks. On the other hand, previous studies have reported that concrete containing the RA requires to more supplementary materials with high fineness to act as pore fillers. Thus, this will reduce the diffusion of carbon dioxide and slow down the speed of carbonation [154,155] (See Figures 14 and 15). However, more research is needed to analyze the effect of RCA on concrete carbonation resistance.

8.5. Resistance to Alkalis and Acid

Resistance of RCA concrete to alkali-silica reactions depends on the RCA sources [12,139,156]. It is noted that RCA obtained from old concretes containing alkalis affects the resistance of newly created RCA concretes to alkali-silica reactions [12]. The results of the researcher showed that the amount of new concretes made using RCA increased disproportionately due to alkaline-silica reactions. The use of low-lime grade F fly ash can significantly minimize expansion due to alkaline silica reactions in RCA concretes. Ziyi Peng and others suggested a method in their study to improve RCA quality by absorbing silica fume slurry into residual mortar for RCA [157] (See Figures 16 and 17).

There is insufficient information available to draw specific conclusions about the alkali-silica reactivity of RCA concrete. Several studies recommend the use of SCMs to dilute AAR in recycled and RCI concrete [157].

It is noted that the acid resistance of RCA concrete is similar or slightly lower than that of NCA concrete [86]. Scientists immersed concrete samples in test baths containing sulfuric acid at pH = 2 to test the acid degradation rate of RCA concrete. Compared to NCA concrete, the penetration of acids into RCA concrete was slightly higher. This may be due to the correlation of acid penetration with high of the porosity and absorption of the RAC and residual mortar compared to that of natural aggregates in concrete [158]. This caused a higher penetration of the acid ions into the concrete, thus caused the dissolution of calcium hydroxide and the destruction of the gel calcification (CSH) in the concrete 8However, researchers have shown that the acid attack resistance of RCA concrete can be improved by adding cement supplementary materials. Kazmi et al. demonstrate the possibility of treating RAC by immersion in the lime with accelerated carbonization and immersion in acetic acid with friction techniques [92] (See Figure 18). This method can be used to improve the acid resistance of RAC-containing concrete in chemically aggressive environments. More studies of RCA concrete under acidic conditions are needed to test their durability due to the small amount of research.

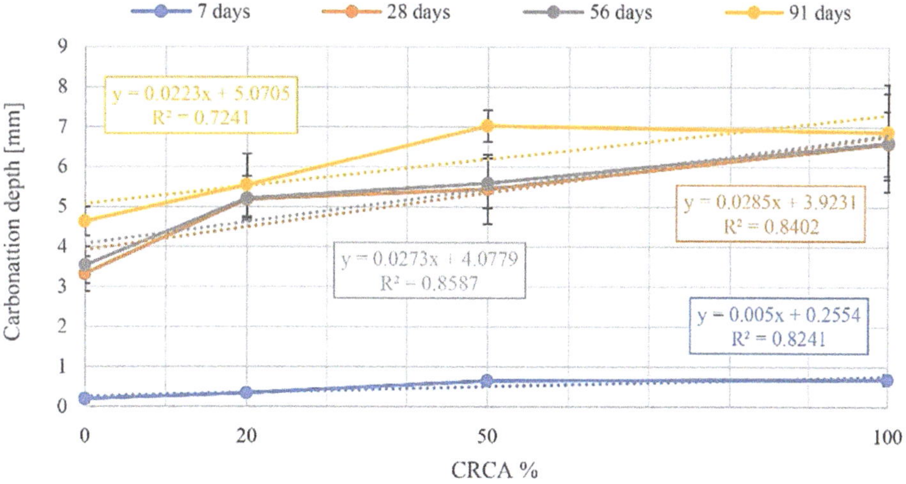

Figure 14. Relation between carbonation depth and coarse recycled concrete aggregates (%) [155]. Reprinted with permission from Elsevier [155].

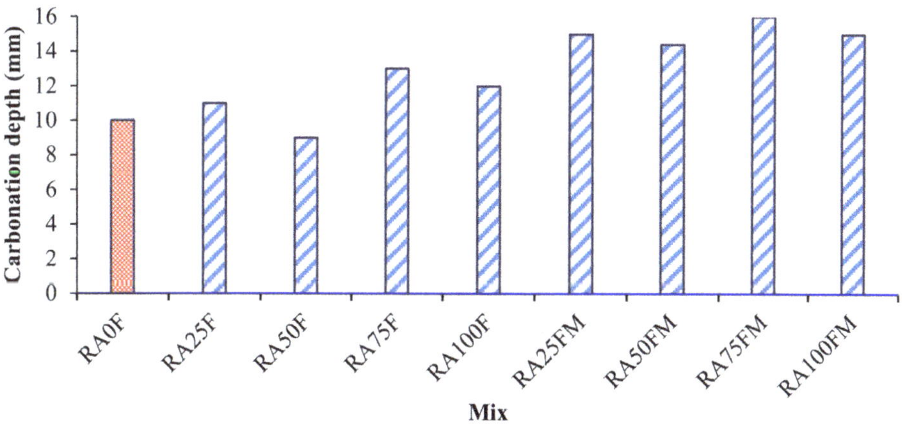

Figure 15. Effect of addition of RCA, FA and MK on concrete carbonation depths an exposure period of 4 weeks [154]. RA0F: without RCA with 30% of FA, RA25F: 25% of RCA with 30%of FA. RA50F: 50% of RCA with 30%FA, RA100F: 100% of RCA with 30%FA, RA25FM: 25% of RCA with 30% of FA & 10% of MK, RA50FM: 50% of RCA with 30% of FA & 10% of MK, RA100FM:100% of RCA with 30% of FA & 10% of MK. Reprinted with permission from Elsevier [154].

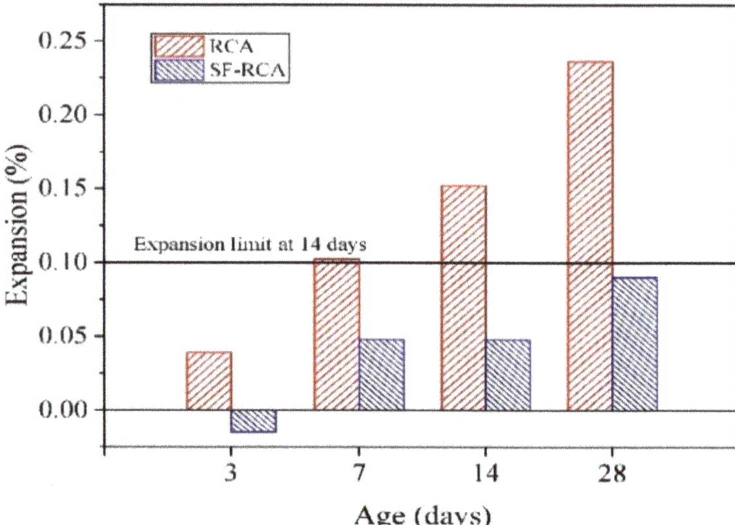

Figure 16. Mitigation of AAR expansion of RCA by sucking fine pozzolanic material into residual mortar [157]. Reprinted with permission from Elsevier [157].

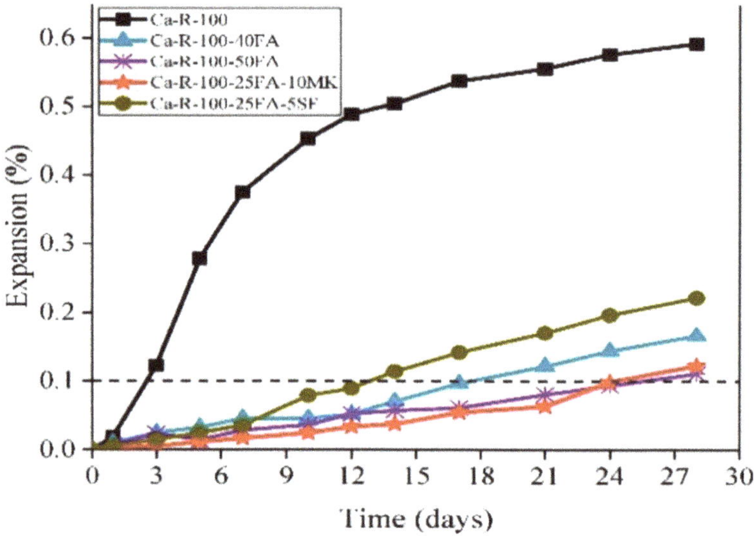

Figure 17. Expansions of RCA with various cementitious material blends [157]. Reprinted with permission from Elsevier [157].

8.6. Freeze-Thaw Resistance

Various researchers have noted that RCA concrete has sufficient resistance to freeze-thaw cycles [141,159,160]. There is evidence that repeated recycling of RCA concrete further improves frost resistance [161]. For RCA concretes, some researchers have reported similar or slightly reduced frost resistance compared to NCA concretes 10 In addition, concretes made from dry and water-saturated RCA have a lower resistance to freeze-thaw. RCA Concretes prepared using water-saturated concretes have shown better results due to improved anchoring at joints between pastes and aggregates. Previous studies showed that the effect of freeze-thaw was greater in mixtures with higher w/b ratio and higher RCA content. However, the effect of w/b ratio on freeze-thaw through percentage of weight change is more significant. The increase in the w/b ratio increased the number and size of the capillary pores as well as the freeze water in the cement paste, causing mainly the extended internal pressure during freezing. In addition, a strong relationship was observed between freeze-thaw damage and water absorption of mixtures [159,162] (See Figures 19–21). However, more research is needed to confirm the effect of RCA on concrete resistance to freezing and thawing.

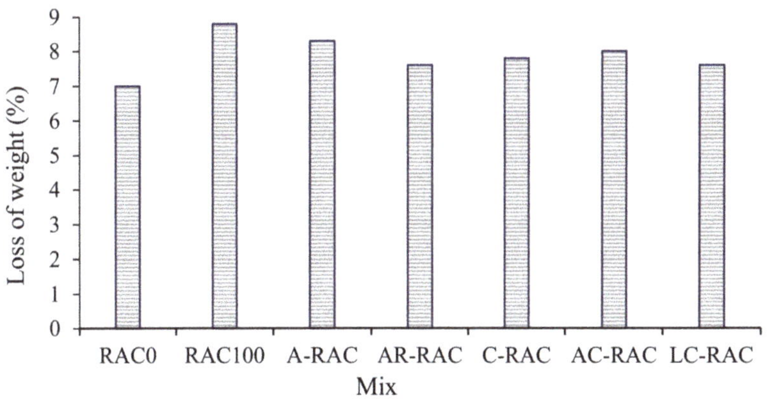

Porosity of CA (%)	NA	RA	A-RA	AR-RA	C-RA	AC-RA	LC-RA
	3.19	19.46	15.94	15.31	14.39	15.79	11.65

Figure 18. Loss in weight of NAC and RAC having untreated and treated RA after acid immersion [92]. Reprinted with permission from Elsevier [92].

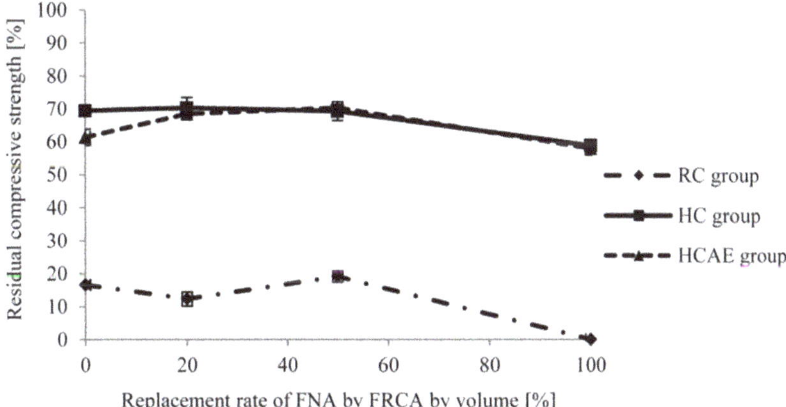

Figure 19. Residual compressive strengths of the mixes after 300 freeze-thaw cycles [159]. FNA: Fine aggregate, RC: Normal strength concrete, HC: High-strength concrete and HCAE: HC with air entraining agent. Reprinted with permission from Elsevier [159].

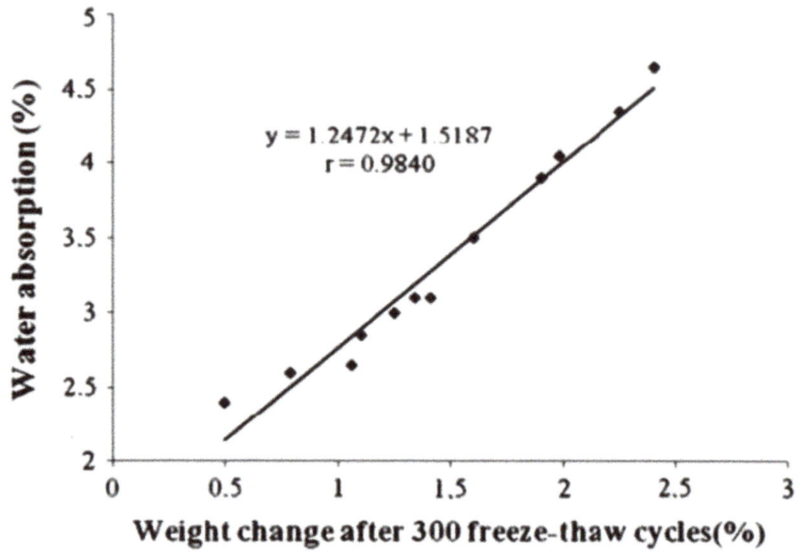

Figure 20. Relationship between water absorption and weight change percentages of after 300 freeze-thaw cycles [162]. Reprinted with permission from Elsevier [162].

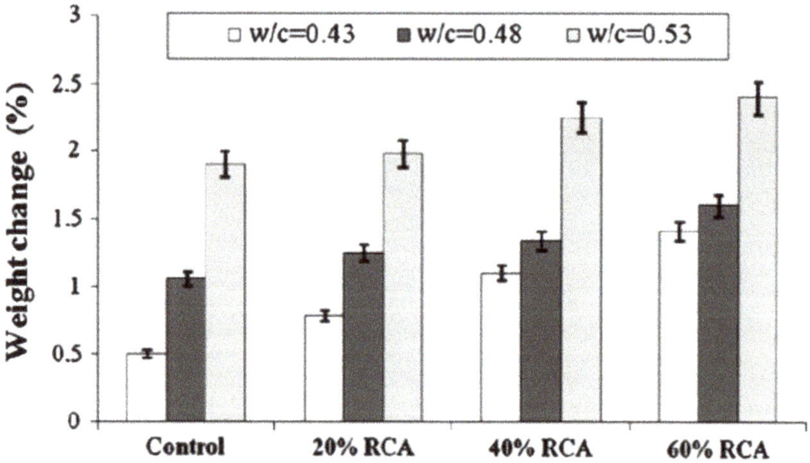

Figure 21. Weight change percentage after 300 freeze-thaw cycles [162]. Reprinted with permission from Elsevier [162].

9. Improvement Methods for RCA Concretes

9.1. RCA Quality Improvements and Adjusting the Ratio of Water and Cement

It is tested heat treatment methods to improve RCA quality [71]. It was noted that RCAs are rationally comparable to the traditional used aggregate removed from the river after heat treatment at 800 °C. It is found that the strength of concrete was adversely affected by unwashed RCA used in concrete mixes [79]. However, the reduction in strength was offset by the use of rinsed RCAs. It is reported that the correct adjustment of the water balance of cement mixes for concrete mixes can increase the strength of RCA concrete [6]. It is have proposed a higher cement content in RCA concrete and a lower water to cement ratio than NCA concrete to achieve similar compressive strength [11]. It is similarly noted that lowering the water-to-cement ratio to certain levels was very beneficial for RCA concretes to establish an equivalent thaw and freeze resistance to those NCA concretes [43].

9.2. Pozzolanic Materials Integration and Soaking of RCA in Pozzolanic Liquids

The strength and durability of RCA concrete can be increased through the use of appropriate pozzolanic substances [87,98,99]. It is showed that the use of 65% fine granular blast furnace slag and 30% pulverized fly ash improved the RCA compressive strength of concrete to control levels of concrete pouring with clean granite gravel [98]. In addition, fine-grained blast-furnace slags and pulverized fuel ash have been successful in increasing the resistance to chloride ion penetration into RCA concrete. Also, silica fumes have been found to significantly improve chloride permeation resistance for RCA concrete. The RCA concrete strengths may be boosted by either allowing reclaimed aggregates to soak up parts of mixing water without or with pozzolanic liquids during mixings or soaking the RCAs in mixes of pozzolanic liquids such as colloidal silicas or water before mixing of concretes. The micro-cracks in RCAs are expected to be filled up by the absorbed pozzolanic liquids or absorbed water with cement gels during pozzolanic reactions or cement hydrations. Therefore, the RCA concrete strengths can be increased.

9.3. Uses of New Mixing and Curing Techniques

It is used two-phase mixing approaches to obtain better quality RCA concrete (Table 5). These scientists used reclaimed aggregates treated with pozzolanic powders to improve the properties of RCA concrete [99]. In addition, the authors [87] have developed a two-phase mixing technique to ensure high-quality use of RCA concretes. These researchers found that, compared to NCA concrete, 100% replacement of virgin concrete aggregate is possible through their mixing approaches to create RCA concrete with suitable characteristics, although the optimal scenario is with 20% NCA replacement. The results of the determination of strength and slump showed that the new mixing methods significantly contributed to the achievement of high values of flexural and compressive strengths, as well as better workability. In addition, using scanning electron microscopes, interfacial transition zones between RCA concrete surfaces were realized. The scanning electron microscopes results established that the new mixing techniques promoted dense microstructures. Internal leaks in additives can be minimized by mixing techniques. The use of long-term cure in a humid environment is another approach to improving the performance of RCA concrete [54,100,102]. One of the most widely used approaches to reducing the carbonization rate of RCA concrete is the long-term cure. In RCA concretes, the carbonization depth is almost half when the concrete is cured with water. It is also reported that the common assumption that RCA is more sensitive to different curing conditions and therefore removes another obstacle to its massive use. [163]. Furthermore, Table 6 shows the ratio of RCA replacement criteria for making structural grades [34].

9.4. Microstructure of RAC

Recycled aggregate used to produce concrete differs from natural aggregate in that it contains old mortar attached to the surface of the aggregate. The RCA recycling process often requires crushing of the concrete parts, so the microstructure of the RCA is exposed to many defects such as micro-cracks, porosity, as well as weakening of the ITZ. Damage to the aggregate microstructure will damage the engineering properties and durability of the concrete containing the recycled aggregate.

In fact, all types of aggregate produced of broken concrete are a composite material consisting of natural aggregate and cement mortar (See Figures 22–26). These broken parts use to completely or partially replace natural aggregates after crushing to small particles [164]. The RCA consists of NA and an old cement mortar, so the RAC consists of three ITZ regions, as follows [164]. First, between NA and old cement matrix, Second, between NA and the new cement matrix, third, between new and old cement matrix.

Several studies focus on studying the transition zone and its relationship to concrete degradation and weakness. The degradation of concrete often depends on the entity of a filtering path across the interface (See Figures 22–26). The transition region, i.e., the interface between the particle assembly and the bulk cement matrix [165]. To reduce this problem, suggestions are made regarding the durability adopted in National Concrete Standards [165–167] that include limits on the minimum cement content as well as on the maximum water-to-cement ratio used for structural concrete. Further, it is also reported that the incorporation of pozzolanic materials, such as silica fume, fly ash and slag, with different mixing methods and accelerated carbonation can improve the microstructure of RCA [168].

Table 6. Ratio of RCA replacement criteria for making structural grades [34].

Country	Standard	Exposure Aggregate or Class	Type 1 (%)		Type 2 (%)	Source/Codes or Standards
Germany	DAfStb	XC1 to XC4 [coarse]	≤35		≤45	DAfStb
		X0 [coarse]	≤35		≤45	
		XA1 [coarse]	≤25		≤25	
		XF1 and XF3 [coarse]	≤25		≤35	
		[Fine]		Not allowed		
			Type 1	Type 2	Type 3	
International	RILEM	Fine (<4 mm)		Not allowed		RILEM, 1994
		Coarse (≥4 mm)	≤100%	≤100%	≤20%	
Netherlands	NEN 5950:1995 (VBT 1995) (NEN 5950:1995 nl	Fine		≤20%		MEN 5950
		Coarse		≤20%		
Belgium	CRIC	Fine		Allowed with restriction		TRA 550, 2004
		Coarse		≤100%		
Denmark	DS 481	Fine		≤20%		DS 481, 1998
		Coarse		≤100%		
UK	BS EN 12620:2002 + A1 (BS)	Fine		Not allowed		BC, 2002
		Coarse		≤20%		

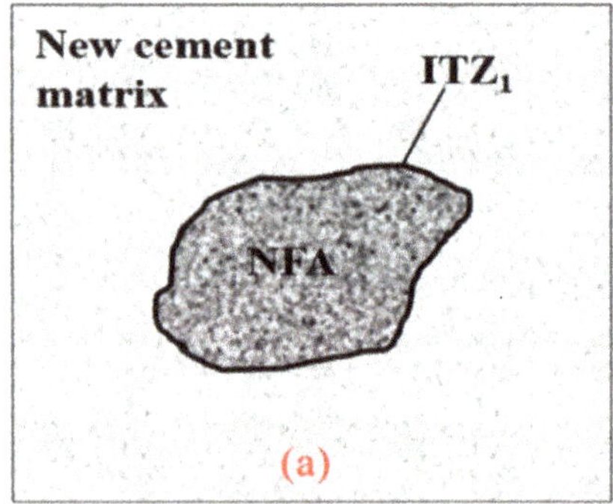

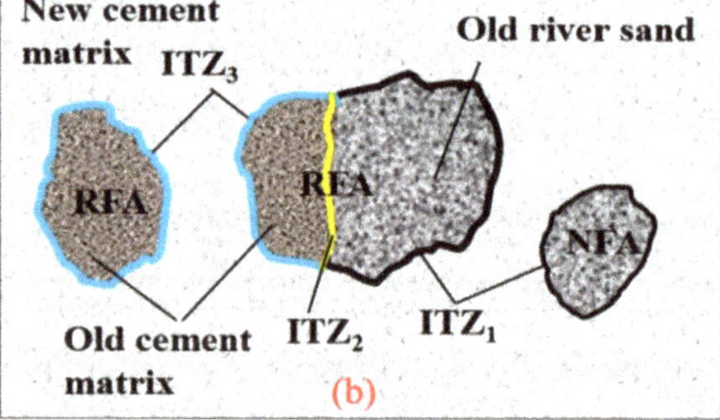

Figure 22. The different types of ITZs contained in concrete prepared with different fine aggregate: (**a**) NFA (quartz or river sand), and (**b**) in concrete with NFA partially replaced by RFA [167]. Reprinted with permission from Elsevier [167].

Figure 23. A low magnification image showing adhered cement in a RAC [87]. Reprinted with permission from Elsevier [87].

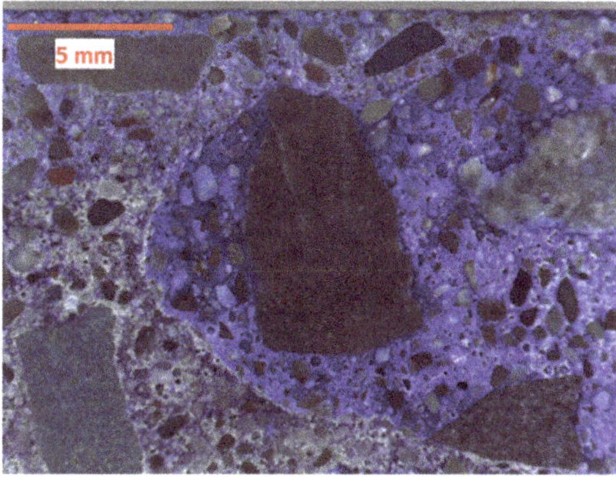

Figure 24. SEM images for polished section in concrete with RCA [87]. Reprinted with permission from Elsevier [87].

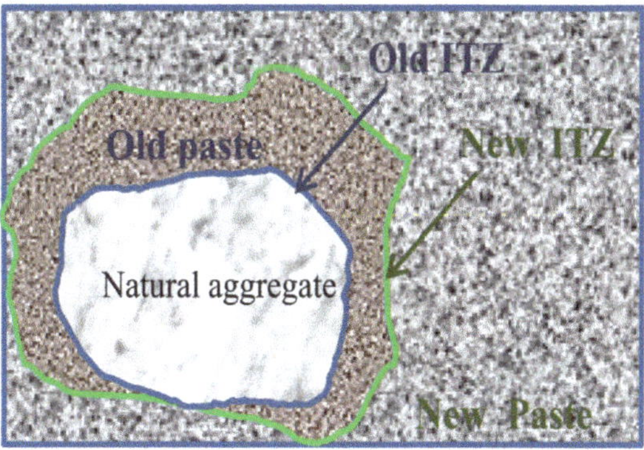

Figure 25. Sectional view of RCA in concrete including the old and new ITZ [169]. Reprinted with permission from [169].

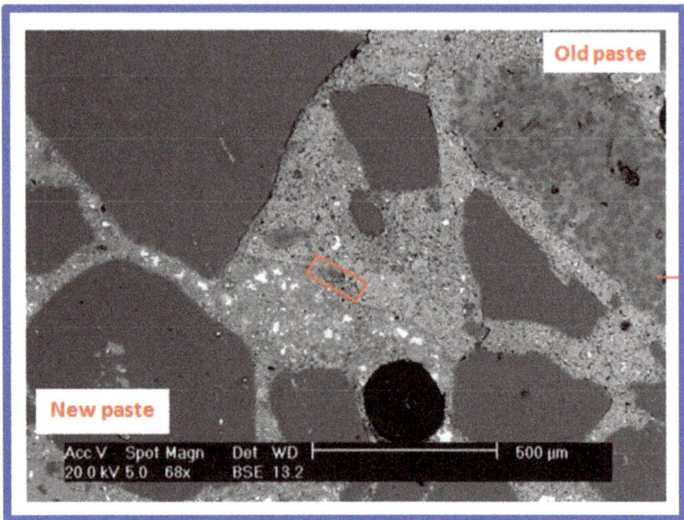

Figure 26. SEM images of interfacial transition zones in RAC [169]. Reprinted with permission from Elsevier [169].

On the other hand, the micro-pores in the mortar attached to the old rubble can contribute to retaining additional quantities of water, which may help to provide a self-treatment that leads to the promotion of filling the pores with moisturizing products (gel) [170]. Thus RA behaves as an internal curing agent, which also improves the ITZ between particles of aggregate and cement paste afterwards reduces the size of the pores within the microstructure of the concrete [171,172] (See Figure 27). This improvement is caused by a better inter-bond between the cement paste and the aggregate particles providing cross-linking sites for the cement paste resulting in better cement wetting and denser ITZ more uniform. The fracture value of aggregate, size, and porosity of ITZ has paradoxical effects on concrete's transport properties.

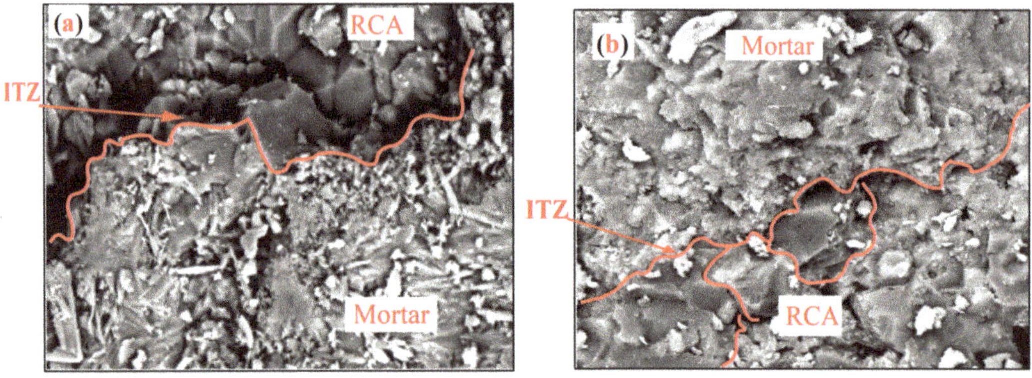

Figure 27. The ITZ between RCA and New Mortar Mixes; (**a**) 100% natural aggregates and (**b**) 75% natural aggregates + 25% recycled aggregates [171]. Reprinted with permission from Sadek and El-Attar [171].

9.5. Treatment of RCA

Due to the varying performance of RCA and its negative impact on most of the properties and concrete durability, several studies have suggested methods for treating RCAs in order to improve their properties and promote its application in the concrete industry safely [172]. The treatment methods depend on several factors, including the source, type of aggregate, and applications of RCA-containing concrete. The studies have suggested different approaches.

- Accelerated carbonization This method has been suggested by researchers [173,174]. This method is based on obtaining carbonated recycled aggregates. For this purpose, first, the RCA was stored at 25 °C and RH of 50% for 72 h days in order to provide the RA with a basic RH between 40%–70% that accommodates the accelerated carbonation [174]. After that, the RCA was placed in a sealed chamber under pressure of −1.0 bar. After the pressure was stabilized, carbon dioxide was applied at a concentration of 100% with a pressure of +0.8 bar for 24 h [175].
- Immersion in saturated lime water In order to improve the RCA performance through accelerated carbonization, the reactive components in the old adhesive slurries of RA play a vital role [174]. So, this method recommends submerging the RCA in lime-saturated water for 24 h, and then placing it in a room at 25 °C and 50% relative humidity for 72 h. This step aims to introduce some calcium into the pores of the old mortar attached to the RCA. After this step was completed, the carbonization process for the RCA treated by immersion in saturated lime water was carried out. The same accelerated carbonization technique described above was used in order to obtain recycled aggregates carbonized with lime [175].
- Treatment with acetic acid This method uses acetic acid to remove the old mortar attached to the RCA. First, the RCAs are washed to remove the dirt and then directly immersed in a 3% acetic acid solution for 24 h. Next, the RCAs are washed again and dried to obtain recycled aggregates treated with acetic acid. After that, mechanical rubbing was performed for 5 minutes using an empty concrete mixer, based on the previous study [175,176].
- Silica fume impregnation In this method, a solution of raw silica fume was prepared 1 kg per 10 liters of water, and a superplasticizer (1% of the mass of silica fume) was added to help disperse the silica fume particles and ensure their distribution on the surface of the RAC [176]. Then, it was placed in a 24 h drying oven, followed by cooling to room temperature and soaking in a silica fume solution for 24 h. In the final stage, the saturated RAC was dried again in the oven to ensure proper penetration of silica fume particles into the surface of the RAC [53].
- Ultrasonic cleaning The crushed concrete was cleaned in an ultrasound bath to remove residual slurry on the surface of the untreated aggregate. The RAC was immersed in an ultrasound bath and treated for 10 minutes, after which the water was replaced with clean water, and the debris was cleaned for an additional 10 minutes. These operations are repeated several times until a clean RAC was obtained [53].
- Pre-soaking treatment methods This method is based on soaking the RCA in an acidic environment at about 20 °C for 24 h, followed by washing the RCA with distilled water to remove the acidic solvents. After that, the RCA was soaked in water for 24 h to ensure the disposal of acids or sticky mortar residues. In this method, three acidic solvents with a concentration of 0.1 mol can be used, which is an acidic environment suitable for removing the old mortar attached to the RCA, as it will not reduce the quality of the RCA. The acids applied in this method are phosphoric acid (H_3PO_4), sulfuric acid (H_2SO_4), or hydrochloric acid (HCl) [177].

10. Practical, Economical, and Environmental Issues of RCA Concretes

Obviously, for the economic feasibility of RCA production, ready-mix concrete plants can be used to obtain production scale for production. Economic impact assessment is effective when it is considered that the action of the case does not require additional payment. He looks at the recycling of waste concrete in ready-mixed concrete plants as infrastructure and construction resources. The prototype developed is based on information obtained from various industrial shops as well as concrete processing shops, which are used as a scenario study to estimate the additional commercial price of a product. The current problems facing Thailand, where there are so many landfills for concrete waste as a result of destruction plans, and inappropriate concrete is being disposed of, will be addressed using RCA production [72]. The growth in RCA replacement is primarily associated with an increase in additional unit costs of resources. In particular, the replacement of the cement content is increased as a result of the increase in the RCA replacement ratio. The increase in RCA cost is associated with additional quality control and pre-processing costs. However, upfront costs can be reduced by going directly to a high quality concrete plant.

Compared to a recycling approach where construction waste is used in recycling plants to obtain new materials, the traditional costs are very high [178]. The first stage of the recycling process includes construction waste that is disposed of at low cost recycling facilities. Additional costs include energy consumption and transportation costs. The second stage is stockpiling, in which lone workers are employed at a rate of $18 an hour. The third step is sorting with machines, such as a shredder and excavator. Also accrued are additional costs for equipment maintenance, labor, fuel, fixed indirect costs, working capital, operating costs, equipment costs, and capital costs [75]. Crushing processes in the fourth phase include magnetic separation, primary crushing, and secondary crushing. This includes fuel costs, fixed indirect costs, operating costs, working capital, and capital costs. With magnetic separation, the process also includes sorting steel scrap and sells for about $100 per ton. The casting phase includes manual removal processes in which pieces of paper, wood, and plastic are separated from the grounded concrete. In this process, the wages are about $18 an hour. The sixth stage involves sitting in the air, watching, and doing laundry, which is identical to the traditional approach. Recycled fuel and water are used to deposit dust particles at this stage [178]. The final stages of the refining process include finished products that sell for between $14 and $22 per ton.

Thus, compared to the traditional approach, RCA has the most cost-effective gain [75]. The financial benefits of recycling concrete lead to longer-term benefits than using natural concrete. Sensitivity analysis is performed, taking into account the main parameter due to the expected uncertainties with the main variables used in the model [153]. The purpose of the analysis is to determine the main variable impact of modification on one unit of output. The main factors to be taken into account are transport distance, additional costs for RCA pretreatment, additional cement required to produce RCA, and RCA composition [138]. Due to depleted natural resources, crushed concrete for demolition is produced in large quantities and must be used in an environmentally friendly and economical manner. Deterministic values are defined as means because the parameters have normal distribution characteristics. The range of input variables can be estimated to determine upper and lower bounds. The RCA price cap range is determined by breaking down changes in the value of a product by changes in quantities [148].

11. Comments and Further Researches

The literature in this paper is summarized to provide comprehensive insights into the potential applications of RCA to produce green and sustainable concrete composites that contribute to environmentally friendly buildings. Therefore, differences in the current state of knowledge between RCAs and NCAs are highlighted, and some suggestions for future research are provided. In addition, this study also contributed to uncovering deficiencies in previous research and studies, which requires further research and investigation into the following points:

1. Estimating the cost of recycled aggregates based on supply and demand, as well as, to many different challenges that must be considered.
2. Utilizing recycled aggregates to obtain a regime of self-curing (internal curing) for concrete, especially for aggregates that have a high pore ratio.

3. The role of chemical components of cement paste adhering on aggregates in the stability and soundness of concrete.
4. Study the role of methods for treating recycled aggregates in order to improve the transport properties of concrete.
5. More understanding of the relationship between microstructures and properties of concrete containing RCA.

12. Conclusions

Reclaimed destroyed concrete has a significant potential for value-added applications to maximize environmental and economic benefits. Significant savings can be achieved by making RCA a valuable resource in the new generation of concrete. In the current study, a critical analysis of a recycled concrete aggregate for the production of heavy-duty concrete structures was carried out. The following conclusions were drawn from this article:

1. As a substitute for natural aggregate, RCA are useful for the production of concretes of standard strength and properties. The main problem with using RCA in new concretes, however, is their incompatible qualities, especially when they come from the demolition of old concrete buildings.
2. By complying with standard virgin aggregate specifications, RCA can be effectively used in new concrete. However, RCA requires new guidelines and specifications.
3. The physical characteristics of RCA strongly influence the properties of hardened and fresh concrete. The magnitude of crushing, aggregate abrasion, and impact viscosity affect the strength characteristics of the concrete. In addition, the negative chemical characteristics of RCA can affect the durability of RCA concrete and, therefore, their performance in service.
4. Total 100% RCA can produce standard quality concrete. Concretes produced with RCA generally have about 81% of that of NCA concrete. Insufficiently dense transition zones between bulk cement pastes and RCA and unfavorable RCA properties cause a decrease in the strength of RCA concrete.
5. The use of crushed RCA in new concrete mixes requires careful research, since the reclaimed fines further reduce the strength of the concrete. Replacement of natural sand with RCAs is usually up to 20%.
6. The performance of RCA concrete can be improved by long-term curing, new mixing methods, the addition of pozzolanic substances, and changing the ratio of water to cement.
7. In high quality concretes such as self-compacting, high strength, and high performance concretes, RCA can be used with proper mix design and material selection.
8. Concretes having RCA can be properly designed and balanced to reduce the effect of RCA on its hardened and fresh strength and characteristics, regardless of the differences in performance between natural concrete aggregates and re-concrete aggregates. The chemical impurities of RCA must be reduced to advance its use in the concrete sector.

The following are the summaries of recommendations:

1. The idea of using existing production facilities to produce recycled concrete aggregate is misplaced as the construction manager may initiate excessive additional costs as a result of the specific conditions of the halls. Nanotechnology can be used to advance RCA manufacturing because of its enormous potential to improve production and quality in high-efficiency apprenticeship manufacturing. In addition, the use of nanotechnology is aimed at changing the operating mode in construction using nanomaterials. The examination describes several properties of RCA at the nanoscale using a variety of state of the art equipment. Therefore, it is worth focusing on nanosciences, which is critical when investigating the microstructure of RCA. In addition, the strength characteristics of RCA can be improved through the use of nanomaterials to obtain greater strength than conventional concretes.

2. Incorporating high performance pozzolans into recycled concrete aggregate to improve RCA performance is likely to increase its cost of use. However, an assessment of the cost-benefit of increasing the amount of fly ash shows that this is necessary due to the increase in strength. Poor quality concrete is one of the problems associated with using recycled resources.
3. In order to accelerate the implementation of the production of secondary concrete aggregate, its use should be included in government projects. The integration of advanced manufacturing technologies that improve quality is essential to reduce the cost of using recycled concrete aggregate. These technologies can be used through client funding and government, including incentives that can support this process. To support this idea, all workers must complete homeschooling programs to facilitate a process that is likely to improve attention to the environment.

Author Contributions: Conceptualization, N.M., R.F. and M.A.; methodology, N.M., R.F., M.A. and A.M.Z.; validation, N.M., R.F., M.A., A.M.Z., G.M., N.V., S.K., T.O. and Y.V.; resources, N.M., R.F., M.A., A.M.Z., G.M., N.V., S.K., T.O. and Y.V.; data curation, N.M., R.F., M.A., A.M.Z., G.M., N.V., S.K., T.O. and Y.V.; writing—original draft preparation, N.M., R.F., M.A. and A.M.Z.; writing-review and editing, N.M., R.F., M.A., A.M.Z., G.M., N.V., S.K., T.O. and Y.V.; supervision, M.A. and R.F.; project administration, R.F., M.A. and N.V.; funding acquisition, M.A., N.V., S.K. and Y.V. All authors have read and agreed to the published version of the manuscript.

Funding: No fund available.

Institutional Review Board Statement: Not applicable.

Informed Consent Statement: Not applicable.

Data Availability Statement: Data sharing not applicable.

Acknowledgments: The authors gratefully acknowledge the support given by Deanship of Scientific Research at Prince Sattam bin Abdulaziz University, Alkharj, Saudi Arabia; Moscow Automobile and Road Construction University, Moscow, Russia; and the Department of Civil Engineering, Faculty of Engineering and IT, Amran University, Yemen, for this research.

Conflicts of Interest: The authors declare no conflict of interest.

References

1. Fediuk, R.; Pak, A.; Kuzmin, D. Fine-Grained Concrete of Composite Binder. *IOP Conf. Ser. Mater. Sci. Eng.* **2017**, *262*, 012025. [CrossRef]
2. Lesovik, V.S.; Glagolev, E.S.; Popov, D.Y.; Lesovik, G.A.; Ageeva, M.S. Textile-reinforced concrete using composite binder based on new types of mineral raw materials. *IOP Conf. Ser. Mater. Sci. Eng.* **2018**, *327*, 032033. [CrossRef]
3. Klyuev, S.V.; Klyuev, A.V.; Khezhev, T.A.; Pukharenko, Y.V. High-strength fine-grained fiber concrete with combined reinforcement by fiber. *J. Eng. Appl. Sci.* **2018**, *13*, 6407–6412. [CrossRef]
4. Fediuk, R.S.; Lesovik, V.S.; Svintsov, A.P.; Mochalov, A.V.; Kulichkov, S.V.; Stoyushko, N.Y.; Gladkova, N.A.; Timokhin, R.A. Self-compacting concrete using pretreatmented rice husk ash. *Mag. Civ. Eng.* **2018**, *79*, 66–76. [CrossRef]
5. de Brito, J.; Agrela, F.; Silva, R.V. Construction and demolition waste. In *New Trends in Eco-Efficient and Recycled Concrete*; Woodhead Publishing: Cambridge, UK, 2018. [CrossRef]
6. Abdel-Hay, A.S. Properties of recycled concrete aggregate under different curing conditions. *HBRC J.* **2017**, *13*, 271–276. [CrossRef]
7. Pavlu, T.; Kočí, V.; Hájek, P. Environmental assessment of two use cycles of recycled aggregate concrete. *Sustainability* **2019**, *11*, 6185. [CrossRef]
8. Abdulmatin, A.; Tangchirapat, W.; Jaturapitakkul, C. Environmentally friendly interlocking concrete paving block containing new cementing material and recycled concrete aggregate. *Eur. J. Environ. Civ. Eng.* **2019**, *23*, 1467–1484. [CrossRef]
9. Rizvi, R.; Tighe, S.L.; Henderson, V.; Norris, J. Incorporating recycled concrete aggregate in pervious concrete pavements. In Proceedings of the 2009 Annual Conference and Exhibition of the Transportation Association of Canada-Transportation in a Climate of Change, Vancouver, BC, Canada, 18–21 October 2009.
10. Shin, M.; Kim, K.; Gwon, S.W.; Cha, S. Durability of sustainable sulfur concrete with fly ash and recycled aggregate against chemical and weathering environments. *Constr. Build. Mater.* **2014**, *69*, 167–176. [CrossRef]
11. Abdel-Shafy, H.I.; Mansour, M.S.M. Solid waste issue: Sources, composition, disposal, recycling, and valorization. *Egypt. J. Pet.* **2018**, *27*, 1275–1290. [CrossRef]

12. Kenai, S. Recycled aggregates. In *Waste and Supplementary Cementitious Materials in Concrete*; Woodhead Publishing: Cambridge, UK, 2018. [CrossRef]
13. Marinković, S.; Radonjanin, V.; Malešev, M.; Ignjatović, I. Comparative environmental assessment of natural and recycled aggregate concrete. *Waste Manag.* **2010**, *30*, 2255–2264. [CrossRef]
14. Pacheco-Torgal, F.; Tam, V.W.Y.; Labrincha, J.A.; Ding, Y.; de Brito, J. *Handbook of Recycled Concrete and Demolition Waste*; Woodhead Publishing: Cambridge, UK, 2013. [CrossRef]
15. Al-Bayati, H.K.A.; Tighe, S.L.; Achebe, J. Influence of recycled concrete aggregate on volumetric properties of hot mix asphalt. *Resour. Conserv. Recycl.* **2018**, *130*, 200–214. [CrossRef]
16. Lee, C.H.; Du, J.C.; Shen, D.H. Evaluation of pre-coated recycled concrete aggregate for hot mix asphalt. *Constr. Build. Mater.* **2012**, *28*, 66–71. [CrossRef]
17. Cho, Y.H.; Yun, T.; Kim, I.T.; Choi, N.R. The application of Recycled Concrete Aggregate (RCA) for Hot Mix Asphalt (HMA) base layer aggregate. *KSCE J. Civ. Eng.* **2011**, *15*, 473–478. [CrossRef]
18. Zulkati, A.; Wong, Y.D.; Sun, D.D. Mechanistic performance of asphalt-concrete mixture incorporating coarse recycled concrete aggregate. *J. Mater. Civ. Eng.* **2013**, *25*, 1299–1305. [CrossRef]
19. Carpenter, S.H.; Wolosick, J.R. Modifier Influence in the Characterization of Hot-Mix Recycled Material. *Transp. Res. Rec. J. Transp. Res. Board.* **1980**, *777*, 15–22.
20. Gálvez-Martos, J.L.; Styles, D.; Schoenberger, H.; Zeschmar-Lahl, B. Construction and demolition waste best management practice in Europe. *Resour. Conserv. Recycl.* **2018**, *136*, 166–178. [CrossRef]
21. Yuan, H.; Shen, L. Trend of the research on construction and demolition waste management. *Waste Manag.* **2011**, *31*, 670–679. [CrossRef] [PubMed]
22. Yeheyis, M.; Hewage, K.; Alam, M.S.; Eskicioglu, C.; Sadiq, R. An overview of construction and demolition waste management in Canada: A lifecycle analysis approach to sustainability. *Clean Technol. Environ. Policy* **2013**, *15*, 81–91. [CrossRef]
23. Mália, M.; de Brito, J.; Pinheiro, M.D.; Bravo, M. Construction and demolition waste indicators. *Waste Manag. Res.* **2013**, *31*, 241–255. [CrossRef]
24. Wu, Z.; Yu, A.T.W.; Shen, L.; Liu, G. Quantifying construction and demolition waste: An analytical review. *Waste Manag.* **2014**, *34*, 1683–1692. [CrossRef]
25. Akhtar, A.; Sarmah, A.K. Construction and demolition waste generation and properties of recycled aggregate concrete: A global perspective. *J. Clean. Prod.* **2018**, *186*, 262–281. [CrossRef]
26. Blengini, G.A. Life cycle of buildings, demolition and recycling potential: A case study in Turin, Italy. *Build. Environ.* **2009**, *44*, 319–330. [CrossRef]
27. Butera, S.; Christensen, T.H.; Astrup, T.F. Life cycle assessment of construction and demolition waste management. *Waste Manag.* **2015**, *44*, 196–205. [CrossRef] [PubMed]
28. Poon, C.S. Management of construction and demolition waste. *Waste Manag.* **2007**, *27*, 159–160. [CrossRef]
29. S Lockrey, H.N.; Crossin, K.V. Recycling the construction and demolition waste in Vietnam: Opportunities and challenges in practice. *J. Clean. Prod.* **2016**, *133*, 757–766. [CrossRef]
30. Al-Swaidani, A.M.; Khwies, W.T. Applicability of Artificial Neural Networks to Predict Mechanical and Permeability Properties of Volcanic Scoria-Based Concrete. *Adv. Civ. Eng.* **2018**, *2018*, 5207962. [CrossRef]
31. Dahlbo, H.; Bachér, J.; Lähtinen, K.; Jouttijärvi, T.; Suoheimo, P.; Mattila, T.; Sironen, S.; Myllymaa, T.; Saramäki, K. Construction and demolition waste management—A holistic evaluation of environmental performance. *J. Clean. Prod.* **2015**, *107*, 333–341. [CrossRef]
32. Coelho, A.; de Brito, J. Influence of construction and demolition waste management on the environmental impact of buildings. *Waste Manag.* **2012**, *32*, 532–541. [CrossRef]
33. AS 1141.6.2. *Methods for Sampling and Testing Aggregates Method 6.2: Particle Density and Water Absorption of Coarse Aggregate—Pycnometer Method*; Standards Australia: Sydney, Australia, 1996.
34. McNeil, K.; Kang, T.H.K. Recycled Concrete Aggregates: A Review. *Int. J. Concr. Struct. Mater.* **2013**, *7*, 61–69. [CrossRef]
35. JIS-5022. *Recycled Concrete Using Recycled Aggregate Class M*; Japanese Standards Association: Tokyo, Japan, 2012.
36. Kim, K.H.; Ahn, J.W.; Lee, D.J.; Cho, H.C. Development of recycled aggregate producing process from waste concrete using autogenous mill and density separation. In Proceedings of the REWAS 2008: Global Symposium on Recycling, Waste Treatment and Clean Technology, Cancun, Mexico, 12–15 October 2008.
37. De Schutter, G.; Ye, G.; Audenaert, K.; Bager, D.; Baroghel-Bouny, V.; Bellmann, F.; Boel, V.; Bonen, D.; Boström, L.; Corradi, M.; et al. Final report of RILEM TC 205-DSC: Durability of self-compacting concrete. *Mater. Struct. Constr.* **2008**, *41*, 225–233. [CrossRef]
38. EHE-08. *Code on Structural Concrete*; Ministerio de Transportes, Movilidad y Agenda Urbana: Madrid, Spanish, 2010.
39. Shi, C.; Li, Y.; Zhang, J.; Li, W.; Chong, L.; Xie, Z. Performance enhancement of recycled concrete aggregate—A review. *J. Clean. Prod.* **2016**, *112*, 466–472. [CrossRef]
40. Li, W.; Xiao, J.; Sun, Z.; Kawashima, S.; Shah, S.P. Interfacial transition zones in recycled aggregate concrete with different mixing approaches. *Constr. Build. Mater.* **2012**, *35*, 1045–1055. [CrossRef]
41. Poon, C.S.; Shui, Z.H.; Lam, L. Effect of microstructure of ITZ on compressive strength of concrete prepared with recycled aggregates. *Constr. Build. Mater.* **2004**, *18*, 461–468. [CrossRef]

42. Etxeberria, M.; Vázquez, E.; Marí, A.; Barra, M. Influence of amount of recycled coarse aggregates and production process on properties of recycled aggregate concrete. *Cem. Concr. Res.* **2007**, *11*, 10094–10101. [CrossRef]
43. Almeida, A.; Cunha, J. The implementation of an Activity-Based Costing (ABC) system in a manufacturing company. *Procedia Manuf.* **2017**, *13*, 932–939. [CrossRef]
44. Malešev, M.; Radonjanin, V.; Marinković, S. Recycled concrete as aggregate for structural concrete production. *Sustainability* **2010**, *2*, 1204–1225. [CrossRef]
45. Tabsh, S.W.; Abdelfatah, A.S. Influence of recycled concrete aggregates on strength properties of concrete. *Constr. Build. Mater.* **2009**, *81*, 179–186. [CrossRef]
46. Guo, H.; Shi, C.; Guan, X.; Zhu, J.; Ding, Y.; Ling, T.C.; Zhang, H.; Wang, Y. Durability of recycled aggregate concrete—A review. *Cem. Concr. Compos.* **2018**, *89*, 251–259. [CrossRef]
47. Tam, V.W.Y.; Gao, X.F.; Tam, C.M. Microstructural analysis of recycled aggregate concrete produced from two-stage mixing approach. *Cem. Concr. Res.* **2005**, *35*, 1195–1203. [CrossRef]
48. Otsuki, N.; Miyazato, S.I.; Yodsudjai, W. Influence of recycled aggregate on interfacial transition zone, strength, chloride penetration and carbonation of concrete. *J. Mater. Civ. Eng.* **2003**, *15*, 443–451. [CrossRef]
49. Silva, R.V.; Neves, R.; de Brito, J.; Dhir, R.K. Carbonation behaviour of recycled aggregate concrete. *Cem. Concr. Compos.* **2015**, *62*, 22–32. [CrossRef]
50. Radonjanin, V.; Malešev, M.; Marinković, S.; al Malty, A.E.S. Green recycled aggregate concrete. *Constr. Build. Mater.* **2013**, *47*, 1503–1511. [CrossRef]
51. Domingo-Cabo, A.; Lázaro, C.; López-Gayarre, F.; Serrano-López, M.A.; Serna, P.; Castaño-Tabares, J.O. Creep and shrinkage of recycled aggregate concrete. *Constr. Build. Mater.* **2009**, *23*, 2545–2553. [CrossRef]
52. Gómez-Soberón, J.M.V. Porosity of recycled concrete with substitution of recycled concrete aggregate: An experimental study. *Cem. Concr. Res.* **2002**, *32*, 1301–1311. [CrossRef]
53. Katz, A. Treatments for the improvement of recycled aggregate. *J. Mater. Civ. Eng.* **2004**, *16*, 597–603. [CrossRef]
54. Bui, N.K.; Satomi, T.; Takahashi, H. Mechanical properties of concrete containing 100% treated coarse recycled concrete aggregate. *Constr. Build. Mater.* **2018**, *163*, 496–507. [CrossRef]
55. Chatterjee, A.; Sui, T. Alternative fuels—Effects on clinker process and properties. *Cem. Concr. Res.* **2019**, *123*, 105777. [CrossRef]
56. Etxeberria, M.; Marí, A.R.; Vázquez, E. Recycled aggregate concrete as structural material. *Mater. Struct. Constr.* **2007**, *40*, 529–541. [CrossRef]
57. Sagoe-Crentsil, K.K.; Brown, T.; Taylor, A.H. Performance of concrete made with commercially produced coarse recycled concrete aggregate. *Cem. Concr. Res.* **2001**, *31*, 707–712. [CrossRef]
58. Tam, V.W.Y.; Soomro, M.; Evangelista, A.C.J. A review of recycled aggregate in concrete applications (2000–2017). *Constr. Build. Mater.* **2018**, *172*, 272–292. [CrossRef]
59. Duan, Z.H.; Kou, S.C.; Poon, C.S. Prediction of compressive strength of recycled aggregate concrete using artificial neural networks. *Constr. Build. Mater.* **2013**, *40*, 1200–1206. [CrossRef]
60. Bui, N.K.; Satomi, T.; Takahashi, H. Improvement of mechanical properties of recycled aggregate concrete basing on a new combination method between recycled aggregate and natural aggregate. *Constr. Build. Mater.* **2017**, *148*, 376–385. [CrossRef]
61. Corinaldesi, V. Mechanical and elastic behaviour of concretes made of recycled-concrete coarse aggregates. *Constr. Build. Mater.* **2010**, *24*, 1616–1620. [CrossRef]
62. Al-Bayati, H.K.A.; Das, P.K.; Tighe, S.L.; Baaj, H. Evaluation of various treatment methods for enhancing the physical and morphological properties of coarse recycled concrete aggregate. *Constr. Build. Mater.* **2016**, *112*, 284–298. [CrossRef]
63. Al-Bayati, H.K.A.; Tighe, S.L. Effect of Recycled Concrete Aggregate on Rutting and Stiffness Characteristics of Asphalt Mixtures. *J. Mater. Civ. Eng.* **2019**, *31*, 04019219. [CrossRef]
64. Elistratkin, M.Y.; Lesovik, V.S.; Zagorodnjuk, L.H.; Pospelova, E.A.; Shatalova, S.V. New point of view on materials development. *IOP Conf. Ser. Mater. Sci. Eng.* **2018**, *327*, 3–032020. [CrossRef]
65. Feduik, R. Reducing permeability of fiber concrete using composite binders. *Spec. Top. Rev. Porous Media.* **2018**, *9*, 79–89. [CrossRef]
66. Klyuev, S.V.; Klyuev, A.V.; Shorstova, E.S. The micro silicon additive effects on the fine-grassed concrete properties for 3-D additive technologies. *Mater. Sci. Forum.* **2019**, *974*, 131–135. [CrossRef]
67. Limbachiya, M.C.; Leelawat, T.; Dhir, R.K. Use of recycled concrete aggregate in high-strength concrete. *Mater. Struct. Constr.* **2000**, *33*, 574–580. [CrossRef]
68. Poon, C.S.; Shui, Z.H.; Lam, L.; Fok, H.; Kou, S.C. Influence of moisture states of natural and recycled aggregates on the slump and compressive strength of concrete. *Cem. Concr. Res.* **2004**, *34*, 31–36. [CrossRef]
69. Poon, C.S.; Kou, S.C.; Lam, L. Use of recycled aggregates in molded concrete bricks and blocks. *Constr. Build. Mater.* **2002**, *16*, 281–289. [CrossRef]
70. Ratnayake, R.M.C.; Samarakoon, S.M.S. Structural integrity assessment and control of ageing onshore and offshore structures. In *Modeling and Simulation Techniques in Structural Engineering*; Samui, P., Chakraborty, S., Kim, D., Eds.; IGI Global: Pennsylvania, PA, USA, 2016. [CrossRef]
71. Thongkamsuk, P.; Sudasna, K.; Tondee, T. Waste generated in high-rise buildings construction: A current situation in Thailand. *Energy Procedia* **2017**, *138*, 411–416. [CrossRef]

72. Xuan, D.; Zhan, B.; Poon, C.S. Durability of recycled aggregate concrete prepared with carbonated recycled concrete aggregates. *Cem. Concr. Compos.* **2017**, *84*, 214–221. [CrossRef]
73. Wang, H.; Sun, X.; Wang, J.; Monteiro, P.J.M. Permeability of concrete with recycled concrete aggregate and pozzolanic materials under stress. *Materials* **2016**, *9*, 252. [CrossRef] [PubMed]
74. Wijayasundara, M.; Mendis, P.; Crawford, R.H. Methodology for the integrated assessment on the use of recycled concrete aggregate replacing natural aggregate in structural concrete. *J. Clean. Prod.* **2017**, *166*, 321–334. [CrossRef]
75. Wijayasundara, M.; Mendis, P.; Crawford, R.H. Integrated assessment of the use of recycled concrete aggregate replacing natural aggregate in structural concrete. *J. Clean. Prod.* **2018**. [CrossRef]
76. Ajdukiewicz, A.; Kliszczewicz, A. Influence of recycled aggregates on mechanical properties of HS/HPC. *Cem. Concr. Compos.* **2002**, *174*, 591–604. [CrossRef]
77. Evangelista, L.; de Brito, J. Mechanical behaviour of concrete made with fine recycled concrete aggregates. *Cem. Concr. Compos.* **2007**, *29*, 397–401. [CrossRef]
78. Tam, V.W.Y. Economic comparison of concrete recycling: A case study approach. *Resour. Conserv. Recycl.* **2008**, *52*, 821–828. [CrossRef]
79. Verian, K.P.; Ashraf, W.; Cao, Y. Properties of recycled concrete aggregate and their influence in new concrete production. *Resour. Conserv. Recycl.* **2018**, *133*, 30–49. [CrossRef]
80. Oikonomou, N.D. Recycled concrete aggregates. *Cem. Concr. Compos.* **2005**, *27*, 315–318. [CrossRef]
81. Teh, S.H.; Wiedmann, T.; Moore, S. Mixed-unit hybrid life cycle assessment applied to the recycling of construction materials. *J. Econ. Struct.* **2018**, *7*, 1–25. [CrossRef]
82. Kisku, N.; Joshi, H.; Ansari, M.; Panda, S.K.; Nayak, S.; Dutta, S.C. A critical review and assessment for usage of recycled aggregate as sustainable construction material. *Constr. Build. Mater.* **2017**, *131*, 721–740. [CrossRef]
83. Tošić, N.; Marinković, S.; Dašić, T.; Stanić, M. Multicriteria optimization of natural and recycled aggregate concrete for structural use. *J. Clean. Prod.* **2015**, *87*, 766–776. [CrossRef]
84. Mas, B.; Cladera, A.; del Olmo, T.; Pitarch, F. Influence of the amount of mixed recycled aggregates on the properties of concrete for non-structural use. *Constr. Build. Mater.* **2012**, *27*, 612–622. [CrossRef]
85. Lye, C.Q.; Dhir, R.K.; Ghataora, G.S.; Li, H. Creep strain of recycled aggregate concrete. *Constr. Build. Mater.* **2016**, *102*, 244–259. [CrossRef]
86. Poon, C.S.; Chan, D. Effects of contaminants on the properties of concrete paving blocks prepared with recycled concrete aggregates. *Constr. Build. Mater.* **2007**, *21*, 164–175. [CrossRef]
87. Andal, J.; Shehata, M.; Zacarias, P. Properties of concrete containing recycled concrete aggregate of preserved quality. *Constr. Build. Mater.* **2016**, *125*, 842–855. [CrossRef]
88. Poon, C.S.; Lam, C.S. The effect of aggregate-to-cement ratio and types of aggregates on the properties of pre-cast concrete blocks. *Cem. Concr. Compos.* **2008**, *30*, 283–289. [CrossRef]
89. Khatib, J.M. Properties of concrete incorporating fine recycled aggregate. *Cem. Concr. Res.* **2005**, *35*, 763–769. [CrossRef]
90. Kou, S.C.; Poon, C.S. Enhancing the durability properties of concrete prepared with coarse recycled aggregate. *Constr. Build. Mater.* **2012**, *35*, 69–76. [CrossRef]
91. Zaetang, Y.; Sata, V.; Wongsa, A.; Chindaprasirt, P. Properties of pervious concrete containing recycled concrete block aggregate and recycled concrete aggregate. *Constr. Build. Mater.* **2016**, *111*, 15–21. [CrossRef]
92. Kazmi, S.M.S.; Munir, M.J.; Wu, Y.F.; Patnaikuni, I.; Zhou, Y.; Xing, F. Effect of recycled aggregate treatment techniques on the durability of concrete: A comparative evaluation. *Constr. Build. Mater.* **2020**, *264*, 120284. [CrossRef]
93. Al-Bayati, H.K.A.; Tighe, S.L. Utilizing a different technique for improving micro and macro characteristics of coarse recycled concrete aggregate. In Proceedings of the TAC 2016: Efficient Transportation-Managing the Demand-2016 Conference and Exhibition of the Transportation Association of Canada, Toronto, ON, Canada, 25–28 September 2016.
94. Limbachiya, M.; Meddah, M.S.; Ouchagour, Y. Use of recycled concrete aggregate in fly-ash concrete. *Constr. Build. Mater.* **2012**, *27*, 439–449. [CrossRef]
95. Padmini, A.K.; Ramamurthy, K.; Mathews, M.S. Influence of parent concrete on the properties of recycled aggregate concrete. *Constr. Build. Mater.* **2009**, *23*, 829–836. [CrossRef]
96. Yang, J.; Du, Q.; Bao, Y. Concrete with recycled concrete aggregate and crushed clay bricks. *Constr. Build. Mater.* **2011**, *25*, 1935–1945. [CrossRef]
97. Rahal, K. Mechanical properties of concrete with recycled coarse aggregate. *Build. Environ.* **2007**, *35*, 763–769. [CrossRef]
98. Arredondo-Rea, S.P.; Corral-Higuera, R.; Gómez-Soberón, J.M.; Gámez-García, D.C.; Bernal-Camacho, J.M.; Rosas-Casarez, C.A.; Ungsson-Nieblas, M.J. Durability parameters of reinforced recycled aggregate concrete: Case study. *Appl. Sci.* **2019**, *9*, 617. [CrossRef]
99. Aliabdo, A.A.; Elmoaty, A.E.M.A.; Fawzy, A.M. Experimental investigation on permeability indices and strength of modified pervious concrete with recycled concrete aggregate. *Constr. Build. Mater.* **2018**, *193*, 105–127. [CrossRef]
100. Dimitriou, G.; Savva, P.; Petrou, M.F. Enhancing mechanical and durability properties of recycled aggregate concrete. *Constr. Build. Mater.* **2018**, *158*, 228–235. [CrossRef]
101. Katz, A. Properties of concrete made with recycled aggregate from partially hydrated old concrete. *Cem. Concr. Res.* **2003**, *33*, 703–711. [CrossRef]

102. Mohammed, S.I.; Najim, K.B. Mechanical strength, flexural behavior and fracture energy of Recycled Concrete Aggregate self-compacting concrete. *Structures* **2020**, *23*, 34–43. [CrossRef]
103. Martín-Morales, M.; Zamorano, M.; Ruiz-Moyano, A.; Valverde-Espinosa, I. Characterization of recycled aggregates construction and demolition waste for concrete production following the Spanish Structural Concrete Code EHE-08. *Constr. Build. Mater.* **2011**, *25*, 742–748. [CrossRef]
104. James, Use of Recycled Aggregate and Fly Ash in Concrete Pavement. *Am. J. Eng. Appl. Sci.* **2011**, *4*, 201–208. [CrossRef]
105. Ceia, F.; Raposo, J.; Guerra, M.; Júlio, E.; de Brito, J. Shear strength of recycled aggregate concrete to natural aggregate concrete interfaces. *Constr. Build. Mater.* **2016**, *109*, 139–145. [CrossRef]
106. Ajdukiewicz, A.B.; Kliszczewicz, A.T. Comparative Tests of Beams and Columns Made of Recycled Aggregate Concrete and Natural Aggregate Concrete. *J. Adv. Concr. Technol.* **2007**, *5*, 259–273. [CrossRef]
107. Lye, C.Q.; Dhir, R.K.; Ghataora, G.S. Shrinkage of recycled aggregate concrete. *Proc. Inst. Civ. Eng. Struct. Build.* **2016**, *169*, 867–891. [CrossRef]
108. Zega, C.J.; Villagrán-Zaccardi, Y.A.; di Maio, A.A. Effect of natural coarse aggregate type on the physical and mechanical properties of recycled coarse aggregates. *Mater. Struct. Constr.* **2010**, *43*, 195–202. [CrossRef]
109. Kleijer, A.L.; Lasvaux, S.; Citherlet, S.; Viviani, M. Product-specific Life Cycle Assessment of ready mix concrete: Comparison between a recycled and an ordinary concrete. *Resour. Conserv. Recycl.* **2017**, *122*, 210–218. [CrossRef]
110. Yazdanbakhsh, A.; Bank, L.C.; Chen, C. Use of recycled FRP reinforcing bar in concrete as coarse aggregate and its impact on the mechanical properties of concrete. *Constr. Build. Mater.* **2016**, *121*, 278–284. [CrossRef]
111. Mohseni, E.; Saadati, R.; Kordbacheh, N.; Parpinchi, Z.S.; Tang, W. Engineering and microstructural assessment of fibre-reinforced self-compacting concrete containing recycled coarse aggregate. *J. Clean. Prod.* **2017**, *168*, 605–613. [CrossRef]
112. Wijayasundara, M.; Mendis, P.; Crawford, R.H. Net incremental indirect external benefit of manufacturing recycled aggregate concrete. *Waste Manag.* **2018**, *78*, 279–291. [CrossRef]
113. Shaikh, F.U.A.; Nath, P.; Hosan, A.; John, M.; Biswas, W.K. Sustainability assessment of recycled aggregates concrete mixes containing industrial by-products. *Mater. Today Sustain.* **2019**, *5*, 100013. [CrossRef]
114. Hiete, M.; Stengel, J.; Ludwig, J.; Schultmann, F. Matching construction and demolition waste supply to recycling demand: A regional management chain model. *Build. Res. Inf.* **2011**, *39*, 333–351. [CrossRef]
115. Rodríguez, C.; Sánchez, I.; Miñano, I.; Benito, F.; Cabeza, M.; Parra, C. On the possibility of using recycled mixed aggregates and GICC thermal plant wastes in non-structural concrete elements. *Sustainability* **2019**, *11*, 633. [CrossRef]
116. Wagih, A.M.; El-Karmoty, H.Z.; Ebid, M.; Okba, S.H. Recycled construction and demolition concrete waste as aggregate for structural concrete. *HBRC J.* **2013**, *9*, 193–200. [CrossRef]
117. Sim, J.; Park, C. Compressive strength and resistance to chloride ion penetration and carbonation of recycled aggregate concrete with varying amount of fly ash and fine recycled aggregate. *Waste Manag.* **2011**, *31*, 2352–2360. [CrossRef] [PubMed]
118. Zega, C.J.; di Maio, Á.A. Use of recycled fine aggregate in concretes with durable requirements. *Waste Manag.* **2011**, *31*, 2336–2340. [CrossRef]
119. Eddine, B.T.; Salah, M.M. Solid waste as renewable source of energy: Current and future possibility in Algeria. *Int. J. Energy Environ. Eng.* **2012**, *3*, 1–12. [CrossRef]
120. Cheng, A.; Hsu, H.M.; Chao, S.J.; Lin, K.L. Experimental study on properties of pervious concrete made with recycled aggregate. *Int. J. Pavement Res. Technol.* **2011**, *4*, 104. [CrossRef]
121. Ravindrarajah, R.S.; Neo, H.W.; Lai, J.E. Performance of pervious recycled aggregate concrete with reduced cement content. In Proceedings of the International Conference on Structural Engineering, Construction and Management, Kandy, Central, Sri Lanka, 16–18 December 2011.
122. Suganthan, J.; Vignesh, K.; Sudhakar, A.; Ashraf, S. An Experimental Investigation on Pervious Concrete by Using Furnace Slag and Recycled Coarse Aggregate. *SSRG Int. J. Civ. Eng.* **2017**, *ICETM-2017*, 108–113.
123. Silva, R.V.; de Brito, J.; Dhir, R.K. Properties and composition of recycled aggregates from construction and demolition waste suitable for concrete production. *Constr. Build. Mater.* **2014**, *65*, 201–217. [CrossRef]
124. Hamad, B.S.; Dawi, A.H. Sustainable normal and high strength recycled aggregate concretes using crushed tested cylinders as coarse aggregates. *Case Stud. Constr. Mater.* **2017**, *7*, 228–239. [CrossRef]
125. Duan, J.; Asteris, P.G.; Nguyen, H.; Bui, X.N.; Moayedi, H. A novel artificial intelligence technique to predict compressive strength of recycled aggregate concrete using ICA-XGBoost model. *Eng. Comput.* **2020**, 1–18. [CrossRef]
126. Aslani, F.; Ma, G.; Wan, D.L.Y.; Muselin, G. Development of high-performance self-compacting concrete using waste recycled concrete aggregates and rubber granules. *J. Clean. Prod.* **2018**, *182*, 553–566. [CrossRef]
127. Seethapathi, M.; Senthilkumar, S.R.R.; Chinnaraju, K. Experimental study on high performance selfcompacting concrete using recycled aggregate. *J. Theor. Appl. Inf. Technol.* **2014**, *67*, 84–90.
128. Nitesh, K.J.N.S.; Rao, S.V.; Kumar, P.R. An experimental investigation on torsional behaviour of recycled aggregate based steel fiber reinforced self compacting concrete. *J. Build. Eng.* **2019**, *22*, 242–251. [CrossRef]
129. Bartolacci, F.; Paolini, A.; Quaranta, A.G.; Soverchia, M. Assessing factors that influence waste management financial sustainability. *Waste Manag.* **2018**, *79*, 571–579. [CrossRef]
130. di Maria, A.; Eyckmans, J.; van Acker, K. Downcycling versus recycling of construction and demolition waste: Combining LCA and LCC to support sustainable policy making. *Waste Manag.* **2018**, *75*, 3–21. [CrossRef] [PubMed]

131. Ding, Z.; Yi, G.; Tam, V.W.Y.; Huang, T. A system dynamics-based environmental performance simulation of construction waste reduction management in China. *Waste Manag.* **2016**, *51*, 130–141. [CrossRef]
132. Fraile-Garcia, E.; Ferreiro-Cabello, J.; López-Ochoa, L.M.; López-González, L.M. Study of the technical feasibility of increasing the amount of recycled concrete waste used in ready-mix concrete production. *Materials* **2017**, *10*, 817. [CrossRef] [PubMed]
133. Durdyev, S.; Omarov, M.; Ismail, S. Causes of delay in residential construction projects in Cambodia. *Cogent Eng.* **2017**, *4*, 1291117. [CrossRef]
134. Du, Z.; Lin, B. Analysis of carbon emissions reduction of China's metallurgical industry. *J. Clean. Prod.* **2018**, *176*, 1177–1184. [CrossRef]
135. Menegaki, M.; Damigos, D. A review on current situation and challenges of construction and demolition waste management. *Curr. Opin. Green Sustain. Chem.* **2018**, *13*, 8–15. [CrossRef]
136. Jacintho, Ana Elisabete Paganelli Guimarães de Avila, Ivanny Soares Gomes Cavaliere, Lia Lorena Pimentel, and Nádia Cazarim Silva Forti. Modulus and Strength of Concretes with Alternative Materials. *Materials* **2020**, *13*, 4378. [CrossRef]
137. Elhakam, A.A.; Mohamed, A.E.; Awad, E. Influence of self-healing, mixing method and adding silica fume on mechanical properties of recycled aggregates concrete. *Constr. Build. Mater.* **2012**, *35*, 421–427. [CrossRef]
138. Ma, J.; Sun, D.; Pang, Q.; Sun, G.; Hu, M.; Lu, T. Potential of recycled concrete aggregate pretreated with waste cooking oil residue for hot mix asphalt. *J. Clean. Prod.* **2019**, *221*, 469–479. [CrossRef]
139. Cabral, A.E.B.; Schalch, V.; Molin, D.C.C.D.; Ribeiro, J.L.D. Mechanical properties modeling of recycled aggregate concrete. *Constr. Build. Mater.* **2010**, *24*, 421–430. [CrossRef]
140. Yang, S. Effect of different types of recycled concrete aggregates on equivalent concrete strength and drying shrinkage properties. *Appl. Sci.* **2018**, *8*, 2190. [CrossRef]
141. Omary, S.; Ghorbel, E.; Wardeh, G. Relationships between recycled concrete aggregates characteristics and recycled aggregates concretes properties. *Constr. Build. Mater.* **2016**, *108*, 163–174. [CrossRef]
142. Deakins, D.; Bensemann, J.O. Achieving Innovation in a Lean Environment: How Innovative Small Firms Overcome Resource Constraints. *Int. J. Innov. Manag.* **2019**, *23*, 1950037. [CrossRef]
143. Jin, R.; Li, B.; Zhou, T.; Wanatowski, D.; Piroozfar, P. An empirical study of perceptions towards construction and demolition waste recycling and reuse in China. *Resour. Conserv. Recycl.* **2017**, *126*, 86–98. [CrossRef]
144. Dhir, R.K.; de Brito, J.; Silva, R.V.; Lye, C.Q. Recycled Aggregate Concrete: Durability Properties. *Sustain. Constr. Mater.* **2019**, 365–418. [CrossRef]
145. Ozbakkaloglu, T.; Gholampour, A.; Xie, T. Mechanical and Durability Properties of Recycled Aggregate Concrete: Effect of Recycled Aggregate Properties and Content. *J. Mater. Civ. Eng.* **2018**, *30*, 04017275. [CrossRef]
146. Jo, B.W.; Park, S.K.; Park, J.C. Mechanical properties of polymer concrete made with recycled PET and recycled concrete aggregates. *Constr. Build. Mater.* **2008**, *22*, 2281–2291. [CrossRef]
147. Júnior, N.S.A.; Silva, G.A.O.; Dias, C.M.R.; Ribeiro, D.V. Concrete containing recycled aggregates: Estimated lifetime using chloride migration test. *Constr. Build. Mater.* **2019**, *222*, 108–118. [CrossRef]
148. Meneses, E.J.; Gaussens, M.; Jakobsen, C.; Mikkelsen, P.S.; Grum, M.; Vezzaro, L. Coordinating rule-based and system-wide model predictive control strategies to reduce storage expansion of combined urban drainage systems: The case study of Lundtofte, Denmark. *Water* **2018**, *10*, 76. [CrossRef]
149. Jain, J.A.; Olek, J.; Verian, K.P.; Whiting, N. Chloride penetration resistance of concrete mixtures with recycled concrete aggregates. In *Brittle Matrix Composites 10*; Woodhead Publishing: Cambridge, UK, 2012; pp. 377–386.
150. Kou, S.C.; Poon, C.S.; Wan, H.W. Properties of concrete prepared with low-grade recycled aggregates. *Constr. Build. Mater.* **2012**, *36*, 881–889. [CrossRef]
151. Silva, R.V.; de Brito, J.; Neves, R.; Dhir, R. Prediction of chloride ion penetration of recycled aggregate concrete. *Mater. Res.* **2015**, *18*, 427–440. [CrossRef]
152. Kou, S.C.; Poon, C.S.; Chan, D. Influence of fly ash as a cement addition on the hardened properties of recycled aggregate concrete. *Mater. Struct. Constr.* **2008**, *41*, 1191–1201. [CrossRef]
153. O'Donnell, B.T.; Ives, C.J.; Mohiuddin, O.A.; Bunnell, B.A. Beyond the Present Constraints That Prevent a Wide Spread of Tissue Engineering and Regenerative Medicine Approaches. *Front. Bioeng. Biotechnol.* **2019**, *7*, 95. [CrossRef]
154. Singh, N.; Singh, S.P. Carbonation and electrical resistance of self compacting concrete made with recycled concrete aggregates and metakaolin. *Constr. Build. Mater.* **2016**, *121*, 400–409. [CrossRef]
155. Levy, S.M.; Helene, P. Durability of recycled aggregates concrete: A safe way to sustainable development. *Cem. Concr. Res.* **2004**, *34*, 1975–1980. [CrossRef]
156. Grilli, A.; Bocci, M.; Tarantino, A.M. Experimental investigation on fibre-reinforced cement-treated materials using reclaimed asphalt. *Constr. Build. Mater.* **2013**, *38*, 491–496. [CrossRef]
157. Peng, Z.; Shi, C.; Shi, Z.; Lu, B.; Wan, S.; Zhang, Z.; Chang, J.; Zhang, T. Alkali-aggregate reaction in recycled aggregate concrete. *J. Clean. Prod.* **2020**, *255*, 120238. [CrossRef]
158. Chi, M. Effects of dosage of alkali-activated solution and curing conditions on the properties and durability of alkali-activated slag concrete. *Constr. Build. Mater.* **2012**, *35*, 240–245. [CrossRef]
159. Bogas, J.A.; de Brito, J.; Ramos, D. Freeze-thaw resistance of concrete produced with fine recycled concrete aggregates. *J. Clean. Prod.* **2016**, *115*, 294–306. [CrossRef]

160. Richardson, A.; Coventry, K.; Bacon, J. Freeze/thaw durability of concrete with recycled demolition aggregate compared to virgin aggregate concrete. *J. Clean. Prod.* **2011**, *19*, 272–277. [CrossRef]
161. Topçu, I.B.; Şengel, S. Properties of concretes produced with waste concrete aggregate. *Cem. Concr. Res.* **2004**, *34*, 1307–1312. [CrossRef]
162. Tuyan, M.; Mardani-Aghabaglou, A.; Ramyar, K. Freeze-thaw resistance, mechanical and transport properties of self-consolidating concrete incorporating coarse recycled concrete aggregate. *Mater. Des.* **2014**, *53*, 983–991. [CrossRef]
163. Fonseca, N.; De Brito, J.; Evangelista, L. The influence of curing conditions on the mechanical performance of concrete made with recycled concrete waste. *Cem. Concr. Compos.* **2011**, *33*, 637–643. [CrossRef]
164. Mardani-Aghabaglou, A.; Andiç-Çakir, Ö.; Ramyar, K. Freeze-thaw resistance and transport properties of high-volume fly ash roller compacted concrete designed by maximum density method. *Cem. Concr. Compos.* **2013**, *37*, 259–266. [CrossRef]
165. Princigallo, A.; van Breugel, K.; Levita, G. Influence of the aggregate on the electrical conductivity of Portland cement concretes. *Cem. Concr. Res.* **2003**, *33*, 1755–1763. [CrossRef]
166. Kong, D.; Lei, T.; Zheng, J.; Ma, C.; Jiang, J.; Jiang, J. Effect and mechanism of surface-coating pozzalanics materials around aggregate on properties and ITZ microstructure of recycled aggregate concrete. *Constr. Build. Mater.* **2010**, *24*, 701–708. [CrossRef]
167. Zhang, H.; Ji, T.; Zeng, X.; Yang, Z.; Lin, X.; Liang, Y. Mechanical behavior of ultra-high performance concrete (UHPC) using recycled fine aggregate cured under different conditions and the mechanism based on integrated microstructural parameters. *Constr. Build. Mater.* **2018**, *192*, 489–507. [CrossRef]
168. Wang, R.; Yu, N.; Li, Y. Methods for improving the microstructure of recycled concrete aggregate: A review. *Constr. Build. Mater.* **2020**, *242*, 118164. [CrossRef]
169. Assia, D. Effect of recycled coarse aggregate on the new interfacial transition zone concrete. *Constr. Build. Mater.* **2018**, *190*, 1023–1033. [CrossRef]
170. Bentz, D.P. Influence of internal curing using lightweight aggregates on interfacial transition zone percolation and chloride ingress in mortars. *Cem. Concr. Compos.* **2009**, *31*, 285–289. [CrossRef]
171. Sadek, D.M.; El-Attar, M.M. Development of high-performance green concrete using demolition and industrial wastes for sustainable construction. *J. Am. Sci.* **2012**, *8*, 120–131.
172. Babu, V.S.; Mullick, A.K.; Jain, K.K.; Singh, P.K. Strength and durability characteristics of high-strength concrete with recycled aggregate-influence of processing. *J. Sustain. Cem. Mater.* **2014**, *4*, 54–71. [CrossRef]
173. Xuan, D.; Zhan, B.; Poon, C.S. Assessment of mechanical properties of concrete incorporating carbonated recycled concrete aggregates. *Cem. Concr. Compos.* **2016**, *65*, 67–74. [CrossRef]
174. Zhan, B.J.; Xuan, D.X.; Poon, C.S. Enhancement of recycled aggregate properties by accelerated CO2 curing coupled with limewater soaking process. *Cem. Concr. Compos.* **2018**, *89*, 230–237. [CrossRef]
175. Kazmi, S.M.S.; Munir, M.J.; Wu, Y.F.; Patnaikuni, I.; Zhou, Y.; Xing, F. Influence of different treatment methods on the mechanical behavior of recycled aggregate concrete: A comparative study. *Cem. Concr. Compos.* **2019**, *104*, 103398. [CrossRef]
176. Wang, L.; Wang, J.; Qian, X.; Chen, P.; Xu, Y.; Guo, J. An environmentally friendly method to improve the quality of recycled concrete aggregates. *Constr. Build. Mater.* **2017**, *144*, 432–441. [CrossRef]
177. Tam, V.W.Y.; Tam, C.M.; Le, K.N. Removal of cement mortar remains from recycled aggregate using pre-soaking approaches. *Resour. Conserv. Recycl.* **2007**, *50*, 82–101. [CrossRef]
178. Wijayasundara, M.; Mendis, P.; Zhang, L.; Sofi, M. Financial assessment of manufacturing recycled aggregate concrete in ready-mix concrete plants. *Resour. Conserv. Recycl.* **2016**, *109*, 187–201. [CrossRef]

MDPI

Review

A State-of-the-Art Review on Suitability of Granite Dust as a Sustainable Additive for Geotechnical Applications

Gudla Amulya [1], Arif Ali Baig Moghal [1,*] and Abdullah Almajed [2,*]

[1] Department of Civil Engineering, National Institute of Technology Warangal, Warangal 506004, India; gamulya015@student.nitw.ac.in
[2] Department of Civil Engineering, College of Engineering, King Saud University, Riyadh 11421, Saudi Arabia
* Correspondence: baig@nitw.ac.in or reach2arif@gmail.com (A.A.B.M.); alabduallah@ksu.edu.sa (A.A.)

Abstract: The increase in infrastructure requirement drives people to use all types of soils, including poor soils. These poor soils, which are weak at construction, must be improved using different techniques. The extinction of natural resources and the increase in cost of available materials require us to think of alternate resources. The usage of industry by-products and related methods for improving the properties of different soils has been studied for several years. Granite dust is an industrial by-product originating from the primary crushing of aggregates. The production of huge quantities of granite dust in the industry causes severe problems from the handling to the disposal stage. Accordingly, in the civil engineering field, the massive utilization of granite dust has been proposed for various applications to resolve these issues. In this context, the present review provides precise and valuable content on granite dust characterization, its effect as a stabilizer on the behavior of different soils, and its interaction mechanisms. The efficacy of the granite dust in replacing sand in concrete is explored followed by its ability to improve the geotechnical characteristics of clays of varying plasticity are explored. The review is even extended to study the effect of binary stabilization on clays with granite dust in the presence of calcium-based binders. The practical limitations encountered and its efficiency over other stabilizers are also assessed. This review is further extended to analyze the effect of the granite dust dosage for various field applications.

Keywords: granite dust; stabilizer; particle size; plasticity; unconfined compression strength

Citation: Amulya, G.; Moghal, A.A.B.; Almajed, A. A State-of-the-Art Review on Suitability of Granite Dust as a Sustainable Additive for Geotechnical Applications. *Crystals* **2021**, *11*, 1526. https://doi.org/10.3390/cryst11121526

Academic Editors: Antonella Sola, Cesare Signorini, Sumit Chakraborty and Valentina Volpini

Received: 8 November 2021
Accepted: 28 November 2021
Published: 7 December 2021

1. Introduction

Soil stabilization is a technique used to improve the geotechnical properties of soil, either physically or chemically. Different types of stabilization methods exist, and each process varies with the type of additive used. Additives include lime, cement, bitumen geosynthetics, and some industrial by-products like flyash, slag, coal, and stone dust, chemicals, reagents, and recycled materials like rubber tire chips, waste plastics, and crushed glass that follows recent advanced bio-stabilization techniques like microbial-induced calcite precipitation and enzyme-induced calcite precipitation [1]. Among these stabilization methods, the most commonly adopted process is the addition of calcium-based materials like lime, cement, and flyash [2,3]. However, these additives have their own limitations in terms of carbon emissions [4,5]. Stabilization with lime and cement also causes a problematic expansion in the presence of sulphate [6]. In silica-rich soils, adding lime decreases the soil performance beyond its optimum level because the soil develops silica gel that withholds water and retains the soil plasticity [7]. Expansive semi-arid soils have been treated with lime and tested for their lime leachability. At 4% lime content, the lime leachability was minimized with increase in the curing period due to pozzolanic reactions [8]. The laterite soil when stabilized with lime caused a decrease in the unconfined compressive strength (UCS) and California bearing ratio (CBR) with the increase in the delay of compaction in hours [9,10]. High-plastic clays stabilize with lime, and the UCS and the coefficient of permeability increase with the increase in the delay of compaction due to the formation of pozzolanic reactions [11]. The stabilization of sand with cement induces cohesion, but is not effective in developing interfacial friction [12]. Previous studies have shown that brittleness is associated with cement-stabilized sands [13]. Alkaline materials, such as cement, lime, and gypsum, make the treated soil brittle with concentration increase and alter the soil pH [14]. Polypropylene fibers and flyash–cement mixtures are used to improve the unconfined compressive strength of clay. The presence of composite cement turns the treated soil to brittle; however, due to the presence of fibers, the treated soil exhibits a plastic behavior upon load application [15]. In order curb the usage of cement and respective alkaline materials, researchers are working on disposal wastes that act as a sustainable supplementary cementitious material. Recently, biopolymer stabilization of plastic and non-plastic fines is gaining attention owing to its sustainable approach and associated low carbon footprint emissions [16]. Coal gangue is another sustainable generated from the coal production process. Coal gangue utilization can reduce ecological issues, but due care should be given to associated leaching of trace elements [4,5,17]. Enzyme-mediated calcite precipitation is a technique used to improve the compressive strength of sand. The substitution of magnesium sulphate in enzymes yields a better improvement compared to conventional calcite precipitation [18,19]. Many other additives like polypropylene fibers, cement kiln dust, ground-granulated blast-furnace slag, and slag play a major role in stabilizing specific soils [20]. In the previous works, every stabilizer has been limited to some aspects (e.g., carbon emission, production cost, groundwater chemistry, change in soil pH, UV radiation, reactivity, etc.) that drive sustainable stabilizers. This review explores the potential applications of granite dust, a waste by-product, as an efficient stabilizer for improving the geotechnical properties of problematic soils, including Atterberg's limits, compaction properties, unconfined compressive strength, permeability, and California bearing ratio, among others. The mechanisms responsible are described based on the physicochemical characteristics of granite dust. The optimum dosage of granite dust in various soils for different applications is proposed to facilitate practical utilization.

Granite dust is an industrial by-product with an ever-increasing demand in the construction industry. It is deposited in huge amounts at quarry sites and crushing industries [21]. Granite dust is a non-plastic material that exhibits high shear strength with zero carbon emissions. The fine state of stone dust results in a large specific surface area. The physical properties, chemical composition, and mineralogy of stone dust vary with the type of parent rock, but is consistent with the quarry at site [22]. Granite dust is an industrial by-product originating from the primary crushing stage of aggregates [23]. These are fine aggregates produced with particle diameters less than 4 mm [24]. The quality of a stone dust depends on the rock type, origin, and processing method. The global production of stone dusts from different plants is approximately 1.48 billion tons and produced by 1430 companies. On an average, a typical rock produces roughly approximately 400–500,000 tons of aggregate every year [25]. Approximately, 20–25% of this goes as unused material [26]. In India, approximately 200 million tons of quarry by-products is produced annually [27]. Mined boulders and blasted rocks from quarry sites are hauled into a crusher bin and fed to crushers [25]. Crushing can be done in three to four stages (i.e., primary, secondary, tertiary, and quaternary). In the primary and secondary stages, two major crusher units are fed with quarried rock to produce aggregates of different sizes determined by demand [25]. An overview of production of quarry fines is described in the Figure 1. Screening is done in each crushing stage to obtain a usable end product.

Table 1 shows that granite dust is an industrial by-product that has high density and zero carbon emission and is abundant and chemically inert with water. The specific gravity of granite dust is greater than the specific gravity of soils [21], which ranges from 2.6 to 2.8.

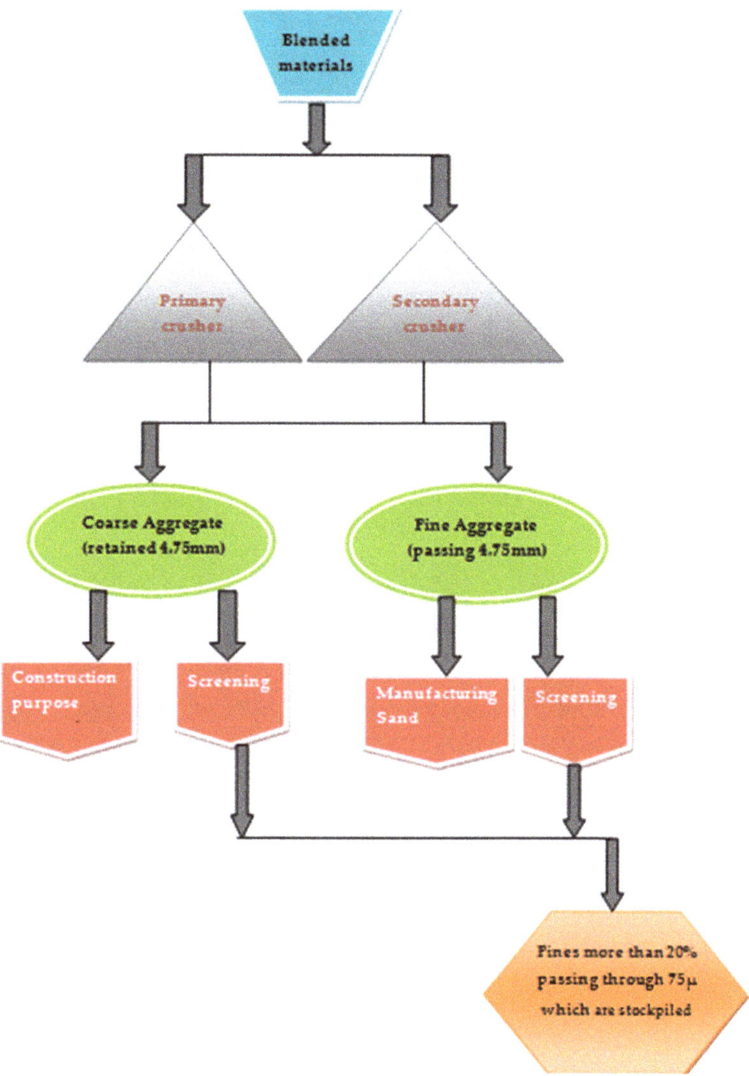

Figure 1. An overview of stages involved in the production of quarry fines.

Table 1. Applications of different types of stone dust.

Name of Rock Dust	Predominant Constituent	Specific Gravity	Civil Engineering Applications	Reference
Dolomite, metamorphic	Calcium magnesium carbonate	2.84	Aggregate, ballast, base material	[27]
Limestone, sedimentary	Calcium carbonate	2.7	Road base and railroad ballast	[28]
Shale, sedimentary	Silica	2.62	Fills and embankment	[29]
Sandstone, sedimentary	Silica	2.5	Replacement of natural sand	[30]
Granite, igneous	Silica	2.7	Filler, subgrade, replacement of natural sand and sub-base	[31]
Marble, metamorphic	Dolomite, quartz, and calcite	2.71	Filler in concrete production	[32]

2. Physical and Chemical Properties of Granite Dust

2.1. Morphology and Mineralogy of Granite Dust

The physical appearance of granite dust changes with topography. The X-ray diffraction and scanning electron microscopy (SEM) of granite dust show mineralogy and morphological variations. Granite is an intrusive igneous rock formed from magma. It is predominantly white, pink, or gray in color. These rocks mainly comprise feldspar, quartz, mica, and amphibole minerals. Dust formed out of granite quarry and aggregate crushing plants varies in the physical appearance of granite dust relevant to location. Table 2 presents petrological details of different granite dusts sourced from different parts of the world.

Table 2. Petrological details of granite dust(s) from different regions of the world.

Location	Petrographic Description	Mineralogy	Reference
Thane, Maharashtra, India	Irregular shaped angular particles	Quartz from XRD	[33]
Local quarry in Kedah, Malaysia	Granular, irregular and angular geometry	Quartz, Microcline, Calcium Aluminium Silicate, Kaolinite, Magnesium Sulphate Hydrate from XRD	[31]
Quarry dust from local crushing plants, Guwahati, Assam, India	Sub-angular to angular	Quartz and feldspar	[34]
Local marble crushing plants, Pakistan	Angular and flaky in shape and bearing rough texture	Quartz, Crystobalite, Zeolite, Wollastonite from XRD	[35]
Garchuk quarry, Guwahati, Assam, India	Sub-angular to angular	Quartz, Feldspar, Biotite, Muscovite and others from petrographic analysis from XRD	[36]

Quartz, granite, limestone, dolomite, and sandstone are the major rock types used by the crushed stone industry. Granite dust is produced from aggregate crushing plants. Most parts of fines are passing the No. 200 sieve and defined as fine aggregate with a particle size less than 4 mm in diameter. The chemical composition of granite dust is an important material characteristic which plays a key role in stabilization. It differs with location, formation and the type of rock available.

2.2. Granite Dust and Composition

Table 3 provides the composition of a granite dust which give a rough estimate of various chemical elements in support of the content provided in Table 2.

Table 3. Chemical composition of granite dust (Sourced from [36–39]).

Element	Composition Range (%)
SiO_2	45–75
Al_2O_3	15–19
CaO	3–14
Fe_2O_3	6–17
K_2O	3–4.5
MgO	1–3.6
Na_2O	0–3.7
P_2O_3	0–0.02
TiO_2	0–2.65

3. Granite Dust as a Sustainable Material

Sand mining is the process of removing sand from the foreshore. Approximately 47 to 59 billion tons of material is mined globally. Sand utilization in construction leads to unjustifiable sand mining caused by the increment in development activities, which are unacceptable. The available sources of characteristic sand are draining. High-class sand can be moved from a significant distance, causing an economical constraint. Therefore, the structure quality relies on a partial or complete material replacement. Granite dust discarded in a huge amount creates a financial and ecological expense to the industry [40]. Granite dust can avoid detrimental effects on the environment, which are caused by the excessive mining of river sand [41]. Some granite dust applications are in geotechnical aspects like embankment, backfills, road-paving materials, underground cavity fillers, barrier wall materials and sub-base.

4. Effect of Granite Dust Addition on the Geotechnical Properties

4.1. Atterberg Limits

Granite dust is a non-plastic material that cannot be influenced by water. Hence, adding granite dust to plastic soils reduces the plasticity index by breaking the particle–water–particle bond and the liquid and plastic limits. Works have been performed on red earth, kaolinite, and sun-dried marine clay, where the Atterberg limits decreased with the dosage increase [21]. The sun-dried marine clay comparatively gave a better response with granite dust addition compared to the other two because the marine clay is a high-plastic soil with a poor gradation curve [21] (Table 4).

Table 4. Response of Atterberg limits with an increase in the dosage of granite dust (Modified after [21]).

Soil Type	% Granite Dust	Specific Gravity	Liquid Limit (%)	Plastic Limit (%)
Red earth	0	2.70	40	25
	20	2.72	35	Non-plastic
	40	2.74	27	-
	60	2.76	25	-
	80	2.78	24	-
Kaolinite	0	2.6	55	30
	20	2.64	47	19
	40	2.68	37	Non-plastic
	60	2.72	30	-
	80	2.76	26	-
Sundried marine clay	0	2.62	73	36
	20	2.66	57	28
	40	2.69	44	21
	60	2.72	35	Non-plastic
	80	2.76	27	-

The high plasticity of soil decreased with the increase in the amount of added granite dust. The liquid limit decreased to 52%, with 60% granite dust addition. Similar works [42,43] have investigated the low-strength/weak soil and concluded that adding granite dust to plastic and high-plastic soils decreases the liquid and plastic limits. Work has also been performed on the granite dust–black cotton soil mixtures and observed the decreasing behavior of Atterberg limits with an increase in granite dust addition [44]. Table 5 lists the summary of the attempts made to improve the Atterberg's limits of different soils with the addition of granite dust.

Table 5. Summary of the attempts to improve the Atterberg's limits of different soils with the addition of granite dust.

Soil Type	Outcome	Reference
Red earth	Decreased gradually	[21]
Lithomargic clay	Significant decrease of liquid limit and plastic limit	[43]
Black cotton soil	Liquid limit is decreased by 42% at 40% addition of granite dust	[44]

Decrease in Atterberg's limits of the soil is due to the decrease in finer fraction of the heterogeneous mix. Change in the finer fraction affects the water absorbing capacity of the soil.

4.2. Compaction Attributes

Maximum dry density (MDD) and optimum moisture content (OMC) are two significant parameters used to assess the field capacity of soil. Adding granite dust to soil increases the MDD and reduces the OMC due to the increase in coarser fraction and the specific gravity of soil–granite dust mixes [43]. Moreover, the increase in the MDD was due to the shift in the gradation curve from a poor to a well-graded mix. In a work on quarry reclamation, granite dust was mixed with silty soil. A decrease in the OMC and an increase in the MDD were observed with the increments in the presence of granite dust (Figure 2) [45]. Generally high-plastic silts show an improved MDD and a decreased OMC with the gradual increase in granite dust substitution [42]. Nwaiwu [43] observed an increase in the MDD of black cotton soil–granite dust mixes and a decrease in the OMC at higher granite dust contents. Irrespective of compaction energy adopted, the addition of granite dust improved OMC and MDD relatively for several soils. Similar observations [22,46] were identified in the case of clays and red earth soils, where an increase in the MDD of mixes were observed with a simultaneous reduction in the OMC values at higher percentages of granite dust dosages was noted. The compaction characteristics of residual soils improved with the addition of granite dust, which consequently led to an increase in the compaction energy [37].

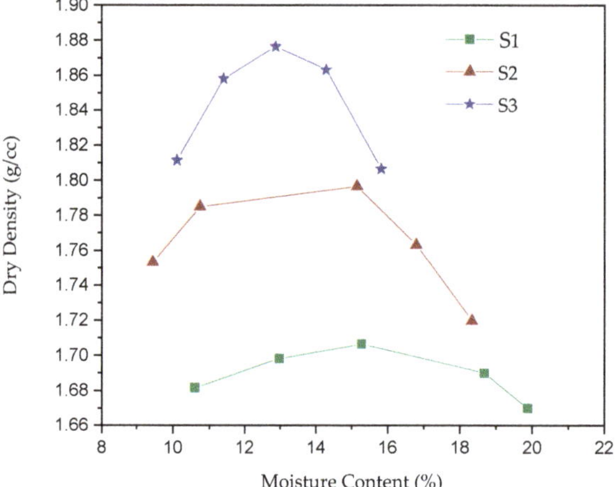

Figure 2. Influence of the granite dust dosage on the compaction characteristics S1: 75% soil + 25% granite dust; S2: 50% soil + 50% granite dust; and S3: 25% soil + 75% granite dust (Modified after [45]).

The maximum dry density of mixed soils improves because of the substitution of dust particles in the clay voids and, to some extent, in silts. This will ensure that macro and micro-voids are minimized at higher compactive efforts. The MDD of marine clay increased by approximately 88%, which is higher compared to other soils as seen in Figure 3 (Adding granite dust to soils allows less water to absorb due to the increase in the coarser fraction compared to fines, which is particularly observed in clays, silts and clayey soils). These changes in the soil–granite mix help to improve the engineering properties of the soil.

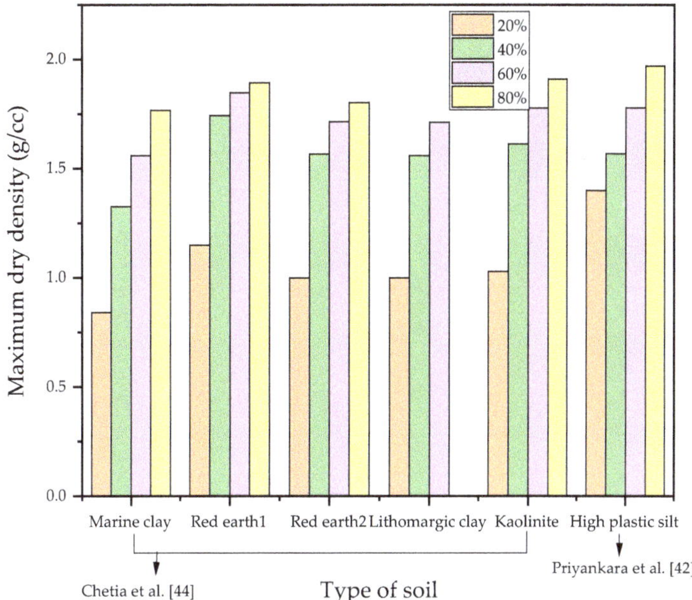

Figure 3. Variation of MDD of different soils with granite dust (Modified after [42,44]).

The change in compaction characteristics is due to the change in particle size distribution of the granite dust mixed soil. Formation of a well graded mix offers greater density and presence of granite dust breaks the water film around the clay particles.

4.3. California Bearing Ratio

To obtain the CBR, a static penetration test is performed to obtain the susceptibility of soil against wheel load penetration. According to [42], the CBR value recommended by the RDA is 15% for subgrade in highway construction. In the case of kaolinite, red earth, and sun-dried marine clay, the CBR values increased with an increase in the percentage of granite dust [21]. However, the percentage increase in CBR for a particular dosage of granite dust is more for clay soil due to the change in grain size distribution (Table 6). Soils with high clay content showed a high improvement with granite dust addition.

Table 6. CBR characteristics of Granite dust amended soils (Modified after [21]).

Soil Type	% Granite Dust	Soaked CBR (%)	Unsoaked CBR (%)
Red earth	0	8.8	9.9
	20	9.8	10.5
	40	10.8	12.0
	60	13.0	14.3
	80	14.7	15.5
Kaolinite	0	5.2	7.8
	20	6.7	8.8
	40	8.8	10.8
	60	18.7	20.8
	80	20.8	22.8
Sun-dried marine clay	0	3.8	4.7
	20	4.3	5.2
	40	5.1	6.2
	60	8.9	9.4
	80	11.2	11.8

This increase in the CBR value is also influenced by the shear strength of the particular soil. In spite of the presence of weak soil, stabilization with granite dust brings the CBR value to the field requirement that helps to reduce the pavement thickness. The increase in the CBR values with the increase in the percentage of granite dust was observed until 50% granite dust addition in the case of black cotton soil [44]. The CBR values of the residual soils increased with an increase in granite dust in soaked and unsoaked conditions [37]. Table 7 shows summary of earlier works related to improvement in CBR characteristics of different soils.

Table 7. Summary of works to improve the CBR of the soils with granite dust.

Soil Type	Outcome	Reference
Marine clay	Good improvement at less amount of granite dust	[21]
Residual soil	Soaked CBR value is comparatively higher than un soaked	[37]
Black cotton soil	Significant increase at 50% addition of granite dust	[44]

4.4. Shear Strength

Sun-dried marine clay highly responded to the granite dust addition compared to red earth. The presence of granite dust in clays filled the voids and developed friction among the mixed particles [22]. A significant improvement in the shear strength with the increase in internal friction and a corresponding decrease in the cohesive nature of high-plastic silt was observed up to 60% addition of granite dust as seen in Figure 4.

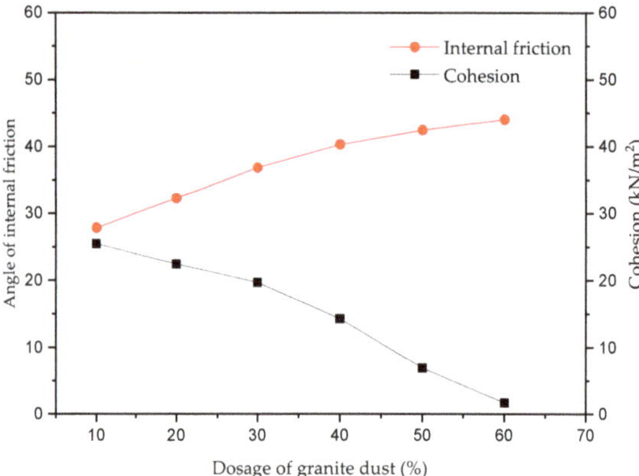

Figure 4. Variation of shear parameters of high-plastic silt with granite dust addition (Modified after [42]).

4.5. UCS and Permeability

The unconfined compressive strength of lithomargic clay is improved with addition of granite dust content up to 20% and decreased with a further increase in granite dust addition. The coefficient of permeability of lithomargic clay proportionally increased with the addition of granite dust [46].

The wealth of literature summarized reveals that, granite dust enhances the geotechnical properties of silts and clays. High-plastic clays and clayey soils hold poor gradation; thus, the granite dust addition turns the mix into a well-graded complex that yields a high confinement as seen from Figure 5.

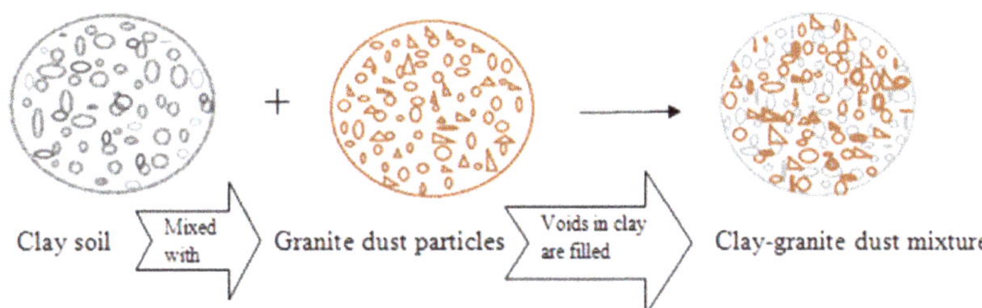

Figure 5. Change in the particle gradation of clay soil with granite dust addition.

5. Granite Dust as a Substitute for Sand

Kumar [47] worked on different sub-base materials like flyash, coarse sand, granite dust, and river bed material (RBM). Granite dust was found to have the least resistance to rutting compared to the other three materials used in this study. The RBM has the maximum resistance to rutting. The static and resilient moduli were both higher for the RBM, implying that it had a better performance than the other three materials in the field. The internal friction angle ranged from 26° to 39°, and the specific gravity was nearly close to the specific gravity of river sand, which is required for the fractional sand replacement [36]. From the chemical composition of granite dust, silica/quartz (SiO_2) is the predominant mineral that helps give a high shear strength similar to sand. Therefore, practically, granite dust can be used as a substitute for sand. Some previous investigation(s) [31,35,48–50] proved that sand can be partially replaced with granite dust without changing the workability and durability of concrete. Furthermore, the shear strength of sandy soil increases with an increase in the percentage of the granite dust content in sand–granite dust mixes to a certain limit. This work concludes that granite dust is the best source of alternative to save sand availability.

6. Alternate Treatment Methods

Reinforced Granite Dust

Sand was earlier referred to as the best backfill material because of its shear strength and permeability characteristics. In view of sustainability, being a cohesion less inert material that can also be used as a backfill due to its bulk utilization, granite dust is also the best substitute for sand [33]. A backfill material should not possess any lateral displacement of facia walls and should be able to resist the settlement due to loading. Granite dust is limited in cohesion property and high density; hence, the concept of granite dust reinforcement has been explored by certain authors. Among several reinforced materials, geosynthetics are considered as the best reinforcing materials due to their workability. Rama Subbarao [51] stated that geo grid reinforcement reduces the shear deformations of granular materials. Reinforced granite dust exhibits a ductile behavior and improves apparent cohesion, but is insignificant in the case of friction. The deviatory stress is the governing factor of shear strength in the case of reinforced stone dust, especially in ductile reinforcements. The EPS geofoam was introduced in granite dust as the load-reduction key. Geofoam is more noticeable for gravelly and sandy fills. The interface shear strength of geofoam–granite dust is highly influenced by normal stress applied [52]. In addition, the presence of geofoam reduces the backfill weight. Reinforced granite dust could be used as a backfill material, even at a lower relative density that reduces facia displacement and vertical settlements [53]. The change in the dimensions of the reinforcement and its location also greatly influence the backfill behavior. Waste plastic strips serve as a reinforcing material for improving the penetration resistance of granite dust. Granite dust is highly influenced by the increase in the density of intruded plastic strips [54]. Earlier works have stated that approximately 1% of plastic strip addition with an aspect ratio of 3 increases the soil CBR. The CBR was improved by the particle interlocking in reinforced layers under the dry condition (unsoaked) and the sedimentation of fines, in which the coarser particles to the top led to a confinement in the wet condition(soaked) [51].

Backfills and soil walls are some of the bulk applications in geotechnical engineering. The reinforced granite dust material is the best substitute for sand. Being a high-density material, the reinforcement helps reduce the pressure on the facia walls, which consequently leads to the reduction of the horizontal displacement and the vertical settlement due to the interlocking phenomenon (Figure 6). The concept of reinforced granite dust also helps improve the penetration resistance due to the development of a confinement among particles.

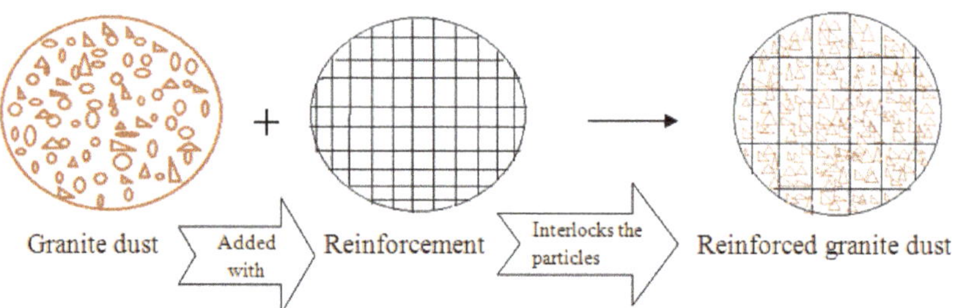

Figure 6. Mechanism of reinforced granite dust.

7. Effect of Granite Dust and Stabilizer(s) on Geotechnical Behavior

Importance of the Stabilizer

Granite dust is a non-plastic cohesion less material with a specific gravity greater than that of soil. Adding granite dust to cohesive soil filled the voids in cohesive soil, which increases the density and the shear strength. However, in some cases, granite dust alone will not be sufficient to fulfil the requirements that may cause sudden drawdown or a slip as seen from Figure 7. The soil-granite mix requires a binding agent to bring an efficient product to work in the field that can withhold the heterogeneous mass.

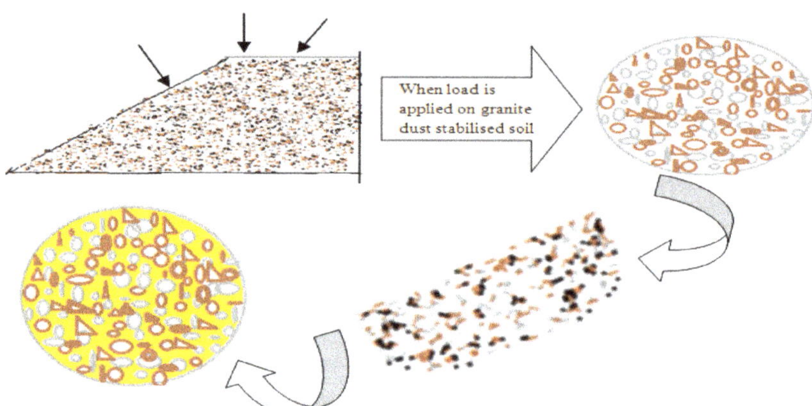

Figure 7. Significance of the secondary stabilizer.

When clay soil was stabilized with granite dust, the plastic nature of the soil decreased with the increase in the granite dust dosage, leading to failure. The presence of a binding material prevented failure and increased the cohesion which enhanced the engineering properties. Black cotton soil when amended with lime and granite dust exhibited better performance compared to untreated scenario [55]. The ettringite formation increased the strength of the soil–granite dust mix (Figure 8). Granite dust, along with calcium carbide residue (CCR) in equal amounts, showed a good influence on problematic silty clay in terms of the CBR (Figure 9) [26]. The presence of the CCR in silty clay led to pozzolanic reactions and increased the chemical bonding between particles.

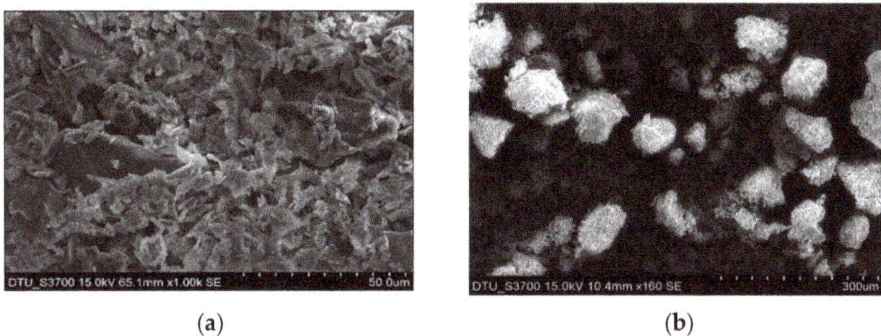

(a) (b)

Figure 8. SEM images (**a**) of black cotton soil and (**b**) black cotton soil mixed with 9% lime and plugged with 25% granite dust (Sourced from [55]).

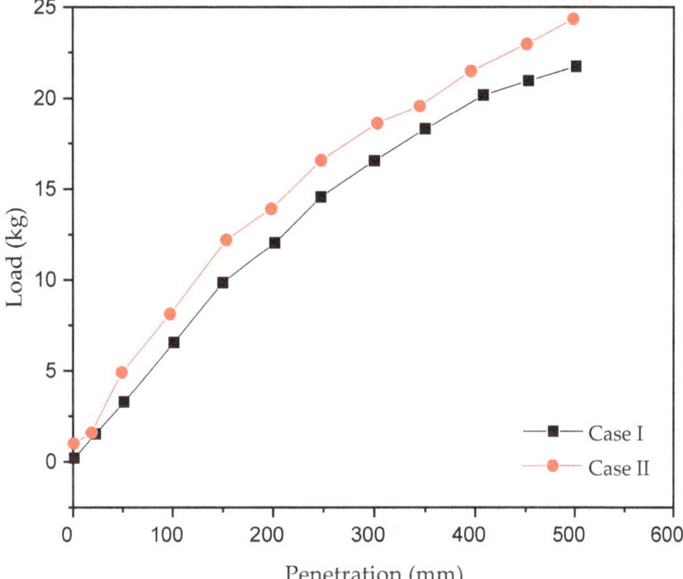

Figure 9. Response for the soaked CBR test after 14 days of curing. Case I-Soil + 5% granite dust + 5% CCR. Case II-Soil + 10% granite dust + 10% CCR (Modified after [26]).

Some works on foundation soil improvement were performed using granite dust and cement, where the soil was improved with the incremental addition of granite dust with 4% cement addition. Granite dust with 10% cement addition improved the shear strength and the hydraulic properties of lithomargic clay [46]. The presence of cement led to ettringite formation, which helps in developing additional strength and cohesion. The presence of cement also helped in decreasing the pore volume. Dutta and Sarda [54] used a waste plastic strip with granite dust/flyash to improve the kaolinite clay properties. The CBR variation in the mix was attributed to the strip intrusion and the strip length in granite dust. The high-plastic silt was added with cement and granite to improve the CBR [42]. Quarry wastes in the form of granite powder and muck could be used as a supplemental subgrade material when added with lime [56]. The effects of quick lime on compacted granite dust were also studied, and the bearing capacity was found to improve with quick lime addition. Table 8 provides the summary of various earlier works which necessitated the inclusion of binder to granite dust.

Table 8. A Summary of works to improve the properties of soil mixed with granite dust and a binder.

Soil Type	Outcome	Reference
Black cotton soil with lime and granite dust	Cohesion increased. Good improvement observed in engineering properties	[55]
Silty clay with CCR and granite dust	Good chemical bonding appeared. CBR increased	[26]
Lithomargic clay with granite dust and cement	Strength and cohesion increased. Pore volume decreased	[46]
Kaolinite clay with granite dust/flyash with waste plastic	CBR increased	[54]
High plastic silt with cement and granite dust	CBR improved	[42]

8. Practical Applications of Granite Dust

An embankment was constructed in Korea using locally available silty material and granite dust sourced from two different quarry sites (biotite granite quarry from Yangju, Gyeonggi province; granitic gneiss from Gongju, South Chung cheong province) [45]. The granite dust is added in multiples of 25% from 0 to 100 and its response to enhancement in targeted geotechnical properties was determined. This case study revealed the fact that, an embankment of silty material stabilized with granite dust should attain a gradient of 1:1.8 for 10 m height and 1:1.5 for 15 m height in order to satisfy the stability analysis as per Korean standards. Up on inclusion of granite dust, the specific gravity of the mix increased whereas the MDD and shear strength of the mix decreased as seen from Table 9.

Table 9. Effect of granite dust on the shear parameters of a local silty soil (Modified after [45]).

Granite Dust: Natural Soil	Yangju, Gyeonggi Province		Gongju, South Chungcheong	
	C (t/m^2)	φ	C(t/m^2)	Φ
100:0	0.52	30.2	0.45	29.3
75:25	0.51	30.8	0.51	30.6
50:50	0.48	32.1	0.48	31.5
25:75	0.44	33.4	0.4	34.2

In Jimma town of Ginjo kebele, Ethiopia, an expansive soil (clayey soil) at subgrade level was stabilized using granite dust [57]. The dosage of granite dust was limited to 50% (added in increments of 5%). CBR requirements for subgrade were met at 30% to 35% of granite dust addition as seen in Figure 10. The thickness of the subgrade was found to reduce by 20.6% compared to an untreated case. Their study concluded that clay-granite dust soil satisfies the requirements for subgrade layer [57].

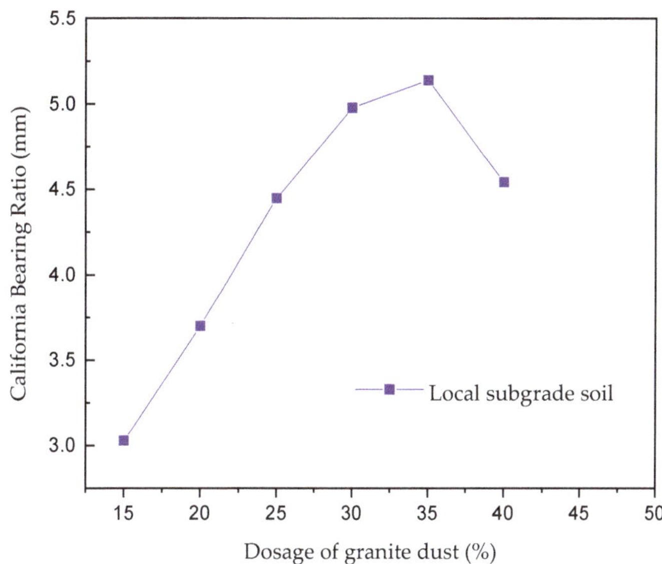

Figure 10. Improvement in the CBR of subgrade soil with addition of granite dust (Modified after [57]).

9. Conclusions

The current review article dealt with the generation of granite dust and shed light on its influence on the engineering properties of soils exhibiting different mineralogy. The workability of granite dust amended soils in the presence of an additional stabilizer is critically reviewed. The performance of granite dust as a backfill and a pavement material is discussed. The mechanism behind the improvement of each soil engineering property with granite dust addition is explained. The major outcomes of this review article are:

- This significant improvement in Atterberg limits and Compaction characteristics is attributed to the increase in the coarser fraction and specific gravity, followed by a decrease in the water absorption capacity of the soil.
- The interfacial friction of cohesive soil increased and cohesion value decreased due to the presence of coarser particles that fills the voids in clays, thereby increasing the friction component between the soil particles.
- Soaked CBR values are increased due to the improvement in corresponding Maximum dry density and Shear strength.
- A small amount of additive (calcium-based stabilizers) (<10%) with granite dust enhances the engineering properties of cohesive soils by causing net reduction in the pore volume, assisting in the rapid formation of ettringite, and substantially enhancing the tensile strength.
- Granite dust is a highly recommended material as a replacement of sand in concrete and geotechnical applications due to its chemical composition and interfacial friction angle.

Granite dust is a sustainable and remarkable material exhibiting relatively low embodied energy levels. For a given scenario, when granite dust is amended with native soil, the amount of CO_2 released due to granite dust addition is compensated by the reduced use of locally available materials. Accordingly, granite dust addition results in reduced carbon footprint values, and this treatment strategy is close to carbon neutral.

Proposed Research Gaps

- To explore the particle size effect of granite dust on the strength characteristics of the soil.

- Dynamic studies on the granite dust stabilized soil can be explored for future rail and roadway applications.
- The stability of embankments and long-term durability of highways constructed with granite dust amended soils may be carried out.

Author Contributions: Conceptualization, G.A. and A.A.B.M.; methodology, G.A. and A.A.B.M.; formal analysis, G.A.; A.A.B.M. and A.A.; resources, A.A.B.M. and A.A.; data curation, G.A. and A.A.B.M.; writing—original draft preparation, G.A., A.A.B.M. and A.A.; writing—review and editing, A.A.B.M. and A.A.; visualization, G.A. and A.A.B.M.; supervision, A.A.B.M.; project administration, A.A.B.M. and A.A.; funding acquisition, A.A. All authors have read and agreed to the published version of the manuscript.

Funding: College of Engineering Research Center and Deanship of Scientific Research at King Saud University in Riyadh, Saudi Arabia.

Data Availability Statement: The content presented here is sourced from existing published literature. Hence, this clause doesn't arise.

Acknowledgments: The authors acknowledge the College of Engineering Research Center and Deanship of Scientific Research at King Saud University in Riyadh, Saudi Arabia, for their financial support for the research work reported in this article. The authors thank the reviewers for their constructive comments which helped the cause of the manuscript.

Conflicts of Interest: The authors declare no conflict of interest.

References

1. Almajed, A.; Lateef, M.A.; Moghal, A.A.B.; Lemboye, K.K. State-of-the-Art Review of the Applicability and Challenges of Microbial-Induced Calcite Precipitation (MICP) and Enzyme-Induced Calcite Precipitation (EICP) Techniques for Geotechnical and Geoenvironmental Applications. *Crystals* **2021**, *11*, 370. [CrossRef]
2. Moghal, A.A.B. State-of-the-Art Review on the Role of Fly Ashes in Geotechnical and Geo environmental Applications. *J. Mater. Civ. Eng.* **2017**, *29*, 04017072. [CrossRef]
3. Moghal, A.A.B. Geotechnical and physico-chemical characterization of low lime fly ashes. *Adv. Mater. Sci. Eng.* **2013**, *2013*, 1. Available online: https://www.hindawi.com/journals/amse/2013/674306 (accessed on 1 July 2021). [CrossRef]
4. Ashfaq, M.; Lal, M.H.; Moghal, A.A.B.; Murthy, V.R. Carbon Footprint Analysis of Coal Gangue in Geotechnical Engineering Applications. *Indian Geotech. J.* **2019**, *50*, 646–654. [CrossRef]
5. Ashfaq, M.; Heeralal, M.; Moghal, A.A.B. Characterization studies on coal gangue for sustainable geotechnics. *Innov. Infrastruct. Solut.* **2020**, *5*, 1–12. [CrossRef]
6. Behnood, A. Soil and clay stabilization with calcium- and non-calcium-based additives: A state-of-the-art review of challenges, approaches and techniques. *Transp. Geotech.* **2018**, *17*, 14–32. [CrossRef]
7. Dash, S.K.; Hussain, M. Lime stabilization of soils: Reappraisal. *J. Mater. Civ. Eng.* **2012**, *24*, 707–714. [CrossRef]
8. Moghal, A.A.B.; Kareem Obaid, A.A.; Al-Refeai, T.O.; Al-Shamrani, M.A. Compressibility and durability characteristics of lime treated expansive semiarid soils. *J. Test. Eval.* **2014**, *43*. [CrossRef]
9. Osinubi, K.J.; Nwaiwu, C.M. Compaction delay effects on properties of lime-treated soil. *J. Mater. Civ. Eng.* **2006**, *18*, 250–258. [CrossRef]
10. Moghal, A.A.B.; Ashfaq, M.; Al-Al-Obaid, A.A.K.H.; Abbas, M.F.; Al-Mahbashi, A.M.; Shaker, A.A. Compaction delay and its effect on the geotechnical properties of lime treated semi-arid soils. *Road Mater. Pavement Des.* **2020**, *22*, 2626–2640. [CrossRef]
11. Ali, H.; Mohamed, M. The effects of compaction delay and environmental temperature on the mechanical and hydraulic properties of lime-stabilized extremely high plastic clays. *Appl. Clay Sci.* **2017**, *150*, 333–341. [CrossRef]
12. Al-Aghbari, M.Y.; Mohamedzein, Y.E.-A.; Taha, R. Stabilisation of desert sands using cement and cement dust. *Proc. Inst. Civ. Eng. -Ground Improv.* **2009**, *162*, 145–151. [CrossRef]
13. Consoli, N.C.; Montardo, J.P.; Prietto, P.D.M.; Pasa, G.S. Engineering behavior of a sand reinforced with plastic waste. *J. Geotech. Geoenviron. Eng.* **2002**, *128*, 462–472. [CrossRef]
14. Chen, Q.; Indraratna, B.; Carter, J.; Rujikiatkamjorn, C. A theoretical and experimental study on the behaviour of lignosulfonate-treated sandy silt. *Comput. Geotech.* **2014**, *61*, 316–327. [CrossRef]
15. Cheng, Q.; Zhang, J.; Zhou, N.; Guo, Y.; Pan, S. Experimental study on unconfined compression strength of polypropylene fiber reinforced composite cemented clay. *Crystals* **2020**, *10*, 247. [CrossRef]
16. Moghal, A.A.B.; Vydehi, K.V. State-of-the-art Review on Efficacy of Xanthan Gum and Guar Gum Inclusion on the Engineering Behavior of Soils. *Innov. Infrastruct. Solut.* **2021**, *6*, 1–14. [CrossRef]

17. Ashfaq, M.; Heera Lal, M.; Moghal, A.A.B. Static and Dynamic Leaching Studies on Coal Gangue. In *Sustainable Environmental Geotechnics 2020*; Lecture Notes in Civil Engineering; Reddy, K.R., Agnihotri, A.K., Yukselen-Aksoy, Y., Dubey, B.K., Bansal, A., Eds.; Springer International Publishing: Cham, Switzerland, 2020; pp. 261–270. [CrossRef]
18. Moghal, A.A.B.; Lateef, M.A.; Mohammed, S.A.S.; Ahmad, M.; Usman, A.R.A.; Almajed, A. Heavy Metal Immobilization Studies and Enhancement in Geotechnical Properties of Cohesive Soils by EICP Technique. *Appl. Sci.* **2020**, *10*, 7568. [CrossRef]
19. Moghal, A.A.B.; Lateef, M.A.; Mohammed, S.A.S.; Lemboye, K.K.; Chittoori, B.C.S.; Abdullah Almajed, A. Efficacy of Enzymatically Induced Calcium Carbonate Precipitation in the Retention of Heavy Metal Ions. *Sustainability* **2020**, *12*, 7019. [CrossRef]
20. Putra, H.; Yasuhara, H.; Kinoshita, N.; Hirata, A. Optimization of enzyme-mediated calcite precipitation as a soil-improvement technique: The effect of aragonite and gypsum on the mechanical properties of treated sand. *Crystals* **2017**, *7*, 59. [CrossRef]
21. David Suits, L.; Sheahan, T.; Soosan, T.; Sridharan, A.; Jose, B.; Abraham, B. Utilization of quarry dust to improve the geotechnical properties of soils in highway construction. *Geotech. Test. J.* **2005**, *28*, 11768. [CrossRef]
22. Sridharan, A.; Soosan, T.G.; Jose, B.T.; Abraham, B.M. Shear strength studies on soil-quarry dust mixtures. *Geotech. Geol. Eng.* **2006**, *24*, 1163–1179. [CrossRef]
23. Birch, W.; Datson, H. Reducing the Environmental Effects of Aggregate Quarrying: Dust, Noise & Vibration. 2008. Available online: https://miningandblasting.files.wordpress.com/2009/09/blast_dust_and_noise_control.pdf (accessed on 5 November 2021).
24. Saghafi, B.; Al Nageim, H.; Atherton, W. Mechanical behavior of a new base material containing high volumes of limestone waste dust, PFA, and APC residues. *J. Mater. Civ. Eng.* **2013**, *25*, 450–461. [CrossRef]
25. Kumar, D.S.; Hudson, W.R. Use of Quarry Fines for Engineering and Environmental Applications. *Natl. Stone Assoc.* **1992**. Available online: https://library.ctr.utexas.edu/digitized/iacreports/1992_specrsrch_natlstoneassoc.pdf (accessed on 1 October 2021).
26. Kumrawat, N.; Ahirwar, S.K. Performance analysis of black cotton soil treated with calcium carbide residue and stone dust. *Int. J. Eng. Res. Sci. Technol.* **2014**, *3*. Available online: http://www.ijerst.com/currentissue.php (accessed on 15 July 2021).
27. Antonov, G.I.; Nedosvitii, V.P.; Semenenko, O.M.; Kulik, A.S.; Gerashchuk, Y.D.; I'chenko, N.V.; Poltavets, L.K. Use of dolomite dust for manufacturing stabilized dolomite refractories. *Refract. Ind. Ceram.* **1997**, *38*, 238–243. [CrossRef]
28. Brooks, R.; Udoeyo, F.F.; Takkalapelli, K.V. Geotechnical properties of problem soils stabilized with fly ash and limestone dust in Philadelphia. *J. Mater. Civ. Eng.* **2011**, *23*, 711–716. [CrossRef]
29. Nweke, O.M.; Okogbue, C.O. The potential of cement stabilized shale quarry dust for possible use as road foundation material. *Int. J. Geo-Eng.* **2017**, *8*, 29. [CrossRef]
30. Haldar, M.K.; Das, S.K. Effect of substitution of sand stone dust for quartz and clay in triaxial porcelain composition. *Bull. Mater. Sci.* **2012**, *35*, 897–904. [CrossRef]
31. Cheah, C.B.; Lim, J.S.; Ramli, M.B. The mechanical strength and durability properties of ternary blended cementitious composites containing granite quarry dust (GQD) as natural sand replacement. *Constr. Build. Mater.* **2019**, *197*, 291–306. [CrossRef]
32. Jain, A.K.; Jha, A.K.; Shivanshi. Improvement in subgrade soils with marble dust for highway construction: A comparative study. *Indian Geotech. J.* **2020**, *50*, 307–317. [CrossRef]
33. Kandolkar, S.S.; Mandal, J.N. Effect of reinforcement on stress–strain behavior of stone dust. *Int. J. Geotech. Eng.* **2014**, *8*, 383–395. [CrossRef]
34. Chetia, M.; Baruah, M.P.; Sridharan, A. Contemporary Issues in Geoenvironmental Engineering, Sustainable Civil Infrastructures. In *Effect of Quarry Dust on Compaction Characteristics of Clay*; Singh, D.N., Galaa, A., Eds.; Springer International Publishing: Cham, Switzerland, 2018; pp. 78–100. [CrossRef]
35. Zafar, M.S.; Javed, U.; Khushnood, R.A.; Nawaz, A.; Zafar, T. Sustainable incorporation of waste granite dust as partial replacement of sand in autoclave aerated concrete. *Constr. Build. Mater.* **2020**, *250*, 118878. [CrossRef]
36. Natasha, K.; Chetia, M. Shear strength of Rock quarry dust and Sand mix. *Emerg. Trends Civ. Eng.* **2020**, 1–13. [CrossRef]
37. Oyediran, I.A.; Idowu, O.D. Performance Analysis of some Quarry dust treated soils. *J. Min. Geol.* **2017**, *53*, 45–53.
38. Bahoria, B.; Parbat, D.K.; Nagarnaik, P.B.; Waghe, U.P. Effect of replacement of natural sand by quarry dust and waste plastic on compressive & split tensile strength of M20 concrete. In Proceedings of the International Conference on Engineering (NUiCONE 2013), Ahmedabad, India, 28–30 November 2013; Available online: www.sciencedirect.com (accessed on 25 November 2021).
39. Etim, R.K.; Ekpo, D.U.; Attah, I.C.; Onyelowe, K.C. Effect of micro sized quarry dust particle on the compaction and strength properties of cement stabilized lateritic soil. *Clean. Mater.* **2021**, *2*, 100023. [CrossRef]
40. Bloodworth, A.J.; Scott, P.W.; McEvoy, F.M. Digging the backyard: Mining and quarrying in the UK and their impact on future land use. *Land Use Policy* **2009**, *26*, S317–S325. [CrossRef]
41. Sanjay, M.; Sindhi, P.R.; Vinay, C.; Ravindra, N.; Vinay, A. Crushed rock sand—An economical and ecological alternative to natural sand to optimize concrete mix. *Eng. Mater. Sci.* **2016**, *8*, 345–347. [CrossRef]
42. Priyankara, N.H.; Wijesooriya, R.M.S.D.; Jayasinghe, S.N.; Wickramasinghe, W.R.M.B.E.; Yapa, S.T.A.J. Suitability of quarry dust in geotechnical applications to improve engineering properties. *Eng. J. Inst. Eng. Sri Lanka* **2009**, *42*. [CrossRef]
43. Nwaiwu, C.; Mshelia, S.; Durkwa, J. Compactive effort influence on properties of quarry dust-black cotton soil mixtures. *Int. J. Geotech. Eng.* **2012**, *6*, 91–101. [CrossRef]
44. Chetia, M.; Sridharan, A. A Review on the Influence of Rock Quarry Dust on Geotechnical Properties of Soil. In *Geo-Chicago 2016*; American Society of Civil Engineers: Reston, VA, USA, 2016; pp. 179–190. [CrossRef]

45. Song, Y.-S.; Kim, K.-S.; Woo, K.-S. Stability of embankments constructed from soil mixed with stone dust in quarry reclamation. *Environ. Earth Sci.* **2011**, *67*, 285–292. [CrossRef]
46. Nayak, S.; Sarvade, P.G. Effect of cement and quarry dust on shear strength and hydraulic characteristics of Lithomargic clay. *Geotech. Geol. Eng.* **2012**, *30*, 419–430. [CrossRef]
47. Kumar, P.; Chandra, S.; Vishal, R. Comparative study of different sub base materials. *J. Mater. Civ. Eng.* **2006**, *18*, 576–580. [CrossRef]
48. Mohammed, S.; Raman, S.N.; Jain, M.F.M. Utilization of Quarry Waste Fine Aggregate in Concrete Mixes. *J. Appl. Sci. Res.* **2007**, *3*, 202–208.
49. Danish, A.; Mosaberpanah, M.A.; Salim, M.U.; Feduik, R.; Rashid, M.F.; Waqas, R.M. Reusing marble and granite dust as cement replacement in cementitious composites: A review on sustainability benefits and critical challenges. *J. Build. Eng.* **2021**, *44*, 102600. [CrossRef]
50. Singh, S.; Nagar, R.; Agarwal, V.; Rana, A.; Tiwari, A. Sustainable Utilization of Granite cutting waste in high strength concrete. *J. Clean. Prod.* **2016**, *116*, 223–235. [CrossRef]
51. Rama, S.R.G.V. Studies on Hexagonal Wire Mesh-Reinforced Crushed Stone Dust. *Slovak J. Civ. Eng.* **2018**, *26*, 20–25. [CrossRef]
52. Beju, Y.Z.; Mandal, J.N. Experimental Investigation of Shear strength behaviour of Stone dust-EPS geofoam Interface. *J. Hazard. Toxic Radioact. Waste* **2018**, *22*, 04018033. [CrossRef]
53. Kandolkar, S.S.; Mandal, J.N. Behavior of reinforced-stone dust walls with backfill at varying relative densities. *J. Hazard. Toxic Radioact. Waste* **2016**, *20*, 04015010. [CrossRef]
54. Dutta, R.K.; Sarda, V.K. CBR behaviour of waste plastic strip-reinforced stone dust/fly ash overlying saturated clay. *Turk. J. Eng. Environ. Sci.* **2007**, *31*, 171–182.
55. Mudgal, A.; Raju, S.; Sahu, A.K. Effect of lime and stone dust in the geotechnical properties of black cotton soil. *Int. J. Geomate* **2014**, *7*, 1033–1039. [CrossRef]
56. Manandhar, S.; Suetsugu, D.; Hara, H.; Hayashi, H. Performance of Waste Quarry By-Products as A Supplementary Recycled Subgrade Material. In Proceedings of the 9th International Symposium on Lowland Technology (ISLT 2014), Saga, Japan, 29 September–1 October 2014; pp. 271–278.
57. Jemal, A.; Agon, E.C.; Geremew, A. Utilization of crushed stone dust as A stabilizer for sub grade soil: A case study in Jimma town. *Int. J. Eng.* **2019**, *17*, 55–63.

MDPI

Review

State-of-the-Art Review of the Applicability and Challenges of Microbial-Induced Calcite Precipitation (MICP) and Enzyme-Induced Calcite Precipitation (EICP) Techniques for Geotechnical and Geoenvironmental Applications

Abdullah Almajed [1,*], Mohammed Abdul Lateef [2], Arif Ali Baig Moghal [3] and Kehinde Lemboye [1]

1 Department of Civil Engineering, College of Engineering, King Saud University, P.O. Box 800, Riyadh 11421, Saudi Arabia; 438105781@student.ksu.edu.sa
2 Department of Civil Engineering, HKBK College of Engineering, Bengaluru 560045, India; abdulmohammed040@gmail.com
3 Department of Civil Engineering, National Institute of Technology, Warangal 506004, India; baig@nitw.ac.in or reach2arif@gmail.com
* Correspondence: alabduallah@ksu.edu.sa

Abstract: The development of alternatives to soil stabilization through mechanical and chemical stabilization has paved the way for the development of biostabilization methods. Since its development, researchers have used different bacteria species for soil treatment. Soil treatment through bioremediation techniques has been used to understand its effect on strength parameters and contaminant remediation. Using a living organism for binding the soil grains to make the soil mass dense and durable is the basic idea of soil biotreatment. Bacteria and enzymes are commonly utilized in biostabilization, which is a common method to encourage ureolysis, leading to calcite precipitation in the soil mass. Microbial-induced calcite precipitation (MICP) and enzyme-induced calcite precipitation (EICP) techniques are emerging trends in soil stabilization. Unlike conventional methods, these techniques are environmentally friendly and sustainable. This review determines the challenges, applicability, advantages, and disadvantages of MICP and EICP in soil treatment and their role in the improvement of the geotechnical and geoenvironmental properties of soil. It further elaborates on their probable mechanism in improving the soil properties in the natural and lab environments. Moreover, it looks into the effectiveness of biostabilization as a remediation of soil contamination. This review intends to present a hands-on adoptable treatment method for in situ implementation depending on specific site conditions.

Keywords: enzyme-induced calcite precipitation; microbial-induced calcite precipitation; geotechnical engineering; geoenvironmental engineering

Citation: Almajed, A.; Lateef, M.A.; Moghal, A.A.B.; Lemboye, K. State-of-the-Art Review of the Applicability and Challenges of Microbial-Induced Calcite Precipitation (MICP) and Enzyme-Induced Calcite Precipitation (EICP) Techniques for Geotechnical and Geoenvironmental Applications. *Crystals* **2021**, *11*, 370. https://doi.org/10.3390/cryst11040370

Academic Editor: Cesare Signorini

Received: 16 February 2021
Accepted: 30 March 2021
Published: 1 April 2021

1. Introduction

Improving soil properties has become inevitable when finding available places with soils of considerable strength is difficult. Instead of finding areas with soil of good geotechnical properties, soil improvement using soil stabilization techniques in the desired location seems preferable. Soil stabilization focuses on the improvement of the soil's bearing capacity and the reduction of settlement and deformation [1–4]. Ground improvement is most effectively addressed through soil stabilization. Researchers have tested various techniques of stabilization, and some have been vastly implemented in the field, specifically in the past four decades. Land with good soil performance is becoming scarce due to population growth. Therefore, using techniques to improve the performance of existing soil has become necessary. Among the soil stabilization methods, mechanical and chemical stabilization are widely acclaimed. Mechanical stabilization involves the process of densifying the soil mass by expelling air voids with nominal variation in water content for better performance, whereas chemical stabilization involves amending the soil with additives to achieve the desired density, reduce permeability, or improve soil strength [5].

Chemical stabilization utilizes cementitious materials like lime, asphalt, and chemicals such as silicates and polymers, and Portland cement. These affect the chemical form of the soil matrix, improving its geotechnical behavior [6]. Chemical stabilization has attracted greater attention due to its effectiveness in soil improvement using traditional binders with a calcium base, like lime, fly ash, and cement, or novel stabilizers, like acids, salts, lignosulfonates, enzymes, petroleum emulsions, resins, and polymers [7]. In this method, the additives must be mechanically mixed with the soil in its natural state. With the influence of a specific chemical stabilizer on the site, the additive mixed must be properly distributed in the soil mass to ensure its effectiveness [8].

Stabilization techniques also include physical methods wherein soil is reinforced to achieve more strength and reduced settlements using reinforcing bars, strips, grids, fibers, and sheets [9]. Due to the rapid population growth, the scarcity of land for construction has increased, leading to construction activities on problematic soils [10]. Using stabilization techniques for problematic soils ensures the safety of structures built on them by improving soil performance against loading. Therefore, soil stabilization serves vital purposes in civil engineering. Apart from increasing the soils' strength, reducing their permeability, improving their bearing capacity, and filling voids, contaminant remediation has also been used to reduce the hazardous effects of pollutants (heavy metals) present in the soils due to anthropogenic activities. Heavy metal contamination in soils is a threat because heavy metals intrude in the food chain and cause hazardous effects [11]. Many techniques are being developed to reduce or recover heavy metals from polluted sites. Physical and chemical methods are proven to be effective in removing a wide spectrum of pollutants. However, the process consumes a lot of energy and may require extra effort to reach the desired level of heavy metal removal [12]. The use of soil stabilization techniques has been proven to be useful in both geotechnical and geoenvironmental applications.

Biologically mediated soil modification is also an emerging trend in soil stabilization. Mitchell and Santamarina [13] explored the possibility of using the biological components of rocks and soils to trigger interests in biological applications in geotechnical engineering. Dejong et al. [14] considered their work as the first of its kind. They also quoted the National Research Council of USA [15] regarding the biogeotechnical field being an important research area in the 21st century. Ants and termites amend soils, making tunnels water-resistant; this shows that biogeotechnical processes happen naturally [16]. Nature has always been an inspiration to humans for exploring possibilities of reaping benefits by replicating natural phenomena. One similar attempt is made in soil stabilization by domesticating microbes to improve soil performance. Mineralization through microbes such as bacteria, fungus, and algae is observed in nature [17], and the mineralization process by using bacteria has various applications in engineering [18]. Bacterial intrusion called microbial-induced calcite precipitation (MICP) in soil treatment improves soils' geotechnical properties through the precipitation of calcium carbonate ($CaCO_3$), binding soil grains together [19]. Soil stabilization through microbes, which precipitate $CaCO_3$, is applied to different soil types like liquefiable soils [20], sand [21–23], sandy soil [24], and tropical residual soils [10]; it is used for the remediation of porous media [25] and the restoration of calcareous stone materials [26]. MICP is even used to seal rock fractures [27], treat wastewater [28], and reduce beach sand erosion [29].

With the understanding of calcite precipitation in the soil through the bacterial method, a similar method of precipitation without bacterial intrusion in the soil was attempted by directly using the urease enzyme, which precipitates calcite. This method of soil stabilization through the precipitation of $CaCO_3$ with the use of enzymes instead of microorganisms is called enzyme-induced calcite precipitation (EICP) [30–33]. This method also has a wide range of engineering applications for soil treatment such as stabilizing slopes, avoiding erosion due to wind and water, reducing the scouring of soil, checking the seepage beneath levees, improving the bearing capacity of soil, tunnelling, and controlling seismic settlement [34] and dust [35]. Further biostabilization of soils has also found its way in the remediation of contaminants. This review considers the available research studies conducted on the EICP and MICP techniques and discusses their applicability and challenges in geotechnical and geoenvironmental applications.

2. Overview of the Microbial-Induced Calcite Precipitation (MICP) and Enzyme-Induced Calcite Precipitation (EICP) Methods

In MICP, precipitates of calcium carbonate are produced by a combination of dissolved calcium ions and urea produced by the urease bacteria after hydrolysis [36]. Due to the complexity of the cultivation of urease-producing microorganisms and the uncontrollability of enzymatic activities, urease activity is incited directly with enzymes, specifically with urease [37–39]. The enzyme-mediated precipitation of calcite is achieved without any bacterial activity. The EICP method is used to improvise the geotechnical properties of soils by using an aqueous chemical solution that precipitates calcite within soil voids. The precipitates help in roughening and binding soil grains and even in pore filling, thereby improving the strength and stiffness of the soils. The EICP method is also distinguished from MICP by its use of free urease instead of bacteria. Enzymes can be derived from microbes, fungi, and agricultural sources [40,41].

Hydrolysis of Urea

The process before the precipitation of $CaCO_3$ in soil voids in biotreatment starts from urea hydrolysis initiated by the urease enzyme. During urea hydrolysis, the decomposition of urea leads to the formation of carbon dioxide and ammonia. The water in the system helps ammonia to dissolve and form hydroxide and ammonia ions. These ions create an environment that allows an increase in the solution's pH. Simultaneously, carbon dioxide dissolves in water and develops ions of bicarbonates and hydrogen due to the increased pH of the environment; carbonates are formed due to the reaction between bicarbonate and hydroxide ions, forming carbonate ions and calcium carbonate in the presence of calcium ions; the calcium carbonate formed is precipitated because of its low dissolution rate in water [42]. Urea hydrolysis can be imitated either through the urease produced by bacteria or directly by using the free urease enzyme. Therefore, the reactions that take place in the soil is common in both MICP and EICP, but these two methods differ in terms of the source that initiates the hydrolysis [43]. Equations (1) and (2) show the chemical reactions that represent urea hydrolysis, leading to the precipitation of $CaCO_3$ [44].

$$CO(NH_2)_2 + 2H_2O \overset{\text{Urease}}{\rightarrow} 2NH_4^+ + CO_3^{2-} \quad (1)$$

$$Ca^{2+} + CO_3^{2-} \rightarrow CaCO_3 \quad (2)$$

Figure 1 presents the EICP and MICP mechanisms. The precipitation of $CaCO_3$ by microbes was tested with and without urea by Golovkina et al. [45], who inferred that two metabolic routes—autotrophic and heterotrophic—were responsible for the precipitation of $CaCO_3$ in the soil. Precipitation with urea was initiated with the usual urea hydrolysis, and the urea free medium also successfully precipitated calcite at low pH values with different bacteria strains.

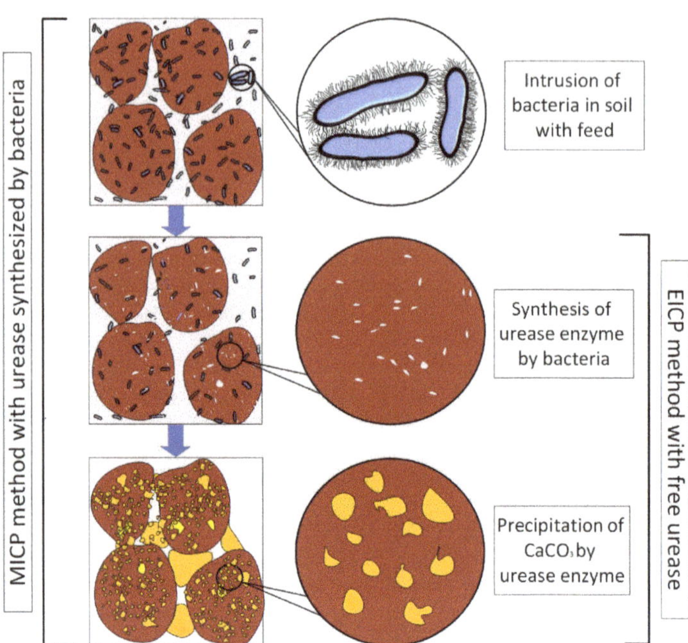

Figure 1. Mechanisms of microbial-induced calcite precipitation (MICP) and enzyme-induced calcite Precipitation (EICP) in the $CaCO_3$ precipitation.

3. Bacterial Precipitation of $CaCO_3$ in Soils

Bacteria can adapt to varying environmental conditions due to their physiology and genetics, which is because they have existed in nature since three and a half billion years ago [46]. Some important features of microorganisms are that they bear cells with a simple structure without an enclosed nucleus. With multiple chromosomes and unique chemical compositions, microbes are characterized and classified by their cell wall, nutrients, RNA, DNA, and type of biochemical changes [47,48]. Bacteria are the most widely found microbes in soils. A bacterial cell has a diameter in the range of 0.5–3.0 μm, with an elongated, spiral, or spherical shape [49]. The bacterial activity in producing calcite for soil treatment involves various bacteria. Burbank et al. [21] studied $CaCO_3$ precipitation through biological mediation using indigenous bacteria and found them effective in increasing the liquefaction resistance of sands. They concluded that using indigenous bacteria is advisable to make soil biomodification more economically feasible. Lee et al. [50] studied the improvement of soil properties using organic materials and found a 1.5–2.5 times increase in soil strength when compared with samples without an organic stabilizer. The stabilizer used was an organic acid material named Con-α, which was developed by Osaki Corp. in Japan. It allows microbe proliferation with aging. The importance of using this organic material is to ensure safety for the environment. pH tests confirmed that the organic acid was eco-friendly. The tests were conducted by preparing samples mixed with 3% and 6% of the organic biostabilizer by weight of the soil, which were tested for different ages. The authors concluded that the pores in the soil were filled with matter produced by the microbes, improving the soil's strength. Although MICP has potential for soil improvement, upscaling this method, optimization and training/educating the technicians on its effective applications are identified as challenges in its implementation [51]. To grade the worth of MICP, the rate of $CaCO_3$ precipitation is said to be 60 kg/m^3 of soil [52]. MICP is carried out by using bacteria such as *Sporosarcina pasteurii* (*S. pasteurii*)/*Bacillus pasteurii* [21–25,50,52–57], *Idiomarina insulisalsae* [24], *Pseudomonas putida* [54], *Bacillus cereus* [58], *Bacillus sphaericus* [59], and indigenous bacteria [60]. MICP is even used for the improvement of the performance of construction materials [54], sediment stabilization [61], and reduction of coastal erosion [62].

Soon et al. [10] used MICP for the improvement of the engineering properties of soils. The species of bacteria used was *Bacillus megaterium*, combined with other cementing reagents. They found that the $CaCO_3$ precipitates were effective in soil stabilization, capable of improving shear strength, and even useful in reducing the hydraulic conductivity of soil and sand. Proto et al. [63] studied the reduction of the permeability of saturated sand through the formation of biofilms on the surface of the sand grains, biofilms being the accumulation of cells and extracellular polymeric substances in an organic process. The process of bioaugmentation was incited by strains of non-native bacteria injected in the sand, and bacteria already present in the sand contributed to the precipitation. The use of biofilms resulted in a considerable decrease in the permeability of the soil. Bioaugmentation is a commonly used method for removing contaminants from soil mass; this process is initiated by allochtonic or autochtonic microbes against nondegradable organic matter from soil [64]. The transformation of harmful compounds into different forms by using bacteria should be possible, showing bioaugmentation [65]. The use of additives, along with microbes, has also been tested for the improvement of soil performance. Zhao et al. [66] used fiber felt scrap (activated carbon) with the MICP technique to treat sand. They observed that unconfined compressive strength (UCS) and tensile strength improved, showing the possibility of using a discarded scrap waste material along with MICP treatment to improve soil strength.

4. Enzyme Usage and Sources for Soil Treatment

The most widely studied enzyme source for soil treatment is the jack bean plant, technically termed *Canavalia ensiformis*. This plant is a draught-resistant species classified in the Fabaceae family [34]. Larsen et al. [67] reported that calcite precipitation increases ten times with the use of the jack bean meal instead of pure urease. The urease enzyme was first crystallized in 1926 by James B. Sumner [68–70]. Oliveira et al. [71] conducted a study on the effect of soil type on the precipitation of calcium carbonate by EICP treatment in the soil by jack bean urease for its improvement. They found that the precipitation of $CaCO_3$ increased the strength of the soil by 40–106%, but the test proved ineffective for organic soils. Renjith et al. [5] used a commercially available enzyme-based additive named "Eko soil" for the construction of unpaved roads in Australia. They found that the treatment methods could be used for cost-effective and sustainable unpaved roads. They also surveyed other enzymes that were commercially available or manufactured from fermented matter, which is converted into a chemical, liquid, or organic form. Javadi et al. [72] used urease enzyme extracted from watermelon seeds. They found that the theoretical maximum precipitation of calcite was around 64%, which was considered promising for soil treatment using urease extracted from watermelon seeds. It is also important to note that using enzymes for soil stabilization is expensive; the cost of enzymes is equal to 90% of the total cost of materials used [73,74].

Use of Additives in the Biotreatment

The use of additives, along with the urease enzyme, in the biotreatment has also been tested to improve the precipitation process for better soil performance. Almajed et al. [75] used non-fat milk powder as an additive to improve the urease activity and obtained surprising results wherein the amount of $CaCO_3$ precipitated and the UCS values were better than those for soils treated with enzyme solutions without non-fat milk powder. It was noted that the amount of calcite precipitated is not regarded as an indicative factor of increase in strength; rather, the precipitation pattern governs the improvement in the geotechnical properties of soils; that is, even low carbonate precipitation with a suitable pattern may lead to higher strength compared to high carbonate precipitation. The enzyme treatment is performed with an aqueous solution consisting of urea, calcium chloride, and urease enzyme in deionized water for mixing/injecting in the soil for calcite precipitation [76,77]. Hamdan and Kavazanjian [78] also used non-fat milk powder as a stabilizer, along with urease, in the enzyme solution to test its effectiveness in fugitive dust control. They observed that the treatment with the enzyme resulted in resistance to wind erosion. Putra et al. [79] used magnesium chloride as a substitute in the enzyme treatment to precipitate $CaCO_3$. They found that the ratio of precipitation was 90% of the theoretical maximum, which was obtained with the addition of a small quantity of magnesium chloride. The use of magnesium resulted in a lower precipitation rate, resulting in the higher injectivity of the enzyme solution. It changed the shape of $CaCO_3$ precipitates, simultaneously precipitating aragonite along with calcite. The use of additives to improve the efficacy of the enzyme treatment on soils has become an important part of research. Yuan et al. [39] used soybean urease for silt improvement in flooded areas and with urease. Additional materials, like glutinous powder of rice, brown sugar, and skim milk in powdered form were used to reinforce the urease activity. Hommel et al. [80] developed a numerical model for the EICP method to simulate the outcomes for different dosages of the enzyme solutions in any experimental setup. They developed a model that could give qualitative outcomes for the experimental setup modeled in the program. Therefore, EICP and MICP can be tested for their proposed outcomes using a numerical model before conducting the experiments physically.

5. Geotechnical Applications of the Biocementation Technique

The use of biostabilization methods wherein $CaCO_3$ precipitation helps in improving soils has attracted the interest of geotechnical engineers substantially [81–83]. The biotreatment of soils further needs suitable environmental conditions to achieve the desired outcomes through the precipitation of $CaCO_3$ [54,84]. However, the use of an enzyme-based stabilization method depends on factors like type of soil, method of construction, curing, and temperature, which may result in poor outcomes unless a suitable adjustment is not made to control the hindrances as per the type of enzymes [5].

5.1. Biotreatment Techniques

Mujah et al. [44] reported that MICP is effectuated by the injection, surface percolation, and premixing methods. In the injection method, the treatment solution is injected in the soil. In the premixing method, soils are mixed with the bacterial solution before dumping the soil in its place to serve the intended purpose. In the surface percolation method, the cementation solution is made to be absorbed in the soil from the surface. Wiffin et al. [81] tested the biotreatment through the injection method on a 5 m long column of sandy soil. They observed that the precipitation of $CaCO_3$ was not even along the length of the column. Sotoudehfar et al. [85] studied the factors influencing MICP applied through the injection method. They used a specially designed pump for injecting the cementation fluid into the soil. They found positive outcomes with their injection method of implementing MICP on soils. The injection method of the biotreatment was carried out by two phase injection procedures wherein initial bacterial strains were injected; later, bacterial feed was injected. Stocks-Fischer et al. [25] reported that injecting bacteria and reagents together may result in clogging at the injection site, especially when the flow rates of the fluids are low in the soil. The injection technique may be suitable when the treatment fluid is of low viscosity. In contrast, injection or biogrouting may need substantial pumping energy to achieve the desired soil strength [86]. The inoculation of bacterial strains in soils is also practiced for contaminant removal from soil mass [65].

Almajed et al. [40] studied the EICP treatment through percolation and premixing methods applied on Ottawa 20–30 sand. They interpreted their findings by comparing both methods of treatment and found that premixing was not effective in maintaining the intactness of the sand specimen, whereas the percolation method portrayed better results. An intact specimen, which could be easily tested for its strength and percolation method, also provides good interparticle bonding. Neupane et al. [87] used percolation for EICP on sand. They observed an almost uniform distribution of calcite precipitates in sand at 5 °C, whereas precipitation was reduced to 5% at a temperature of 23.5 °C. It can be understood that implementation techniques also play a vital role in the biotreatment of soils, and specific methods of implementation can be devised depending on the soil type and the environmental conditions.

5.2. Effect of Biotreatment on the Unconfined Compressive Strength (UCS) Test

The biotreatment of soils is well understood for its degree of effectiveness by UCS values [36,88–90]. Ali et al. [91] stated that the UCS test is most trusted to ascertain the effectiveness of soil stabilization methods. Sharma and Ramkrishnan [92] studied soils (fine grained) treated with MICP and observed the improvement in the UCS value of soil. They also inferred that the particle packing plays an important role in improving soil strength and even leads to improved bearing capacity, reduced settlement and permeability, and diminished shrink-swell characteristics of soils. The development of pore pressure can also be stopped with soil treatment. Strength enhancement in soils treated with biocementation methods is mainly achieved because of the adhesion of soil grains due to calcite precipitates in soil voids. Figure 2 shows a comparison of UCS values obtained for the soils before and after the biotreatment.

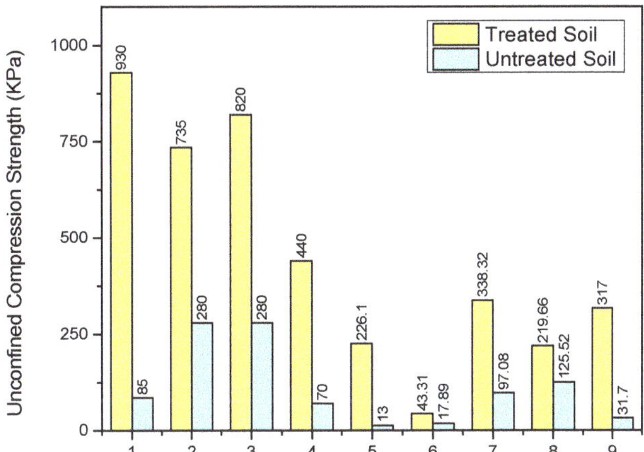

Figure 2. Comparison of unconfined compressive strength (UCS) results for soil samples before and after biotreatment. 1—Sotoudehfar et al. [85], 2—Wani and Mir [93], 3—Wani and Mir [93], 4—Moghal et al. [94], 5—Moghal et al. [94], 6—Xiao et al. [95], 7—Sharma and R [92], 8—Sharma and R [92], 9—Park et al. [96].

Table 1 shows the UCS test results obtained by researchers for different soils treated with MICP/EICP techniques. Yasuhara et al. [31] used free enzyme (urease) supplied by Kishida Chemical to treat sand through $CaCO_3$ precipitation. They found that the experiments showed the effectiveness of the enzyme treatment on UCS samples and permeability. Strength gain in the soils depends on the amount of calcite precipitated. Notably, for the substantial improvement in the stiffness and strength of soils treated with a biostabilization technique, a minimum of 4% of calcite precipitation per mass of treated soils is required [96].

Table 1. Unconfined Compressive Strength (USC) results obtained by different researchers after the biotreatment of soils.

Sl No	Bacteria Type	Type of Soil	Maximum UCS Value Obtained after Treatment (kPa)	UCS Value for Untreated Soil (kPa)	Reference
1	*Sporosarcina pasteurii*	Poorly graded sand	930	85	[85]
2	Jack Bean Urease	Ottawa 20–30 sand	88.8	-	[40]
3	*Bacillus subtilis* *Bacillus pasteurii*	Dredged soils	735 820	280	[93]
4	Pararhodobacter sp.	Fine grained sand Coarse sand Mixed sand	1330 2870 2800	-	[97]
5	Jack Bean Urease	Silica sand	1745	-	[75]
6	Jack Bean Urease	Red soil Black soil	440 226.1	70 13	[94]
8	*Sporosarcina pasteurii*	Fine to medium-grain sand	12,400	-	[98]
9	Urease Enzyme	Silica sand	380	-	[99]
10	Urease Enzyme	Sand	1600	-	[100]
11	Urease Enzyme	Silica sand	600	-	[79]
12	Urease enzyme	F-60 silica sand Ottawa 20–30 sand	529 391	-	[38]
13	*Sporosarcina pasteurii*	Soft Clay	43.31	17.89	[95]
14	*Sporosarcina pasteurii*	Sand (commercially available)	14,000	-	[101]
15	*Sporosarcina pasteurii*	Fine grained soil (CL) Fine grained soil (CH)	338.32 219.66	97.08 125.52	[92]
16	Jack Bean Urease	Poorly graded silica sand	555	-	[102]
17	In-situ soil bacteria	Poorly-graded sands	5300	-	[103]
18	Urease enzyme from watermelon seeds	Mikawa sand	3000	-	[104]
19	Jack-bean extract	Nakdong River sand	317	31.7	[96]
20	Jack Bean urease	Ottawa sand	1700	-	[105]
21	Jack Bean urease	Ottawa 20–30	1600	-	[106]
22	Terrazyme	Clay with low plasticity	1073	-	[107]

5.3. Reduction of Hydraulic Conductivity by Biotreatment

The use of biotreatment methods for permeability reduction in soils has led to effective results [10,63,90,108–112]. One of the reasons that leads to the reduction in soil permeability is the cementation of soil grains with precipitates of $CaCO_3$, leading to the blockage of connected pores in the soil mass. Although permeability reduction due to biotreatment depends on the size of soil grains, finer soils have miniscule flow paths due to the proper packing of soil grains; even the size of precipitates plays an important role in reducing permeability. Sometimes, suspended precipitates also get accommodated in the voids of the soil mass, thereby contributing further in the reduction of flow paths [108,113]. Cuthbert et al. stated that permeability reduction due to the precipitation of $CaCO_3$ depends on the quantity of precipitates; that is, with more precipitates, permeability reduction will be higher [114]. Ferris et al. [115] studied the use of bacterially precipitated calcite as a plugging material in porous media and proved a permeability reduction in the sand tested. They concluded that a 40% increase in bacteria paved the way for a 70% reduction in sand permeability, which can be attributed to greater precipitates from more bacteria, leading to a higher percentage of $CaCO_3$ precipitation in the soil mass. Reduced permeability after the soil treatment by calcite precipitates is also due to the reduction of the pore throat at the points of contact between the soil grains where precipitation occurs [82]. Gui et al. [116] used the MICP method in porous media for bioclogging. They found that permeability reduction was greater than 72%. The main reason they identified for the reduction of permeability was the formation of biofilms on the sand grain surfaces. Figure 3 shows the flow paths in a soil mass and the blockage of the flow paths by biofilms on soil grains, plugging, and sealing of pore throats with calcite after the biotreatment.

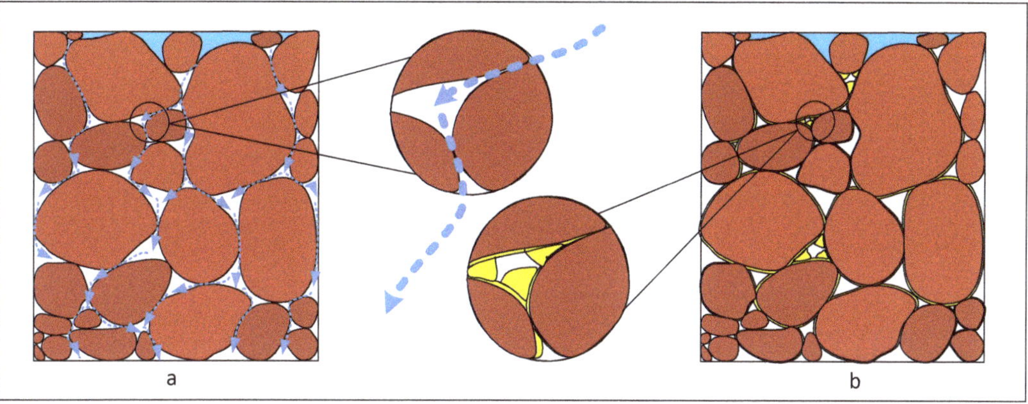

Figure 3. Schematic representation of soil mass: (**a**) Flow paths allowing water to percolate through soil mass. (**b**) Representation of the blockage of pore throats by calcite precipitation and the formation of calcite biofilms on soil grains, leading to narrowing/blockage of flow paths and reducing permeability after biotreatment.

Nemati et al. [109] compared the microbial and enzymatic processes applied for permeability reduction. They found that bacterial precipitation may include a degradable biomass developed as a plugging agent in soil pores, which may dissolve or decompose with time or after exposure to moisture, whereas enzymatically precipitated calcite proves to be a durable plugging agent that contributes to permeability reduction and is, hence, more convenient for geotechnical applications. Rittmann [117] suggested that reduction in permeability is possible because of the formation of biofilms, which reduces pore sizes after coating, clogs flow paths on the units of porous medium, and increases the friction factor of the porous medium after clogging. Chittoori et al. tested the effect of porosity, consolidation, and the unit weight of expansive clays. Their study was important in the wake of soil treatment with microbes because pore size and size of pore throat play important roles in microbial treatment [118]. Figure 4 shows the maximum reduction of permeability in percentage achieved by different researchers in their studies after the biomineralization of soils. It can be inferred from Figure 4 that permeability reduction is achieved through biocementation, which is also promising in applications like lining the base of water bodies and seepage control in water retaining structures.

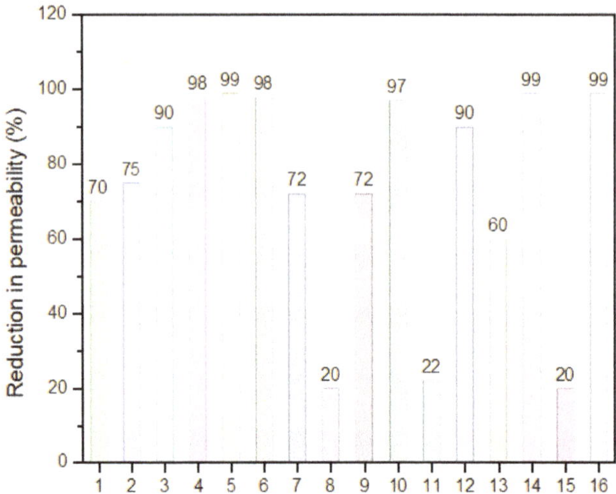

Figure 4. Maximum permeability reductions in percentage achieved by different researchers in their studies after soil biomineralization. 1—Yasuhara et al. [100], 2—Whiffin et al. [81], 3—Soon et al. [119], 4—Nemati and Voordouw [32], 5—Moghal et al. [120], 6—Handley-Sidhu et al. [110], 7—Zamani and Montoya [108], 8—Ferris et al. [115], 9—Gui et al. [116], 10—Proto et al. [63], 11—Ragusa et al. [121], 12—Cunningham et al. [122], 13—Van Paassen [123], 14—Ivanov et al. [124], 15—Al Qabany and Soga [125], 16—Ivanov and Chu [126].

5.4. Liquefaction Control by Biotreatment

Earthquakes and explosions make soils vulnerable and liquefy them, causing serious damage to the structures. Poorly graded and saturated sands are potential targets to liquefaction [127]. Major threats encountered due to the liquefaction of soils are landslides, damaged underground sewage lines and tunnels, and quicksand effects, although liquefaction helps in preventing seismic waves from reaching the Earth's surface since it produces a damping effect to the waves [128]. Liquefaction control is achieved by various techniques. Densification or compaction of existing soil is also among the methods adopted, but this method poses a threat to adjacent structures [129]. Biocementation has been proven as an effective method in controlling liquefaction in soils since calcite precipitation reduces permeability in soil [115]. Burbunk et al. [20,21] quoted that MICP improved resistance to liquefaction. Water in the voids of soils develops pressure transmitted through flow paths and tends to detach the soil grains apart or facilitate in the possible space in the vicinity. This pore water pressure contributes to the factors leading to the liquefaction of soils. Soil grains, when gelled together after biocementation, are less prone to liquefaction because of the disruption of flow paths and, later, due to permeability reduction and reduced pore water pressure [130]. Zamani and Montoya [131] tested the use of MICP on the permeability and shear of sand with silt and found a reduction in permeability depending on the amount of fines in the sample. Figure 5 shows the development of pore pressure in soil voids due to pore water and the development of resistance to the pore pressure through precipitates of calcite.

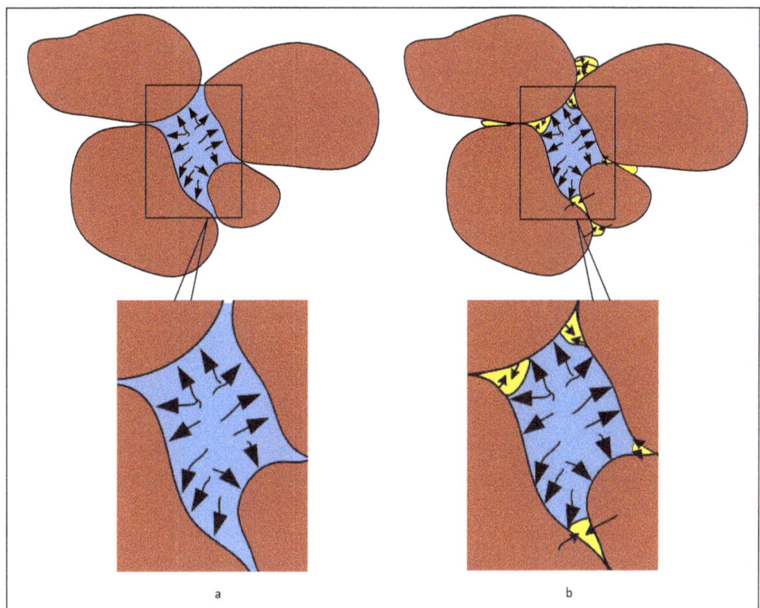

Figure 5. Representation of pore pressure in soil void: (**a**) Pore pressure developed by pore water pushing soil grains away from each other, leading to the loosening of the soil mass. (**b**) Cementation of soil grains by bioprecipitated calcite offering resistance to pore pressure.

6. Biotreatment of Soils for Geoenvironmental Applications

Bacterially precipitated $CaCO_3$ has also been used for capturing heavy metal contaminants to reduce their hazardous effects by converting the heavy metal traces into carbonates [12,132–135]. Bioremediation through urease in soil is effective for contaminant remediation [136]. Although natural calcite is used as an adsorbent for removing ions of heavy metal from contaminated water [137,138], the use of calcium for remediating heavy metal contaminants in soil is also practiced by researchers [139–141]. Natural calcite used as an adsorbent is very rare, and even the quality of naturally available calcite is not feasible for adsorption. Therefore, calcite precipitated by microbes was tested for the purpose of adsorption [142]. Kulczycki et al. [143] used bacterial ferrihydrite for the sorption of cadmium and lead and found that the precipitates of ferrihydrite were effective in providing sites for the heavy metal ions to sorb. Pan et al. [144] studied the microbial strategy for lead remediation. They inferred from their study that use of microbes was cost-efficient and environmentally-friendly as a lead remediation method. Velmurugan et al. [145] studied the kinetics of lead absorption by *Penicillium* sp. MRF-1 in a contaminated mining site in South Korea. Their study covered the use of this metal-resistant fungus stain for the remediation of Pb(II) within the dimensions of time of exposure, pH, and temperature. They concluded that *Penicillium* sp. MRF-1 was an inexpensive and conveniently cultivable fungus for the removal of Pb from contaminated solutions. Moghal et al. [94,120] used the enzyme treatment for adsorption and desorption studies for cadmium, nickel, and lead contaminants and found that the urease enzyme was effective in precipitating the carbonates of cadmium, nickel, and lead. They also obtained encouraging results in the level of desorption of heavy metals even after washing the contaminated soils with harsh extractants like ethylene diamine tetra-acetic acid (EDTA) and citric acid. The sorption studies, on the other hand, depicted better results in providing the sites on the soil grains to sorb. Sorption and desorption studies were conducted for individuals and for cocktail solutions of contaminants. These studies obtained appreciable results, encouraging the application of these techniques in situ. Nathan et al. [146] used the EICP method for heavy metal remediation in paper pulp deinking. They found that the enzymatic bioremediation is effective in reducing the hazardous effects of heavy metals. It also established that the urease enzyme is a nickel-based enzyme [147] and suggested that calcite precipitates by urease enzyme are active in showing affinity to the nickel contaminant, encapsulating them in the precipitates, and converting them into nickel carbonates [94].

Lauchnor et al. [148] studied the co-precipitation of strontium (Sr) in the porous media along with calcite precipitated by the MICP technique. They inferred that Sr precipitation was effective, thereby indicating the effective implementation of this method on site for the remediation of Sr. Mitchell and Ferris [149] tested the use of calcite precipitated with the MICP method to coprecipitate Sr in contaminated water and found that calcite precipitated by the MICP technique was exceptionally effective in the remediation of groundwater. Sr was also remediated by the formation of $SrCO_3$ in microenvironments of soil mass, leading to the reduced effect of radio nucleoids [150]. Wang et al. [151] studied the effect $CaCO_3$ on immobilizing heavy metals and observed that $CaCO_3$ was successful in serving the purpose. Precipitates of calcite in the soil mass contributing to heavy metal immobilization can also be attributed to the number of heavy metal ions in the soil mass. If there are fewer heavy metal ions, then the sites for the ions to settle down will be sufficient, leading to better results in terms of the immobilization of heavy metals. Varenya et al. [152] studied lead retention using the MICP method and found that the precipitates of $CaCO_3$ could be effective in the remediation of lead. They concluded that the MICP method has the potential to be applied in arid areas where phytoremediation cannot be used to remove heavy metals. The MICP method can be effective in reducing the hazardous effects caused by heavy metals like arsenic, cadmium, chromium, copper, and lead [153]. Therefore, the effectiveness of the biomineralization method can be well understood. The bioremediation of contaminants is a promising technique to reduce adverse effects caused by heavy metals. Table 2 shows a brief list of heavy metals remediated by the biostabilization method.

Table 2. List of contaminants remediated by the biostabilization method.

Applied On	Bacteria Used	Reference
Toxic metals	Sporosarcina luteola	[154]
Lead-contaminated mine wastes	Pararhodobacter sp.	[97]
Lead	Bacillus (pumilus and cereus)	[155]
Lead (II)	Rhodococcus opacus	[156]
Zn(II), Ni(II) and Cr(VI)	Trichoderma viride	[157]
Cobalt and copper	Lyngbya putealis	[158]
AA 6061 nuclear alloy	Bacillus cereus RE 10	[159]
Au(III)	Bacillus subtilis	[160]
Lead	Pseudomonas aeruginosa	[161]
Cd, Ni and Pb	Urease Enzyme	[94,120]
Cr^{6+}	Rhodococcus erythropolis	[162]
Cu and Pb	Comamonas testosterone, Enterobacter ludwigii and Zoogloea ramigerais	[163]
Copper	Stenotrophomonas maltophilia	[164]
Nickel	Lysinibacillus sp.	[165]
Pb and Cu	Bacillus thioparans	[166]
Cr(VI)	Bacillus cereus	[167]
Cr(VI)	Cellulosimicrobium funkei	[168]
Cd(II)	Bacillus cereus RC-1	[169]
Polycyclic aromatic hydrocarbons (PAHs)	Pseudomonas plecoglossicida J26	[170]
Cadmium	Exiguobacterium undae	[171]

Effect of Biotreatment on Mine Tailings/Dust Control

The various applications of the biotreatment of soils include the control of permeability, improvement of the bearing capacity, strength and stiffness development, and the control of dust due to erosion [172]. The air quality in the majority of cities worldwide is becoming a grave concern due to the increase in population and urbanization [173,174]. In this scenario, measures to curb the deterioration of air quality are developed to safeguard the environment. Dust contributes to the deterioration of air quality. According to Watson et al. [175], the major sources of air pollution in the major cities of the U.S. are vehicular emissions and dust from roads. With dust's severe impact on the air quality, methods to reduce dust emission are employed. These methods include using dust suppressants, spraying water, and providing wind shield walls against dust emission [176]. Chang et al. [177] reported that water spraying to control dust serves the purpose for a maximum duration of 4 h. Using water for dust suppression impacts water reserves, since water offers a temporary remedy and needs repeated application. Additionally, chemically activated suppressants for dust control are highly corrosive and hamper the environment. Sustainable and eco-friendly dust control techniques are on high demand [176]. Dust control using biotreatment can be achieved in the same way that sand solidification in the desert can be carried out; potential dust sources with dust particles can be dealt with through biocementation [178].

Sun et al. [176] conducted a study on dust near a quarry site in China by developing a simulation of rainfall erosion and by conducting field tests. Their test methods involved ascertaining surface strength, which is most vulnerable to wind erosion. Surface hardening was obtained by spraying biotreatment solutions on the surface; after spraying a thin and hard calcite on the surface, a crust of soil was formed. They confirmed that the cementation of dust particles using $CaCO_3$ precipitates through the enzyme treatment reduced dust pollution and that implementing EICP for dust control could be efficient during sandstorm and rainfall. Meyer et al. [56] used *Sporosarcina pasteurii* to treat two soil types to control air pollution due to dust. The treated soils were made to pass through a wind tunnel, and the amount of reduction in soil mass after exposure in the wind tunnel was observed to express the amount of wind erosion. The results obtained from this work showed that microbially precipitated calcite was very much effective in controlling soil erosion, as proven by wind tunnel experiments. Naeimi and Chu [179] compared the effectiveness of dust control through the biotreatment method and conventional techniques. *Sporosarcina pasteurii* was used in their study to treat sand against dust emission. The comparison was done on the same sand treated with calcium lignosulfonate, water, and calcium chloride. Their results showed that the biotreatment of soil exposed to wind in the wind tunnel improved erosion resistance, and only 1.5% mass loss was observed. Other treatment methods showed greater loss in mass after wind tunnel testing; hence, the use of biomediated soils was the best treatment method used in their study. Table 3 provides different bio-stabilizers adopted for dust control in different non-plastic materials.

Table 3. Bio-stabilizers adopted for dust control in different non-plastic materials.

Specimen Tested	Bio-Stabilizer	Reference
Coal dust	MICP (Urease microbes)	[180]
Coal dust	MICP (Staphylococcus succinus)	[181]
Desert soil	MICP (Indigenous bacteria)	[182]
Sand	MICP (Sporosarcina pasteurii)	[179]
Sand (well-graded)	MICP (Sporosarcina pasteurii)	[56]
F60 silica sand	EICP (Urease enzyme)	[183]
Ottawa F-60 fine grained uniform silica sand Well graded silty fine sand Mine tailings	EICP (Urease enzyme)	[78]

The erosion of deposits of mine tailings caused by wind erosion is one of the most serious environmental concerns [184,185]. Wind carrying mine tailings poses a threat to water bodies nearby and deteriorates air quality, which means serious risks to human and animal health [186]. Controlling dust by suppressants sprayed on the deposits is a common method. Dust suppressants agglomerate the fine particles and check the possibility of dust/finer mine tailings escaping with the wind movement; agglomerated dust particles form dense deposits on the ground, trapping dust sources beneath [187]. Chen et al. [188] conducted a study on the reduction of dust from mine tailings with a biopolymer coating on deposits. They found that the biopolymer coating was effective in mitigating mine tailing dust. Govarthanan et al. [189] used bacteria for the mineralization of lead contaminants found in mine tailing. They found that precipitates of calcium carbonate mineralized by bacteria were effective in lead bioremediation. Bacteria used in their study were effective in changing nitrates of lead to silicon oxides and sulfides of lead, thereby reducing the severity of lead on the atmosphere. Zamani et al. [190] studied the effect of MICP on mine tailing stability and found that microbial precipitates of $CaCO_3$ were effective in improving the stability of slopes of mine tailing materials. It can be observed from the literature that the remediation of heavy metals and dust control from mine tailings are well addressed by the biocementation process.

7. Limitations of Biocementation Techniques

Miftah et al. [43] reviewed the effectiveness of the MICP and EICP techniques in soil improvement and expressed that these methods could be effective in many geotechnical applications. However, certain concerns limit the effectiveness of these techniques. In the MICP method, concerns such as the type of soil, environmental issues, and the uniform treatment of soil mass are factors that create problems for its application. In the EICP method, the cost of enzymes happens to be too high since 57–98% of the cost of enzyme solutions is incurred on the urease enzyme. The soil type also plays an important role in governing the effectiveness of the biotreatment. The MICP method is restricted to the subsoil, and other regions of the soil may not provide a feasible environment to the bacterial growth. MICP does not show good results when used on very fine soils because comparatively larger sizes of bacteria cannot be accommodated in the pores of fine soils. On the other hand, EICP does not pose any hindrance in its application due to its size. Miftah et al. [43] also discussed the environmental concerns related to the use of MICP. This technique leaves microbes in soils after treatment, which means that it may require the permission of the concerned authorities and regular inspection to ensure that the energy of microbes is not hazardous to the surroundings. Furthermore, the release of ammonia through MICP is dangerous to people and the ecology of the area where it is applied, especially to the air and water. Additionally, the increase in pH may develop potential corrosion, and further contamination of groundwater due to chloride may be possible after the precipitation of $CaCO_3$. On the other hand, urease used in the EICP technique may not have a long-term impact on the environment because it becomes degraded after a certain time period. The use of microbes for soil treatment needs a specific environment in the soil mass for their cultivation, and the storage of bacterial strains is an expensive process. With these limitations in the use of alternate means for calcite precipitation, EICP seems to be better than MICP [191,192]. It has also been observed that high concentrations of calcium chloride and urea hinder the bacterial activity, reducing the amount of calcite precipitation. Conversely, using enzymes can very well be possible with high concentrations of calcium chloride and urea, which paves the way for a greater amount of calcite precipitation [109]. Therefore, the EICP technique is preferable over MICP. Yasuhara et al. [100] mentioned that maintaining bacteria for their cultivation requires technical expertise. Controlling bacterial activity also poses a challenge in the MICP method; the EICP technique is free of this constraint.

8. Conclusions

Biostabilization of the soil is an emerging trend with relatively simple onsite applications; the application of the stabilizer of a particular type onsite is carried out by injecting the stabilizer in the work area. Biostabilization of soils through EICP and MICP have the potential to meet the ever-growing demands of setting new infrastructure and remediating contaminants in the soils. Furthermore, the challenge of reducing environmental pollution and developing sustainable techniques can be achieved by the implementation of these techniques. The development of precipitates in the voids of the soil is helpful in reducing hydraulic conductivity. The application of MICP/EICP has stretched to the extent that ponds can be created in regions with soils of high permeability by relying on $CaCO_3$ precipitates. Salient observations made from the existing literature are summarized below.

- The development of biostabilization methods has been proven to be sustainable, eco-friendly, and effective in soil treatment, leading to the improvement in the geotechnical performance of soils such as reduced permeability, reduced porosity of soil mass, improved bearing capacity, control of soil erosion/dust, mitigated liquefaction of soils, seepage control, stabilized slopes, and contaminant remediation.
- The possibility of the intrusion of bacteria in soils for calcite precipitation is limited due to their sizes. Soil pores with sizes less than 0.5 μ cannot accommodate the microbes for the process of calcite precipitation since sizes of microns range from 0.5 μm to 3 μm. Enzyme particulates have sizes of about 12 nm, which can make the precipitation of calcite more convenient, even in finer clays.
- Soil treatment with MICP/EICP may increase chloride and ammonium ion (formed during hydrolysis) concentrations in the groundwater due to the precipitates of $CaCO_3$. It even causes an increase in the pH of the surrounding groundwater, triggering corrosion for structures built on them. The applicability of these techniques also pose challenges such as thee type of soil to be treated and the associated costs. Further studies on biotreatment can address these issues and aid in developing better application methods of biotreatment.
- Between the two methods, EICP is preferred over MICP as it requires less monitoring. The literature suggests that precipitates developed through MICP are vulnerable to moisture and may dissolve. MICP requires the environment to be maintained for proper bacterial growth and the production of urease enzymes. On the other hand, in EICP, the use of free urease is more promising for the calcite precipitation in the voids of soil grains. It provides a convenient approach to soil treatment because of its ease of application and lower maintenance in comparison to the MICP method.

Author Contributions: Conceptualization, A.A.B.M. and A.A.; Methodology, A.A.; Formal analysis, A.A.B.M.; Investigation, K.L.; Resources, A.A.B.M.; Data curation, K.L.; Writing—original draft preparation, M.A.L.; Writing—review and editing, M.A.L., A.A.B.M. and A.A.; Visualization, A.A.; Supervision, A.A.B.M.; Project administration, A.A.B.M.; Funding acquisition, A.A. All authors have read and agreed to the published version of the manuscript.

Funding: This work is funded by College of Engineering Research Center and Deanship of Scientific Research at King Saud University in Riyadh, Saudi Arabia.

Institutional Review Board Statement: Not applicable.

Informed Consent Statement: Not applicable.

Data Availability Statement: The content presented here was sourced from existing published literature, hence, this clause is not applicable.

Acknowledgments: The authors acknowledge the College of Engineering Research Center and Deanship of Scientific Research at King Saud University in Riyadh, Saudi Arabia for their financial support for the research work reported in this article.

Conflicts of Interest: The authors declare no conflict of interest.

References

1. Changizi, F.; Haddad, A. Effect of Nano-SiO_2 on the Geotechnical Properties of Cohesive Soil. *Geotech. Geol. Eng.* **2016**, *34*, 725–733. [CrossRef]
2. Harichane, K.; Ghrici, M.; Kenai, S.; Grine, K. Use of Natural Pozzolana and Lime for Stabilization of Cohesive Soils. *Geotech. Geol. Eng.* **2011**, *29*, 759–769. [CrossRef]
3. Attoh-Okine, N.O. Lime Treatment of Laterite Soils and Gravels—Revisited. *Constr. Build. Mater.* **1995**, *9*, 283–287. [CrossRef]
4. Moghal, A.A.B.; Vydehi, K.V. State-of-the-Art Review on Efficacy of Xanthan Gum and Guar Gum Inclusion on the Engineering Behavior of Soils. *Innov. Infrastruct. Solut.* **2021**, *6*, 108. [CrossRef]
5. Renjith, R.; Robert, D.J.; Gunasekara, C.; Setunge, S.; O'Donnell, B. Optimization of Enzyme-Based Soil Stabilization. *J. Mater. Civ. Eng.* **2020**, *32*. [CrossRef]
6. Calik, U.; Sadoglu, E. Engineering Properties of Expansive Clayey Soil Stabilized with Lime and Perlite. *Geomech. Eng.* **2014**, *6*, 403–418. [CrossRef]
7. Tingle, J.S.; Santoni, R.L. Stabilization of Clay Soils with Nontraditional Additives. *Transp. Res. Rec.* **2003**, *1819*, 72–84. [CrossRef]
8. Correia, A.A.S.; Rasteiro, M.G. Nanotechnology Applied to Chemical Soil Stabilization. *Procedia Eng.* **2016**, *143*, 1252–1259. [CrossRef]
9. Pradhan, P.K.; Kar, R.K.; Naik, A. Effect of Random Inclusion of Polypropylene Fibers on Strength Characteristics of Cohesive Soil. *Geotech. Geol. Eng.* **2012**, *30*, 15–25. [CrossRef]
10. Soon, N.W.; Lee, L.M.; Khun, T.C.; Ling, H.S. Improvements in Engineering Properties of Soils through Microbial-Induced Calcite Precipitation. *KSCE J. Civ. Eng.* **2013**, *17*, 718–728. [CrossRef]
11. Sangeetha, V.; Thenmozhi, A.; Devasena, M. Enhanced Removal of Lead from Soil Using Biosurfactant Derived from Edible Oils. *Soil Sediment Contam. Int. J.* **2021**, *30*, 135–147. [CrossRef]
12. Kang, C.-H.; Kwon, Y.-J.; So, J.-S. Bioremediation of Heavy Metals by Using Bacterial Mixtures. *Ecol. Eng.* **2016**, *89*, 64–69. [CrossRef]
13. Mitchell, J.K.; Santamarina, J.C. Biological Considerations in Geotechnical Engineering. *J. Geotech. Geoenviron. Eng.* **2005**, *131*, 1222–1233. [CrossRef]
14. Dejong, J.T.; Soga, K.; Kavazanjian, E.; Burns, S.; Van Paassen, L.A.; Al Qabany, A.; Aydilek, A.; Bang, S.S.; Burbank, M.; Caslake, L.F.; et al. Biogeochemical Processes and Geotechnical Applications: Progress, Opportunities and Challenges. *Géotechnique* **2013**, *63*, 287–301. [CrossRef]
15. National Research Council. *Geological and Geotechnical Engineering in the New Millennium: Opportunities for Research and Technological Innovation*; National Academies Press: Washington, DC, USA, 2006; ISBN 978-0-309-10009-0.
16. Espinoza, D.N.; Santamarina, J.C. Ant Tunneling—A Granular Media Perspective. *Granul. Matter* **2010**, *12*, 607–616. [CrossRef]
17. Li, M.; Cheng, X.; Guo, H. Heavy Metal Removal by Biomineralization of Urease Producing Bacteria Isolated from Soil. *Int. Biodeterior. Biodegrad.* **2013**, *76*, 81–85. [CrossRef]
18. Han, L.; Li, J.; Xue, Q.; Chen, Z.; Zhou, Y.; Poon, C.S. Bacterial-Induced Mineralization (BIM) for Soil Solidification and Heavy Metal Stabilization: A Critical Review. *Sci. Total Environ.* **2020**, *746*, 140967. [CrossRef]
19. Rahman, M.M.; Hora, R.N.; Ahenkorah, I.; Beecham, S.; Karim, M.R.; Iqbal, A. State-of-the-Art Review of Microbial-Induced Calcite Precipitation and Its Sustainability in Engineering Applications. *Sustainability* **2020**, *12*, 6281. [CrossRef]
20. Burbank, M.B.; Weaver, T.J.; Green, T.L.; Williams, B.C.; Crawford, R.L. Precipitation of Calcite by Indigenous Microorganisms to Strengthen Liquefiable Soils. *Geomicrobiol. J.* **2011**, *28*, 301–312. [CrossRef]
21. Burbank, M.; Weaver, T.; Lewis, R.; Williams, T.; Williams, B.; Crawford, R. Geotechnical Tests of Sands Following Bioinduced Calcite Precipitation Catalyzed by Indigenous Bacteria. *J. Geotech. Geoenviron. Eng.* **2013**, *139*, 928–936. [CrossRef]
22. Chou, C.-W.; Seagren, E.A.; Aydilek, A.H.; Lai, M. Biocalcification of Sand through Ureolysis. *J. Geotech. Geoenviron. Eng.* **2011**, *137*, 1179–1189. [CrossRef]
23. DeJong, J.T.; Fritzges, M.B.; Nüsslein, K. Microbially Induced Cementation to Control Sand Response to Undrained Shear. *J. Geotech. Geoenviron. Eng.* **2006**, *132*, 1381–1392. [CrossRef]
24. Venda Oliveira, P.J.; da Costa, M.S.; Costa, J.N.P.; Nobre, M.F. Comparison of the Ability of Two Bacteria to Improve the Behavior of Sandy Soil. *J. Mater. Civ. Eng.* **2015**, *27*. [CrossRef]
25. Stocks-Fischer, S.; Galinat, J.K.; Bang, S.S. Microbiological Precipitation of $CaCO_3$. *Soil Biol. Biochem.* **1999**, *31*, 1563–1571. [CrossRef]
26. Rodriguez-Navarro, C.; Rodriguez-Gallego, M.; Chekroun, K.B.; Gonzalez-Muñoz, M.T. Conservation of Ornamental Stone by Myxococcus Xanthus-Induced Carbonate Biomineralization. *Appl. Environ. Microbiol.* **2003**, *69*, 2182–2193. [CrossRef]
27. Bucci, N.A.; Ghazanfari, E.; Lu, H. Microbially-Induced Calcite Precipitation for Sealing Rock Fractures. *Geo-Chicago* **2016**, 558–567. [CrossRef]
28. Hammes, F.; Seka, A.; de Knijf, S.; Verstraete, W. A Novel Approach to Calcium Removal from Calcium-Rich Industrial Wastewater. *Water Res.* **2003**. [CrossRef]
29. Chek, A.; Crowley, R.; Ellis, T.N.; Durnin, M.; Wingender, B. Evaluation of Factors Affecting Erodibility Improvement for MICP-Treated Beach Sand. *J. Geotech. Geoenviron. Eng.* **2021**, *147*. [CrossRef]
30. Neupane, D.; Yasuhara, H.; Kinoshita, N.; Unno, T. Applicability of Enzymatic Calcium Carbonate Precipitation as a Soil-Strengthening Technique. *J. Geotech. Geoenviron. Eng.* **2013**, *139*, 2201–2211. [CrossRef]

31. Yasuhara, H.; Hayashi, K.; Okamura, M. Evolution in Mechanical and Hydraulic Properties of Calcite-Cemented Sand Mediated by Biocatalyst. *Adv. Geotech. Eng.* **2012**, 3984–3992. [CrossRef]
32. Nemati, M.; Voordouw, G. Modification of Porous Media Permeability, Using Calcium Carbonate Produced Enzymatically in Situ. *Enzyme Microb. Technol.* **2003**, *33*, 635–642. [CrossRef]
33. Putra, H.; Yasuhara, H.; Erizal; Sutoyo; Fauzan, M. Review of Enzyme-Induced Calcite Precipitation as a Ground-Improvement Technique. *Infrastructures* **2020**, *5*, 66. [CrossRef]
34. Kavazanjian, E.; Almajed, A.; Hamdan, N. Bio-Inspired Soil Improvement Using EICP Soil Columns and Soil Nails. *Grouting* **2017**, 13–22. [CrossRef]
35. Song, J.Y.; Sim, Y.; Jang, J.; Hong, W.-T.; Yun, T.S. Near-Surface Soil Stabilization by Enzyme-Induced Carbonate Precipitation for Fugitive Dust Suppression. *Acta Geotech.* **2020**, *15*, 1967–1980. [CrossRef]
36. Harkes, M.P.; van Paassen, L.A.; Booster, J.L.; Whiffin, V.S.; van Loosdrecht, M.C.M. Fixation and Distribution of Bacterial Activity in Sand to Induce Carbonate Precipitation for Ground Reinforcement. *Ecol. Eng.* **2010**, *36*, 112–117. [CrossRef]
37. Ossai, R.; Rivera, L.; Bandini, P. Experimental Study to Determine an EICP Application Method Feasible for Field Treatment for Soil Erosion Control. *Am. Soc. Civ. Eng.* **2020**, 205–213. [CrossRef]
38. Kavazanjian, E.; Hamdan, N. Enzyme Induced Carbonate Precipitation (EICP) Columns for Ground Improvement. *IFFCEE* **2015**, 2252–2261. [CrossRef]
39. Yuan, H.; Ren, G.; Liu, K.; Zheng, W.; Zhao, Z. Experimental Study of EICP Combined with Organic Materials for Silt Improvement in the Yellow River Flood Area. *Appl. Sci.* **2020**, *10*, 7678. [CrossRef]
40. Almajed, A.; Khodadadi Tirkolaei, H.; Kavazanjian, E. Baseline Investigation on Enzyme-Induced Calcium Carbonate Precipitation. *J. Geotech. Geoenviron. Eng.* **2018**, *144*. [CrossRef]
41. Almajed, A.; Abbas, H.; Arab, M.; Alsabhan, A.; Hamid, W.; Al-Salloum, Y. Enzyme-Induced Carbonate Precipitation (EICP)-Based Methods for Ecofriendly Stabilization of Different Types of Natural Sands. *J. Clean. Prod.* **2020**, *274*, 122627. [CrossRef]
42. Choi, S.-G.; Chang, I.; Lee, M.; Lee, J.-H.; Han, J.-T.; Kwon, T.-H. Review on Geotechnical Engineering Properties of Sands Treated by Microbially Induced Calcium Carbonate Precipitation (MICP) and Biopolymers. *Constr. Build. Mater.* **2020**, *246*, 118415. [CrossRef]
43. Miftah, A.; Tirkolaei, H.K.; Bilsel, H. Bio-Precipitation of CaCO3 for Soil Improvement: A Review. *IOP Conf. Ser. Mater. Sci. Eng.* **2020**, *800*, 012037. [CrossRef]
44. Mujah, D.; Shahin, M.A.; Cheng, L. State-of-the-Art Review of Biocementation by Microbially Induced Calcite Precipitation (MICP) for Soil Stabilization. *Geomicrobiol. J.* **2017**, *34*, 524–537. [CrossRef]
45. Golovkina, D.A.; Zhurishkina, E.V.; Ivanova, L.A.; Baranchikov, A.E.; Sokolov, A.Y.; Bobrov, K.S.; Masharsky, A.E.; Tsvigun, N.V.; Kopitsa, G.P.; Kulminskaya, A.A. Calcifying Bacteria Flexibility in Induction of CaCO3 Mineralization. *Life* **2020**, *10*, 317. [CrossRef]
46. Stotzky, G. Soil as an Environment for Microbial Life. In *Modern Soil Microbiology*, 3rd ed.; van Elsas, J.D., Trevors, J.T., Wellington, E.M.H., Eds.; CRC Press: Boca Raton, FL, USA, 2019; ISBN 9781498763530.
47. Woese, C.R.; Kandler, O.; Wheelis, M.L. Towards a Natural System of Organisms: Proposal for the Domains Archaea, Bacteria, and Eucarya. *Proc. Natl. Acad. Sci. USA* **1990**, *87*, 4576–4579. [CrossRef] [PubMed]
48. Ehrlich, H.L. Geomicrobiology: Its Significance for Geology. *Earth-Sci. Rev.* **1998**, *45*, 45–60. [CrossRef]
49. Brock Biology of Microorganisms | 15th Edition | Pearson. Available online: https://www.pearson.com/store/p/brock-biology-of-microorganisms/P100000185862/9780134261928 (accessed on 16 January 2021).
50. Lee, J.; Kim, K.; Chun, B. Strength Characteristics of Soils Mixed with an Organic Acid Material for Improvement. *J. Mater. Civ. Eng.* **2012**, *24*, 1529–1533. [CrossRef]
51. Umar, M.; Kassim, K.A.; Ping Chiet, K.T. Biological Process of Soil Improvement in Civil Engineering: A Review. *J. Rock Mech. Geotech. Eng.* **2016**, *8*, 767–774. [CrossRef]
52. Van Paassen, L.A.; Daza, C.M.; Staal, M.; Sorokin, D.Y.; van der Zon, W.; van Loosdrecht, M.C.M. Potential Soil Reinforcement by Biological Denitrification. *Ecol. Eng.* **2010**, *36*, 168–175. [CrossRef]
53. Mortensen, B.M.; Haber, M.J.; DeJong, J.T.; Caslake, L.F.; Nelson, D.C. Effects of Environmental Factors on Microbial Induced Calcium Carbonate Precipitation. *J. Appl. Microbiol.* **2011**, *111*, 338–349. [CrossRef]
54. De Muynck, W.; De Belie, N.; Verstraete, W. Microbial Carbonate Precipitation in Construction Materials: A Review. *Ecol. Eng.* **2010**, *36*, 118–136. [CrossRef]
55. Park, S.-S.; Choi, S.-G.; Kim, W.-J.; Lee, J.-C. Effect of Microbially Induced Calcite Precipitation on the Strength of Cemented Sand. *New Front. Geotech. Eng.* **2014**, 47–56. [CrossRef]
56. Meyer, F.D.; Bang, S.; Min, S.; Stetler, L.D.; Bang, S.S. Microbiologically-Induced Soil Stabilization: Application of *Sporosarcina Pasteurii* for Fugitive Dust Control. *Adv. Geotech. Eng.* **2012**, 4002–4011. [CrossRef]
57. Ashraf, M.S.; Azahar, S.B.; Yusof, N.Z. Soil Improvement Using MICP and Biopolymers: A Review. *IOP Conf. Ser. Mater. Sci. Eng.* **2017**, *226*, 012058. [CrossRef]
58. Castanier, S.; Le Métayer-Levrel, G.; Perthuisot, J.-P. Ca-Carbonates Precipitation and Limestone Genesis—The Microbiogeologist Point of View. *Sediment. Geol.* **1999**, *126*, 9–23. [CrossRef]
59. Shirakawa, M.A.; Kaminishikawahara, K.K.; John, V.M.; Kahn, H.; Futai, M.M. Sand Bioconsolidation through the Precipitation of Calcium Carbonate by Two Ureolytic Bacteria. *Mater. Lett.* **2011**, *65*, 1730–1733. [CrossRef]

60. Chittoori, B.C.S.; Rahman, T.; Burbank, M.; Moghal, A.A.B. Evaluating Shallow Mixing Protocols as Application Methods for Microbial Induced Calcite Precipitation Targeting Expansive Soil Treatment. *Am. Soc. Civ. Eng.* **2019**, 250–259. [CrossRef]
61. Stal, L.J. Microphytobenthos as a Biogeomorphological Force in Intertidal Sediment Stabilization. *Ecol. Eng.* **2010**, *36*, 236–245. [CrossRef]
62. Shanahan, C.; Montoya, B.M. Erosion Reduction of Coastal Sands Using Microbial Induced Calcite Precipitation. *Geo-Chicago* **2016**, 42–51. [CrossRef]
63. Proto, C.J.; DeJong, J.T.; Nelson, D.C. Biomediated Permeability Reduction of Saturated Sands. *J. Geotech. Geoenviron. Eng.* **2016**, *142*, 04016073. [CrossRef]
64. Baćmaga, M.; Wyszkowska, J.; Kucharski, J. Bioaugmentation of Soil Contaminated with Azoxystrobin. *Water. Air. Soil Pollut.* **2016**, *228*, 19. [CrossRef] [PubMed]
65. Ma, X.-K.; Ding, N.; Peterson, E.C. Bioaugmentation of Soil Contaminated with High-Level Crude Oil through Inoculation with Mixed Cultures Including Acremonium Sp. *Biodegradation* **2015**, *26*, 259–269. [CrossRef] [PubMed]
66. Zhao, Y.; Fan, C.; Ge, F.; Cheng, X.; Liu, P. Enhancing Strength of MICP-Treated Sand with Scrap of Activated Carbon-Fiber Felt. *J. Mater. Civ. Eng.* **2020**, *32*, 04020061. [CrossRef]
67. Larsen, J.; Poulsen, M.; Lundgaard, T.; Agerbaek, M. Plugging of Fractures in Chalk Reservoirs by Enzyme-Induced Calcium Carbonate Precipitation. *SPE Prod. Oper.* **2008**, *23*, 478–483. [CrossRef]
68. Sumner, J.B. The Isolation and Crystallization of the Enzyme Urease: Preliminary Paper. *J. Biol. Chem.* **1926**, *69*, 435–441. [CrossRef]
69. Balasubramanian, A.; Ponnuraj, K. Crystal Structure of the First Plant Urease from Jack Bean: 83 Years of Journey from Its First Crystal to Molecular Structure. *J. Mol. Biol.* **2010**, *400*, 274–283. [CrossRef]
70. Blakeley, R.L.; Zerner, B. Jack Bean Urease: The First Nickel Enzyme. *J. Mol. Catal.* **1984**, *23*, 263–292. [CrossRef]
71. Oliveira, P.J.V.; Freitas, L.D.; Carmona, J.P.S.F. Effect of Soil Type on the Enzymatic Calcium Carbonate Precipitation Process Used for Soil Improvement. *J. Mater. Civ. Eng.* **2017**, *29*, 04016263. [CrossRef]
72. Javadi, N.; Khodadadi, H.; Hamdan, N.; Kavazanjian, E. EICP Treatment of Soil by Using Urease Enzyme Extracted from Watermelon Seeds. *IFCEE* **2018**, 115–124. [CrossRef]
73. Pratama, G.B.S.; Yasuhara, H.; Kinoshita, N.; Putra, H. Application of Soybean Powder as Urease Enzyme Replacement on EICP Method for Soil Improvement Technique. *IOP Conf. Ser. Earth Environ. Sci.* **2021**, *622*, 012035. [CrossRef]
74. Almajed, A.A. Enzyme Induced Carbonate Precipitation (EICP) for Soil Improvement. Ph.D. Thesis, Arizona State University, Phoenix, AZ, USA, 2017.
75. Almajed, A.; Tirkolaei, H.K.; Kavazanjian, E.; Hamdan, N. Enzyme Induced Biocementated Sand with High Strength at Low Carbonate Content. *Sci. Rep.* **2019**, *9*, 1135. [CrossRef] [PubMed]
76. Almajed, A. Enzyme Induced Cementation of Biochar-Intercalated Soil: Fabrication and Characterization. *Arab. J. Geosci.* **2019**, *12*, 403. [CrossRef]
77. Almajed, A.; Khodadadi, H.; Kavazanjian, E. Sisal Fiber Reinforcement of EICP-Treated Soil. *IFCEE* **2018**, 29–36. [CrossRef]
78. Hamdan, N.; Kavazanjian, E. Enzyme-Induced Carbonate Mineral Precipitation for Fugitive Dust Control. *Géotechnique* **2016**, *66*, 546–555. [CrossRef]
79. Putra, H.; Yasuhara, H.; Kinoshita, N.; Neupane, D.; Lu, C.-W. Effect of Magnesium as Substitute Material in Enzyme-Mediated Calcite Precipitation for Soil-Improvement Technique. *Front. Bioeng. Biotechnol.* **2016**, *4*. [CrossRef]
80. Hommel, J.; Akyel, A.; Frieling, Z.; Phillips, A.J.; Gerlach, R.; Cunningham, A.B.; Class, H. A Numerical Model for Enzymatically Induced Calcium Carbonate Precipitation. *Appl. Sci.* **2020**, *10*, 4538. [CrossRef]
81. Whiffin, V.S.; van Paassen, L.A.; Harkes, M.P. Microbial Carbonate Precipitation as a Soil Improvement Technique. *Geomicrobiol. J.* **2007**, *24*, 417–423. [CrossRef]
82. DeJong, J.T.; Mortensen, B.M.; Martinez, B.C.; Nelson, D.C. Bio-Mediated Soil Improvement. *Ecol. Eng.* **2010**, *36*, 197–210. [CrossRef]
83. Burbank, M.B.; Weaver, T.J.; Williams, B.C.; Crawford, R.L. Urease Activity of Ureolytic Bacteria Isolated from Six Soils in Which Calcite Was Precipitated by Indigenous Bacteria. *Geomicrobiol. J.* **2012**, *29*, 389–395. [CrossRef]
84. Barabesi, C.; Galizzi, A.; Mastromei, G.; Rossi, M.; Tamburini, E.; Perito, B. Bacillus Subtilis Gene Cluster Involved in Calcium Carbonate Biomineralization. *J. Bacteriol.* **2007**, *189*, 228–235. [CrossRef]
85. Sotoudehfar, A.R.; Sadeghi, M.M.; Mokhtari, E.; Shafiei, F. Assessment of the Parameters Influencing Microbial Calcite Precipitation in Injection Experiments Using Taguchi Methodology. *Geomicrobiol. J.* **2016**, *33*, 163–172. [CrossRef]
86. Jiang, N.-J.; Soga, K.; Kuo, M. Microbially Induced Carbonate Precipitation for Seepage-Induced Internal Erosion Control in Sand–Clay Mixtures. *J. Geotech. Geoenviron. Eng.* **2017**, *143*, 04016100. [CrossRef]
87. Neupane, D.; Yasuhara, H.; Kinoshita, N.; Putra, H. Distribution of Grout Material within 1-m Sand Column in Insitu Calcite Precipitation Technique. *Soils Found.* **2015**, *55*, 1512–1518. [CrossRef]
88. Ivanov, V.; Chu, J.; Stabnikov, V.; Li, B. Strengthening of Soft Marine Clay Using Bioencapsulation. *Mar. Georesources Geotechnol.* **2015**, *33*, 320–324. [CrossRef]
89. Zhao, Q.; Li, L.; Li, C.; Li, M.; Amini, F.; Zhang, H. Factors Affecting Improvement of Engineering Properties of MICP-Treated Soil Catalyzed by Bacteria and Urease. *J. Mater. Civ. Eng.* **2014**, *26*, 04014094. [CrossRef]

90. Cheng, L.; Cord-RuwischRalf, A.S. Cementation of Sand Soil by Microbially Induced Calcite Precipitation at Various Degrees of Saturation. *Can. Geotech. J.* **2013**. [CrossRef]
91. Ali, F.H.; Adnan, A.; Choy, C.K. Geotechnical Properties of a Chemically Stabilized Soil from Malaysia with Rice Husk Ash as an Additive. *Geotech. Geol. Eng.* **1992**, *10*, 117–134. [CrossRef]
92. Sharma, A.; Ramkrishnan, R. Study on Effect of Microbial Induced Calcite Precipitates on Strength of Fine Grained Soils. *Perspect. Sci.* **2016**, *8*, 198–202. [CrossRef]
93. Wani, K.M.N.S.; Mir, B.A. Unconfined Compressive Strength Testing of Bio-Cemented Weak Soils: A Comparative Upscale Laboratory Testing. *Arab. J. Sci. Eng.* **2020**, *45*, 8145–8157. [CrossRef]
94. Moghal, A.A.B.; Lateef, M.A.; Mohammed, S.A.S.; Lemboye, K.; Chittoori, B.; Almajed, A. Efficacy of Enzymatically Induced Calcium Carbonate Precipitation in the Retention of Heavy Metal Ions. *Sustainability* **2020**, *12*, 7019. [CrossRef]
95. Xiao, J.Z.; Wei, Y.Q.; Cai, H.; Wang, Z.W.; Yang, T.; Wang, Q.H.; Wu, S.F. Microbial-Induced Carbonate Precipitation for Strengthening Soft Clay. *Adv. Mater. Sci. Eng.* **2020**, *2020*, 8140724. [CrossRef]
96. Park, S.-S.; Choi, S.-G.; Nam, I.-H. Effect of Plant-Induced Calcite Precipitation on the Strength of Sand. *J. Mater. Civ. Eng.* **2014**, *26*, 06014017. [CrossRef]
97. Mwandira, W.; Nakashima, K.; Kawasaki, S. Bioremediation of Lead-Contaminated Mine Waste by Pararhodobacter Sp. Based on the Microbially Induced Calcium Carbonate Precipitation Technique and Its Effects on Strength of Coarse and Fine Grained Sand. *Ecol. Eng.* **2017**, *109*, 57–64. [CrossRef]
98. Van Paassen, L.A.; Ghose, R.; van der Linden, T.J.M.; van der Star, W.R.L.; van Loosdrecht, M.C.M. Quantifying Biomediated Ground Improvement by Ureolysis: Large-Scale Biogrout Experiment. *J. Geotech. Geoenviron. Eng.* **2010**, *136*, 1721–1728. [CrossRef]
99. Neupane, D.; Yasuhara, H.; Kinoshita, N.; Ando, Y. Distribution of Mineralized Carbonate and Its Quantification Method in Enzyme Mediated Calcite Precipitation Technique. *Soils Found.* **2015**, *55*, 447–457. [CrossRef]
100. Yasuhara, H.; Neupane, D.; Hayashi, K.; Okamura, M. Experiments and Predictions of Physical Properties of Sand Cemented by Enzymatically-Induced Carbonate Precipitation. *Soils Found.* **2012**, *52*, 539–549. [CrossRef]
101. Mahawish, A.; Bouazza, A.; Gates, W.P. Unconfined Compressive Strength and Visualization of the Microstructure of Coarse Sand Subjected to Different Biocementation Levels. *J. Geotech. Geoenviron. Eng.* **2019**, *145*, 04019033. [CrossRef]
102. Putra, H.; Yasuhara, H.; Kinoshita, N. Optimum Condition for the Application of Enzyme-Mediated Calcite Precipitation Technique as Soil Improvement Technique. *Int. J. Adv. Sci. Eng. Inf. Technol.* **2017**, *7*, 2145–2151. [CrossRef]
103. Gomez, M.G.; Anderson, C.M.; DeJong, J.T.; Nelson, D.C.; Lau, X.H. Stimulating In Situ Soil Bacteria for Bio-Cementation of Sands. *Geo-Charact. Modeling Sustain.* **2014**, 1674–1682. [CrossRef]
104. Dilrukshi, R.A.N.; Nakashima, K.; Kawasaki, S. Soil Improvement Using Plant-Derived Urease-Induced Calcium Carbonate Precipitation. *Soils Found.* **2018**, *58*, 894–910. [CrossRef]
105. Khodadadi, T.H.; Krishnan, V.; Martin, K.; Hamdan, N.; Kavazanjian, E.; Almajed, A. Variation in Strength of EICP Treated "Standard" Sand. In Proceedings of the International Symposium on Bio-mediated and Bio-inspired Geotechnics, Atlanta, GA, USA, 10–12 September 2018.
106. Krishnan, V.; Khodadadi Tirkolaei, H.; Martin, K.; Hamdan, N.; van Paassen, L.A.; Kavazanjian, E. Variability in the Unconfined Compressive Strength of EICP-Treated "Standard" Sand. *J. Geotech. Geoenviron. Eng.* **2021**, *147*, 06021001. [CrossRef]
107. Thomas, A.; Tripathi, R.K.; Yadu, L.K. Variation in a Shear Modulus of Enzyme-Treated Soil under Cyclic Loading. *Geo-China* **2016**, 88–95. [CrossRef]
108. Zamani, A.; Montoya, B.M. Permeability Reduction Due to Microbial Induced Calcite Precipitation in Sand. *Geo-Chicago* **2016**, 94–103. [CrossRef]
109. Nemati, M.; Greene, E.A.; Voordouw, G. Permeability Profile Modification Using Bacterially Formed Calcium Carbonate: Comparison with Enzymic Option. *Process Biochem.* **2005**, *40*, 925–933. [CrossRef]
110. Handley-Sidhu, S.; Sham, E.; Cuthbert, M.O.; Nougarol, S.; Mantle, M.; Johns, M.L.; Macaskie, L.E.; Renshaw, J.C. Kinetics of Urease Mediated Calcite Precipitation and Permeability Reduction of Porous Media Evidenced by Magnetic Resonance Imaging. *Int. J. Environ. Sci. Technol.* **2013**, *10*, 881–890. [CrossRef]
111. Hataf, N.; Baharifard, A. Reducing Soil Permeability Using Microbial Induced Carbonate Precipitation (MICP) Method: A Case Study of Shiraz Landfill Soil. *Geomicrobiol. J.* **2020**, *37*, 147–158. [CrossRef]
112. Wani, K.M.N.S.; Mir, B.A. Microbial Geo-Technology in Ground Improvement Techniques: A Comprehensive Review. *Innov. Infrastruct. Solut.* **2020**, *5*, 82. [CrossRef]
113. Martinez, B.C.; DeJong, J.T.; Ginn, T.R.; Montoya, B.M.; Barkouki, T.H.; Hunt, C.; Tanyu, B.; Major, D. Experimental Optimization of Microbial-Induced Carbonate Precipitation for Soil Improvement. *J. Geotech. Geoenviron. Eng.* **2013**, *139*, 587–598. [CrossRef]
114. Cuthbert, M.O.; McMillan, L.A.; Handley-Sidhu, S.; Riley, M.S.; Tobler, D.J.; Phoenix, V.R. A Field and Modeling Study of Fractured Rock Permeability Reduction Using Microbially Induced Calcite Precipitation. *Environ. Sci. Technol.* **2013**, *47*, 13637–13643. [CrossRef]
115. Ferris, F.G.; Stehmeier, L.G.; Kantzas, A.; Mourits, F.M. Bacteriogenic Mineral Plugging. *J. Can. Pet. Technol.* **1996**, *35*. [CrossRef]
116. Gui, R.; Pan, Y.; Ding, D.; Liu, Y.; Zhang, Z. Experimental Study on Bioclogging in Porous Media during the Radioactive Effluent Percolation. *Adv. Civ. Eng.* **2018**, *2018*, 9671371. [CrossRef]
117. Rittmann, B.E. The Significance of Biofilms in Porous Media. *Water Resour. Res.* **1993**, *29*, 2195–2202. [CrossRef]

118. Chittoori, B.C.S.; Moghal, A.A.B.; Pedarla, A.; Al-Mahbashi, A.M. Effect of Unit Weight on Porosity and Consolidation Characteristics of Expansive Clays. *J. Test. Eval.* **2017**, *45*, 94–104. [CrossRef]
119. Soon, N.W.; Lee, L.M.; Khun, T.C.; Ling, H.S. Factors Affecting Improvement in Engineering Properties of Residual Soil through Microbial-Induced Calcite Precipitation. *J. Geotech. Geoenviron. Eng.* **2014**, *140*, 04014006. [CrossRef]
120. Moghal, A.A.B.; Lateef, M.A.; Abu Sayeed Mohammed, S.; Ahmad, M.; Usman, A.R.A.; Almajed, A. Heavy Metal Immobilization Studies and Enhancement in Geotechnical Properties of Cohesive Soils by EICP Technique. *Appl. Sci.* **2020**, *10*, 7568. [CrossRef]
121. Ragusa, S.R.; de Zoysa, D.S.; Rengasamy, P. The Effect of Microorganisms, Salinity and Turbidity on Hydraulic Conductivity of Irrigation Channel Soil. *Irrig. Sci.* **1994**, *15*, 159–166. [CrossRef]
122. Cunningham, A.B.; Characklis, W.G.; Abedeen, F.; Crawford, D. Influence of Biofilm Accumulation on Porous Media Hydrodynamics. *Environ. Sci. Technol.* **1991**, *25*, 1305–1311. [CrossRef]
123. Van Paassen, L.A. Biogrout, Ground Improvement by Microbial Induced Carbonate Precipitation. Ph.D. Thesis, Delft University of Technology, Delft, The Netherlands, 2009.
124. Ivanov, V.; Chu, J.; Naeimi, M.; Stabnikov, V.; He, J. Iron-Based Bio-Grout for Soil Improvement and Land Reclamation. In Proceedings of the 2nd international Conference on Sustainable Construction Materials and Technologies, Ancone, Italy, 28–30 June 2010; pp. 415–420.
125. Al Qabany, A.; Soga, K. Effect of Chemical Treatment Used in MICP on Engineering Properties of Cemented Soils. In Proceedings of the Bio- and Chemo-Mechanical Processes in Geotechnical Engineering-Geotechnique Symposium in Print 2013, London, UK, 1 January 2013; pp. 107–115. [CrossRef]
126. Ivanov, V.; Chu, J. Applications of Microorganisms to Geotechnical Engineering for Bioclogging and Biocementation of Soil in Situ. *Rev. Environ. Sci. Biotechnol.* **2008**, *7*, 139–153. [CrossRef]
127. Seed, H.B.; Idriss, I.M. Simplified Procedure for Evaluating Soil Liquefaction Potential. *J. Soil Mech. Found. Div.* **1971**, *97*, 1249–1273. [CrossRef]
128. Sharma, M.; Satyam, N.; Reddy, K.R. State of the Art Review of Emerging and Biogeotechnical Methods for Liquefaction Mitigation in Sands. *J. Hazard. Toxic Radioact. Waste* **2021**, *25*, 03120002. [CrossRef]
129. Wang, Z.; Zhang, N.; Cai, G.; Jin, Y.; Ding, N.; Shen, D. Review of Ground Improvement Using Microbial Induced Carbonate Precipitation (MICP). *Mar. Georesour. Geotechnol.* **2017**, *35*, 1135–1146. [CrossRef]
130. Montoya, B.M.; Dejong, J.T.; Boulanger, R.W. Dynamic Response of Liquefiable Sand Improved by Microbial-Induced Calcite Precipitation. *Géotechnique* **2013**, *63*, 302–312. [CrossRef]
131. Zamani, A.; Montoya, B.M. Shearing and Hydraulic Behavior of MICP Treated Silty Sand. *Geotech. Front.* **2017**, 290–299. [CrossRef]
132. Warren, L.A.; Maurice, P.A.; Parmar, N.; Ferris, F.G. Microbially Mediated Calcium Carbonate Precipitation: Implications for Interpreting Calcite Precipitation and for Solid-Phase Capture of Inorganic Contaminants. *Geomicrobiol. J.* **2001**, *18*, 93–115. [CrossRef]
133. Dixit, R.; Wasiullah; Malaviya, D.; Pandiyan, K.; Singh, U.B.; Sahu, A.; Shukla, R.; Singh, B.P.; Rai, J.P.; Sharma, P.K.; et al. Bioremediation of Heavy Metals from Soil and Aquatic Environment: An Overview of Principles and Criteria of Fundamental Processes. *Sustainability* **2015**, *7*, 2189–2212. [CrossRef]
134. Ali, N.; Hameed, A.; Ahmed, S. Physicochemical Characterization and Bioremediation Perspective of Textile Effluent, Dyes and Metals by Indigenous Bacteria. *J. Hazard. Mater.* **2009**, *164*, 322–328. [CrossRef]
135. Torres-Aravena, Á.E.; Duarte-Nass, C.; Azócar, L.; Mella-Herrera, R.; Rivas, M.; Jeison, D. Can Microbially Induced Calcite Precipitation (MICP) through a Ureolytic Pathway Be Successfully Applied for Removing Heavy Metals from Wastewaters? *Crystals* **2018**, *8*, 438. [CrossRef]
136. Ran, D.; Kawasaki, S. Effective Use of Plant-Derived Urease in the Field of Geoenvironmental/Geotechnical Engineering. *J. Civ. Environ. Eng.* **2016**, *6*, 1–13. [CrossRef]
137. Aziz, H.A.; Adlan, M.N.; Ariffin, K.S. Heavy Metals (Cd, Pb, Zn, Ni, Cu and Cr(III)) Removal from Water in Malaysia: Post Treatment by High Quality Limestone. *Bioresour. Technol.* **2008**, *99*, 1578–1583. [CrossRef]
138. Yavuz, O.; Guzel, R.; Aydin, F.; Tegin, I.; Ziyadanogullari, R. Removal of Cadmium and Lead from Aqueous Solution by Calcite. *Pol. J. Environ. Stud.* **2007**, *16*, 467–471.
139. Moghal, A.A.B.; Reddy, K.R.; Mohammed, S.A.S.; Al-Shamrani, M.A.; Zahid, W.M. Lime-Amended Semi-Arid Soils in Retaining Copper, Lead, and Zinc from Aqueous Solutions. *Water. Air. Soil Pollut.* **2016**, *227*, 372. [CrossRef]
140. Moghal, A.; Mohammed, S.; Al-Shamrani, M.; Zahid, W. Retention Studies on Arsenic from Aqueous Solutions by Lime Treated Semi Arid Soils. *Int. J. Geomate* **2017**, *12*. [CrossRef]
141. Moghal, A.A.B.; Reddy, K.R.; Mohammed, S.A.S.; Al-Shamrani, M.A.; Zahid, W.M. Sorptive Response of Chromium (Cr^{+6}) and Mercury (Hg^{+2}) From Aqueous Solutions Using Chemically Modified Soils. *J. Test. Eval.* **2017**, *45*, 105–119. [CrossRef]
142. Liu, R.; Yu, Y.; Liu, X.; Guan, Y.; Chen, L.; Lian, B. Adsorption of Ni2+ and Cu2+ Using Bio-Mineral: Adsorption Isotherms and Mechanisms. *Geomicrobiol. J.* **2018**, *35*, 742–748. [CrossRef]
143. Kulczycki, E.; Fowle, D.A.; Fortin, D.; Ferris, F.G. Sorption of Cadmium and Lead by Bacteria–Ferrihydrite Composites. *Geomicrobiol. J.* **2005**, *22*, 299–310. [CrossRef]
144. Pan, X.; Chen, Z.; Li, L.; Rao, W.; Xu, Z.; Guan, X. Microbial Strategy for Potential Lead Remediation: A Review Study. *World J. Microbiol. Biotechnol.* **2017**, *33*, 35. [CrossRef]

145. Velmurugan, N.; Hwang, G.; Sathishkumar, M.; Choi, T.K.; Lee, K.-J.; Oh, B.-T.; Lee, Y.-S. Isolation, Identification, Pb(II) Biosorption Isotherms and Kinetics of a Lead Adsorbing Penicillium Sp. MRF-1 from South Korean Mine Soil. *J. Environ. Sci.* **2010**, *22*, 1049–1056. [CrossRef]
146. Nathan, V.K.; Rani, M.E.; Gunaseeli, R.; Kannan, N.D. Enhanced Biobleaching Efficacy and Heavy Metal Remediation through Enzyme Mediated Lab-Scale Paper Pulp Deinking Process. *J. Clean. Prod.* **2018**, *203*, 926–932. [CrossRef]
147. Mazzei, L.; Musiani, F.; Ciurli, S. The Structure-Based Reaction Mechanism of Urease, a Nickel Dependent Enzyme: Tale of a Long Debate. *JBIC J. Biol. Inorg. Chem.* **2020**, *25*, 829–845. [CrossRef]
148. Lauchnor, E.G.; Schultz, L.N.; Bugni, S.; Mitchell, A.C.; Cunningham, A.B.; Gerlach, R. Bacterially Induced Calcium Carbonate Precipitation and Strontium Coprecipitation in a Porous Media Flow System. *Environ. Sci. Technol.* **2013**, *47*, 1557–1564. [CrossRef]
149. Mitchell, A.C.; Ferris, F.G. The Coprecipitation of Sr into Calcite Precipitates Induced by Bacterial Ureolysis in Artificial Groundwater: Temperature and Kinetic Dependence. *Geochim. Cosmochim. Acta* **2005**, *69*, 4199–4210. [CrossRef]
150. Dhami, N.K.; Reddy, M.S.; Mukherjee, A. Biomineralization of Calcium Carbonates and Their Engineered Applications: A Review. *Front. Microbiol.* **2013**, *4*. [CrossRef] [PubMed]
151. Wang, Y.M.; Chen, T.C.; Yeh, K.J.; Shue, M.F. Stabilization of an Elevated Heavy Metal Contaminated Site. *J. Hazard. Mater.* **2001**, *88*, 63–74. [CrossRef]
152. Achal, V.; Pan, X.; Fu, D.Z. Bioremediation of Pb-Contaminated Soil Based on Microbially Induced Calcite Precipitation. *J. Microbiol. Biotechnol.* **2012**, *22*, 244–247. [CrossRef] [PubMed]
153. Kumari, D.; Qian, X.-Y.; Pan, X.; Achal, V.; Li, Q.; Gadd, G.M. Chapter Two-Microbially-induced Carbonate Precipitation for Immobilization of Toxic Metals. In *Advances in Applied Microbiology*; Sariaslani, S., Gadd, G.M., Eds.; Academic Press: Cambridge, MA, USA, 2016; Volume 94, pp. 79–108. [CrossRef]
154. Cuaxinque-Flores, G.; Aguirre-Noyola, J.L.; Hernández-Flores, G.; Martínez-Romero, E.; Romero-Ramírez, Y.; Talavera-Mendoza, O. Bioimmobilization of Toxic Metals by Precipitation of Carbonates Using Sporosarcina Luteola: An in Vitro Study and Application to Sulfide-Bearing Tailings. *Sci. Total Environ.* **2020**, *724*, 138124. [CrossRef] [PubMed]
155. Çolak, F.; Atar, N.; Yazıcıoğlu, D.; Olgun, A. Biosorption of Lead from Aqueous Solutions by Bacillus Strains Possessing Heavy-Metal Resistance. *Chem. Eng. J.* **2011**, *173*, 422–428. [CrossRef]
156. Bueno, B.Y.M.; Torem, M.L.; de Carvalho, R.J.; Pino, G.A.H.; de Mesquita, L.M.S. Fundamental Aspects of Biosorption of Lead (II) Ions onto a Rhodococcus Opacus Strain for Environmental Applications. *Miner. Eng.* **2011**, *24*, 1619–1624. [CrossRef]
157. Kumar, R.; Bhatia, D.; Singh, R.; Rani, S.; Bishnoi, N.R. Sorption of Heavy Metals from Electroplating Effluent Using Immobilized Biomass Trichoderma Viride in a Continuous Packed-Bed Column. *Int. Biodeterior. Biodegrad.* **2011**, *65*, 1133–1139. [CrossRef]
158. Kiran, B.; Thanasekaran, K. Metal Tolerance of an Indigenous Cyanobacterial Strain, Lyngbya Putealis. *Int. Biodeterior. Biodegrad.* **2011**, *65*, 1128–1132. [CrossRef]
159. Forte Giacobone, A.F.; Rodriguez, S.A.; Burkart, A.L.; Pizarro, R.A. Microbiological Induced Corrosion of AA 6061 Nuclear Alloy in Highly Diluted Media by Bacillus Cereus RE 10. *Int. Biodeterior. Biodegrad.* **2011**, *65*, 1161–1168. [CrossRef]
160. Ji, Y.; Gao, H.; Sun, J.; Cai, F. Experimental Probation on the Binding Kinetics and Thermodynamics of Au(III) onto Bacillus Subtilis. *Chem. Eng. J.* **2011**, *172*, 122–128. [CrossRef]
161. Naik, M.M.; Pandey, A.; Dubey, S.K. Pseudomonas Aeruginosa Strain WI-1 from Mandovi Estuary Possesses Metallothionein to Alleviate Lead Toxicity and Promotes Plant Growth. *Ecotoxicol. Environ. Saf.* **2012**, *79*, 129–133. [CrossRef] [PubMed]
162. Banerjee, S.; Joshi, S.R.; Mandal, T.; Halder, G. Insight into Cr^{6+} Reduction Efficiency of Rhodococcus Erythropolis Isolated from Coalmine Waste Water. *Chemosphere* **2017**, *167*, 269–281. [CrossRef] [PubMed]
163. Black, R.; Sartaj, M.; Mohammadian, A.; Qiblawey, H.A.M. Biosorption of Pb and Cu Using Fixed and Suspended Bacteria. *J. Environ. Chem. Eng.* **2014**, *2*, 1663–1671. [CrossRef]
164. Ghosh, A.; Saha, P.D. Optimization of Copper Bioremediation by Stenotrophomonas Maltophilia PD2. *J. Environ. Chem. Eng.* **2013**, *1*, 159–163. [CrossRef]
165. Prithviraj, D.; Deboleena, K.; Neelu, N.; Noor, N.; Aminur, R.; Balasaheb, K.; Abul, M. Biosorption of Nickel by Lysinibacillus Sp. BA2 Native to Bauxite Mine. *Ecotoxicol. Environ. Saf.* **2014**, *107*, 260–268. [CrossRef]
166. Rodríguez-Tirado, V.; Green-Ruiz, C.; Gómez-Gil, B. Cu and Pb Biosorption on Bacillus Thioparans Strain U3 in Aqueous Solution: Kinetic and Equilibrium Studies. *Chem. Eng. J.* **2012**, *181–182*, 352–359. [CrossRef]
167. Chen, Z.; Huang, Z.; Cheng, Y.; Pan, D.; Pan, X.; Yu, M.; Pan, Z.; Lin, Z.; Guan, X.; Wu, Z. Cr(VI) Uptake Mechanism of Bacillus Cereus. *Chemosphere* **2012**, *87*, 211–216. [CrossRef]
168. Karthik, C.; Ramkumar, V.S.; Pugazhendhi, A.; Gopalakrishnan, K.; Arulselvi, P.I. Biosorption and Biotransformation of Cr(VI) by Novel Cellulosimicrobium Funkei Strain AR6. *J. Taiwan Inst. Chem. Eng.* **2017**, *70*, 282–290. [CrossRef]
169. Huang, F.; Dang, Z.; Guo, C.-L.; Lu, G.-N.; Gu, R.R.; Liu, H.-J.; Zhang, H. Biosorption of Cd(II) by Live and Dead Cells of Bacillus Cereus RC-1 Isolated from Cadmium-Contaminated Soil. *Colloids Surf. B Biointerfaces* **2013**, *107*, 11–18. [CrossRef] [PubMed]
170. Parellada, E.A.; Ramos, A.N.; Ferrero, M.; Cartagena, E.; Bardón, A.; Valdez, J.C.; Neske, A. Squamocin Mode of Action to Stimulate Biofilm Formation of Pseudomonas Plecoglossicida J26, a PAHs Degrading Bacterium. *Int. Biodeterior. Biodegrad.* **2011**, *65*, 1066–1072. [CrossRef]
171. Kumari, D.; Pan, X.; Lee, D.-J.; Achal, V. Immobilization of Cadmium in Soil by Microbially Induced Carbonate Precipitation with Exiguobacterium Undae at Low Temperature. *Int. Biodeterior. Biodegrad.* **2014**, *94*, 98–102. [CrossRef]

172. Chandra, A.; Ravi, K. Application of Enzyme-Induced Carbonate Precipitation (EICP) to Improve the Shear Strength of Different Type of Soils. In *Problematic Soils and Geoenvironmental Concerns*; Latha Gali, M., Raghuveer Rao, P., Eds.; Lecture Notes in Civil Engineering; Springer: Singapore, 2021; Volume 88, pp. 617–632. ISBN 9789811562365.
173. Kelly, J.T.; Reff, A.; Gantt, B. A Method to Predict PM2.5 Resulting from Compliance with National Ambient Air Quality Standards. *Atmos. Environ.* **2017**, *162*, 1–10. [CrossRef]
174. Shen, J.; Gao, Z.; Ding, W.; Yu, Y. An Investigation on the Effect of Street Morphology to Ambient Air Quality Using Six Real-World Cases. *Atmos. Environ.* **2017**, *164*, 85–101. [CrossRef]
175. Watson, J.G.; Chow, J.C.; Mathai, C.V. Receptor Models in Air Resources Management: A Summary of the APCA International Specialty Conference. *JAPCA* **1989**, *39*, 419–426. [CrossRef]
176. Sun, X.; Miao, L.; Yuan, J.; Wang, H.; Wu, L. Application of Enzymatic Calcification for Dust Control and Rainfall Erosion Resistance Improvement. *Sci. Total Environ.* **2021**, *759*, 143468. [CrossRef] [PubMed]
177. Chang, Y.-M.; Chou, C.-M.; Su, K.-T.; Tseng, C.-H. Effectiveness of Street Sweeping and Washing for Controlling Ambient TSP. *Atmos. Environ.* **2005**, *39*, 1891–1902. [CrossRef]
178. Zhan, Q.; Qian, C.; Yi, H. Microbial-Induced Mineralization and Cementation of Fugitive Dust and Engineering Application. *Constr. Build. Mater.* **2016**, *121*, 437–444. [CrossRef]
179. Naeimi, M.; Chu, J. Comparison of Conventional and Bio-Treated Methods as a Dust Suppressant. *Environ. Sci. Pollut. Res.* **2017**, *24*. [CrossRef]
180. Fan, Y.; Hu, X.; Zhao, Y.; Wu, M.; Wang, S.; Wang, P.; Xue, Y.; Zhu, S. Urease Producing Microorganisms for Coal Dust Suppression Isolated from Coal: Characterization and Comparative Study. *Adv. Powder Technol.* **2020**, *31*, 4095–4106. [CrossRef]
181. Song, W.; Yang, Y.; Qi, R.; Li, J.; Pan, X. Suppression of Coal Dust by Microbially Induced Carbonate Precipitation Using Staphylococcus Succinus. *Environ. Sci. Pollut. Res.* **2019**, *26*, 35968–35977. [CrossRef] [PubMed]
182. Raveh-Amit, H.; Tsesarsky, M. Biostimulation in Desert Soils for Microbial-Induced Calcite Precipitation. *Appl. Sci.* **2020**, *10*, 2905. [CrossRef]
183. Woolley, M.A.; van Paassen, L.; Kavazanjian, E. Impact on Surface Hydraulic Conductivity of EICP Treatment for Fugitive Dust Mitigation. *Geo-Congress* **2020**, 132–140. [CrossRef]
184. Chen, R.; Lee, I.; Zhang, L. Biopolymer Stabilization of Mine Tailings for Dust Control. *J. Geotech. Geoenviron. Eng.* **2015**, *141*, 04014100. [CrossRef]
185. Chen, R.; Zhang, L.; Budhu, M. Biopolymer Stabilization of Mine Tailings. *J. Geotech. Geoenviron. Eng.* **2013**, *139*, 1802–1807. [CrossRef]
186. Mendez, M.O.; Maier, R.M. Phytostabilization of Mine Tailings in Arid and Semiarid Environments—An Emerging Remediation Technology. *Environ. Health Perspect.* **2008**, *116*, 278–283. [CrossRef]
187. Bolander, P.; Yamada, A. *Dust Palliative Selection and Application Guide*; USDA San Dimas Technology and Development Center: San Dimas, CA, USA, 1999.
188. Chen, R.; Ding, X.; Lai, H.; Zhang, L. Improving Dust Resistance of Mine Tailings Using Green Biopolymer. *Environ. Geotech.* **2020**, 1–10. [CrossRef]
189. Govarthanan, M.; Lee, K.-J.; Cho, M.; Kim, J.S.; Kamala-Kannan, S.; Oh, B.-T. Significance of Autochthonous Bacillus Sp. KK1 on Biomineralization of Lead in Mine Tailings. *Chemosphere* **2013**, *90*, 2267–2272. [CrossRef]
190. Zamani, A.; Liu, Q.; Montoya, B.M. Effect of Microbial Induced Carbonate Precipitation on the Stability of Mine Tailings. *IFCEE* **2018**, 291–300. [CrossRef]
191. Carmona, J.P.S.F.; Venda Oliveira, P.J.; Lemos, L.J.L.; Pedro, A.M.G. Improvement of a Sandy Soil by Enzymatic Calcium Carbonate Precipitation. *Proc. Inst. Civ. Eng. Geotech. Eng.* **2017**, *171*, 3–15. [CrossRef]
192. Hamdan, N.; Zhao, Z.; Mujica, M.; Kavazanjian, E.; He, X. Hydrogel-Assisted Enzyme-Induced Carbonate Mineral Precipitation. *J. Mater. Civ. Eng.* **2016**, *28*, 04016089. [CrossRef]

Article

Physical, Mechanical and Durability Properties of Ecofriendly Ternary Concrete Made with Sugar Cane Bagasse Ash and Silica Fume

Laura Landa-Ruiz [1,2], Aldo Landa-Gómez [2], José M. Mendoza-Rangel [3], Abigail Landa-Sánchez [2], Hilda Ariza-Figueroa [2], Ce Tochtli Méndez-Ramírez [2], Griselda Santiago-Hurtado [4,*], Victor M. Moreno-Landeros [4,*], René Croche [5,*] and Miguel Angel Baltazar-Zamora [2,*]

1 Facultad de Ingeniería Mecánica y Eléctrica, Doctorado en Ingeniería, Universidad Veracruzana, Lomas del Estadio S/N, Zona Universitaria, Xalapa 91000, Veracruz, Mexico; lalanda@uv.mx
2 Facultad de Ingeniería Civil—Xalapa, Universidad Veracruzana, Lomas del Estadio S/N, Zona Universitaria, Xalapa 91000, Veracruz, Mexico; aldolanda_12@hotmail.com (A.L.-G.); liagiba07@hotmail.com (A.L.-S.); hilda_af@hotmail.com (H.A.-F.); cmendez@uv.mx (C.T.M.-R.)
3 Facultad de Ingeniería Civil, Universidad Autónoma de Nuevo León, Ave. Pedro de Alba S/N, Ciudad Universitaria, San Nicolás de los Garza 66450, Nuevo León, Mexico; jmmr.rangel@gmail.com
4 Facultad de Ingeniería Civil—Unidad Torreón, UADEC, Torreón 27276, Coahuila, Mexico
5 Facultad de Ingeniería Mecánica y Eléctrica—Xalapa, Universidad Veracruzana, Lomas del Estadio S/N, Zona Universitaria, Xalapa 91000, Veracruz, Mexico
* Correspondence: grey.shg@gmail.com (G.S.-H.); vmmorlan@gmail.com (V.M.M.-L.); rcroche@uv.mx (R.C.); mbaltazar@uv.mx (M.A.B.-Z.); Tel.: +52-2282-5252-94 (M.A.B.-Z.)

Citation: Landa-Ruiz, L.; Landa-Gómez, A.; Mendoza-Rangel, J.M.; Landa-Sánchez, A.; Ariza-Figueroa, H.; Méndez-Ramírez, C.T.; Santiago-Hurtado, G.; Moreno-Landeros, V.M.; Croche, R.; Baltazar-Zamora, M.A. Physical, Mechanical and Durability Properties of Ecofriendly Ternary Concrete Made with Sugar Cane Bagasse Ash and Silica Fume. *Crystals* **2021**, *11*, 1012. https://doi.org/10.3390/cryst11091012

Academic Editors: Cesare Signorini, Antonella Sola, Sumit Chakraborty and Valentina Volpini

Received: 12 July 2021
Accepted: 21 August 2021
Published: 24 August 2021

Publisher's Note: MDPI stays neutral with regard to jurisdictional claims in published maps and institutional affiliations.

Abstract: In the present investigation, the physical, mechanical and durability properties of six concrete mixtures were evaluated, one of conventional concrete (CC) with 100% Portland cement (PC) and five mixtures of Ecofriendly Ternary Concrete (ETC) made with partial replacement of Portland Cement by combinations of sugar cane bagasse ash (SCBA) and silica fume (SF) at percentages of 10, 20, 30, 40 and 50%. The physical properties of slump, temperature, and unit weight were determined, as well as compressive strength, rebound number, and electrical resistivity as a durability parameter. All tests were carried out according to the ASTM and ONNCCE standards. The obtained results show that the physical properties of ETC concretes are very similar to those of conventional concrete, complying with the corresponding regulations. Compressive strength results of all ETC mixtures showed favorable performances, increasing with aging, presenting values similar to CC at 90 days and greater values at 180 days in the ETC-20 and ETC-30 mixtures. Electrical resistivity results indicated that the five ETC mixtures performed better than conventional concrete throughout the entire monitoring period, increasing in durability almost proportionally to the percentage of substitution of Portland cement by the SCBA–SF combination; the ETC mixture made with 40% replacement had the highest resistivity value, which represents the longest durability. The present electrical resistivity indicates that the durability of the five ETC concretes was greater than conventional concrete. The results show that it is feasible to use ETC, because it meets the standards of quality, mechanical resistance and durability, and offers a very significant and beneficial contribution to the environment due to the use of agro-industrial and industrial waste as partial substitutes up to 50% of CPC, which contributes to reduction in CO_2 emissions due to the production of Portland cement, responsible for 8% of total emissions worldwide.

Keywords: properties; mechanical; electrical resistivity; durability; ecofriendly ternary concrete; SCBA; SF

1. Introduction

Concrete is the most widely used construction material worldwide, due to its great mechanical and physical properties, with a demand that grows every year due to the need for the development of civil infrastructure across all countries in the world [1–8]. Even though concrete is durable, it is compromised when exposed to aggressive media where chloride and sulfate ions may be present, which are considered to be the main responsible agents for the premature deterioration of reinforced concrete structures, in which the main problem is the corrosion of reinforcing steel [9–14]. This compromises sustainable development by not complying with the useful lifetime for which the structures were designed; additionally, it is known that the manufacture of Portland cement, the main component for the development of concrete, is responsible for around 5 to 8% of total CO_2 emissions worldwide [15–17]. This has led the scientific community to look for options to reduce the environmental impact due to the use of concrete, of which the addition of supplementary materials to Portland cement is a very favorable option. These materials are industrial wastes, of which Fly ash is a waste material in the power generation industry, and reusing this highly active pozzolan in the construction industry may bring about several advantages [18]; silica fume (SF) is a byproduct from the production of silicon alloys such as ferro-chromium, ferro-manganese, calcium silicon, etc., which also creates environmental pollution and health hazards [19]; blast furnace slag is a waste product of the steel manufacturing process [20]; and among agro-industrial wastes, the most used as alternative materials to Portland cement are rice husk ash [21] and sugar cane bagasse ash (SCBA) [22–24].

Lua et al. found that fly ash (FA) and blast furnace slag (BFS) with various contents (cement replacement ratio at 0, 20, and 40%) significantly affected the autogenous self-healing ability of early age cracks. The self-healing efficiency of early age cracks decreased with increases in FA and BFS content. BFS mortars exhibited greater recovery in relation to water penetration resistance compared to the reference and FA mortars [25]. Likewise, Anandan et al. determined that the mechanical properties of processed fly ash based concrete with 50% OPC replacement had equal or better strength gain at later ages than unprocessed fly ash based concrete with 25% OPC replacement [26], and in another research work it was shown that binary concretes with 20% fly ash reinforced with AISI 304 Steel presented a higher corrosion resistance than AISI 1018 steel when exposed to a simulated marine environment [27].

Atis et al. showed that the compressive strength of silica fume concrete cured at 65% RH was easier to influence than that of Portland cement concrete. It was found that the compressive strength of silica fume concrete cured at 65% RH was, on average, 13% lower than silica fume concrete cured at 100% RH in concretes with three different water/cement ratios and SF percentages of 10, 15 and 20% [28]. Bhanja et al., based on findings of compressive and tensile strength increases with silica fume incorporation, determined that the optimum replacement percentage is not a constant one but depends on the water–cementitious material (w/cm) ratio of the mix [29]. Ozcan et al. concluded that inclusion of silica fume in concrete increased the compressive strength between 20% and 50% compared to control PC concrete and there was an optimum replacement ratio of silica fume, which could be predicted using artificial neural networks (ANN) and fuzzy logic (FL) [30]. Landa et al. determined that sustainable binary concretes made with 10% SF provided high corrosion resistance to AISI 1018 steel when exposed to sulfates for more than 300 days [31].

Fly ash, silica fume and SCBA have been used in various investigations as supplementary materials to cement with excellent results, such as from Srinivasan et al. who in their studies showed that SCBA in blended concrete had significantly higher compressive strength, tensile strength, and flexural strength compared to concrete without SCBA. It was found that the cement could be advantageously replaced with SCBA up to a maximum limit of 10% [32]. Another study showed that green concretes with substitution of 20% of Portland cement for with SCBA presented a great resistance to corrosion when reinforced with stainless steel [33]. Kawade et al. obtained results showing that SCBA concrete had significantly higher compressive strength compared to concrete without SCBA. The optimal level of SCBA content was achieved with 15.0% replacement and the partial replacement of cement by SCBA increased workability of fresh concrete; therefore, use of super plasticizer was not essential [34]. Castaldelli et al. evaluated different BFS/SCBA mixtures, replacing part of the BFS with SCBA from 0 to 40% by weight; the results of the mechanical resistance values were approximately 60 MPa of compressive strength for BFS/SCBA systems after 270 days of curing at 20 °C. This demonstrated that sugar cane bagasse ash is an interesting source for preparing alkali-activated binders [35]. There are several studies of sustainable concretes, including SCBA, that have shown that corrosion resistance increased compared to that of reinforcing steel when exposed to sulfated media or marine media [22,36–39], and some research has also been reported on the use of SCBA for green road construction [40,41].

Despite the fact that a large number of studies have been carried out worldwide on the benefits of the inclusion of SCBA for the preparation of concretes and mortars, there is still no standardized process for its commercial use as there is for fly ash and silica fume.

When using alternative materials to Portland cement, there are three very important impacts on development in the construction field. The first is the improvement in physical, mechanical and durability properties of the concretes. The second is the reduction in CO_2 emissions when making concrete to build civil infrastructure (bridges, houses, dams, hospitals, roads) by reducing the amount of Portland cement per cubic meter of concrete. The decrease is proportional to the amount in which the Pozzolanic material replaces Portland cement, so that the more volume of Portland cement is replaced, the greater the impact on the environment will be, in accordance with the findings of Dong et al.: when 50% of the cement content was replaced by FA, the embodied CO_2 emissions for the UHPC mixture were reduced by approximately 50% as compared to the CO_2 emissions calculated from conventional normal-strength concrete [42]. The third is the impact on the culture of recycling waste materials. In first-world countries, the use of fly ash and silica fume is already significant compared to emerging countries, such as Mexico, where at the moment there does not exist civil infrastructure where concrete has been used with replacement in large volumes by this type of material.

Therefore, in this research work, physical, mechanical and durability tests were carried out on Ecological Ternary Concretes (ETC), made with substitution of Portland cement in 10, 20, 30, 40 and 50% of combinations of SCBA and SF, in order to determine the most suitable substitution percentage for the fabrication of ETC that provides better performance than a conventional mixture. Six concrete mixes were produced with a water–cement ratio of 0.65. The physical properties of the concrete in the fresh state, such as slump, volumetric weight, and temperature, were determined according to ASTM and ONNCCE standards. For the mechanical properties, compressive strength tests were carried out as well as rebound number tests, and for the durability parameter of all the study mixtures, the electrical resistivity was determined.

2. Materials and Methods

2.1. Materials

For the elaboration of the study specimens, Portland cement type CPC 30R was used according to the NMX-C-414-ONNCCE standard [43], sugar cane bagasse ash (SCBA) was obtained from a sugar mill located in the town of Mahuixtlán, Veracruz, México, and silica fume (SF) was acquired commercially. Six concrete mixtures were made for the present research, the first of conventional concrete, denoted the control mix (MC), and the remaining five of Ecofriendly Ternary Concrete (ETC), made by substituting the CPC 30R for combinations of SCBA and SF at percentages of 10, 20, 30, 40 and 50%. SCBA and SF were used because they are agro-industrial and industrial wastes with pozzolanic properties due to their chemical composition. The results of the chemical characterization of the cementitious materials used, obtained by X-ray fluorescence (XRF) analysis, are presented in Table 1.

Table 1. Chemical composition of the cementitious materials obtained by XRF.

	Chemical Composition (% by Mass)							
	Fe_2O_3	Al_2O_3	SiO_2	CaO	Na_2O	K_2O	MgO	SO_3
Cement Portland	3.872	5.478	21.187	63.346	0.564	0.83	2.068	2.157
Sugar Cane Bagasse Ash	5.105	3.150	77.739	3.995	0.569	6.672	0.563	0.406
Silica Fume	1.574	0.792	92.261	0.436	0.383	1.314	0.292	0.335

The coarse and fine aggregates used for the preparation of the study mixtures were from banks of the Xalapa region. Table 2 summarizes the physical characteristics of the materials used; the tests were carried out according to ASTM standards [44–47].

Table 2. Physical characteristics of the aggregates.

Aggregates	Relative Density (Specific Gravity)	Bulk Density ("Unit Weight") (kg/m^3)	Absorption (%)	Fineness Modulus	Maximum Aggregate Size (mm)
Coarse (Gravel)	2.38	1381	5.10	-	19
Fine (Sand)	2.60	1764	1.56	3.40	-

2.2. Proportioning of the Mixtures MC and ETC

For the design and proportioning of the concrete mixtures, the ACI 211.1 method [48] was used; a water/cement ratio = 0.65 and a slump of 10 cm were measured for all concrete mixes. Table 3 presents the dosing of the six studied mixtures, the control mix (MC) and the five Ecofriendly Ternary Concrete (ETC) mixtures made with substitution of CPC 30 with combinations of SCBA-SF at 10, 20, 30, 40 and 50% (ETC-10, ETC-20, ETC-30, ETC-40, ETC-50).

Table 3. Dosage of ternary concrete mixtures (Kg/m^3).

Mixture	CPC 30R	SCBA	SF	Water	Aggregate Fine	Aggregate Coarse
MC	315.00	-	-	205.00	746.00	881.00
ETC-10	283.50	15.75	15.75	205.00	746.00	881.00
ETC-20	252.00	31.50	31.50	205.00	746.00	881.00
ETC-30	220.5	47.25	47.25	205.00	746.00	881.00
ETC-40	189.00	63.00	63.00	205.00	746.00	881.00
ETC-50	157.50	78.75	78.75	205.00	746.00	881.00

2.3. Physical Properties of Concrete Mixtures

To determine the physical properties of the six studied mixtures (MC, ETC-10, ETC-20, ETC-30, ETC-40, ETC-50), slump, temperature and unit weight tests were carried out. All tests were carried out in accordance with the ASTM and ONNCCE.

According to the NMX-C-156-ONNCCE-2010 standard [49] for determining slump, a truncated conical mold was used where the fresh concrete was poured and compacted. The mold was placed over a base and raised upwards. The measure of the consistency or workability of the concrete was provided by the amount of concrete slumped and the distance slumped, see Figure 1a.

The temperature was determined according to the ASTM C 1064/C1064M-08 standard [50], which indicates that the concrete must be placed in a non-absorbent container with at least 75 mm of concrete in all directions from the temperature sensor, which must have a resolution of ±1.0 °C or smaller with an interval of 0 °C to 50 °C. The thermometer was submerged in fresh concrete to a minimum depth of 75 mm, leaving it for over 2 min until the reading was established (see Figure 1b). The unitary mixture was calculated according to the NMX-C-162-ONNCCE-2014 standard [51]; the equipment used for this test was a balance with a precision of 50 gr., maze of gum, ruler plate, verification plate, measuring container and compaction rod. The concrete was placed in three layers inside the container, and each one was compacted via 25 penetrations with the compaction rod. When compression was complete, the mold was made flush with the ruler plate; finally, the container with the compacted concrete was weighed (see Figure 1c,d).

Figure 1. Tests of (**a**) Slump, (**b**) Temperature, (**c**,**d**) Unit weight.

2.4. Mechanical and Durability Properties of Eco-Friendly Ternary Concrete Mixtures

2.4.1. Compressive Strength

Compressive strength is the parameter or property of the mechanical behavior of hydraulic concrete most necessary for the structural design of civil infrastructure built on the basis of reinforced concrete. Compressive strength testing was carried out according to the NMX-C-083 ONNCCE standard [52], for which specimens were manufactured using cylindrical steel molds of 100 × 200 mm. After 24 h they were removed from the molds and placed in a curing tank according to the NMX-C-ONNCCE standard [53]. The specimens of the six study mixtures were tested at the ages of 7, 14, 28, 90 and 180 days using a loading rate of 0.3 MPa/s (see Figure 2). The compressive strength values analyzed in the results section are the average of the values of three specimens of each mix of concrete.

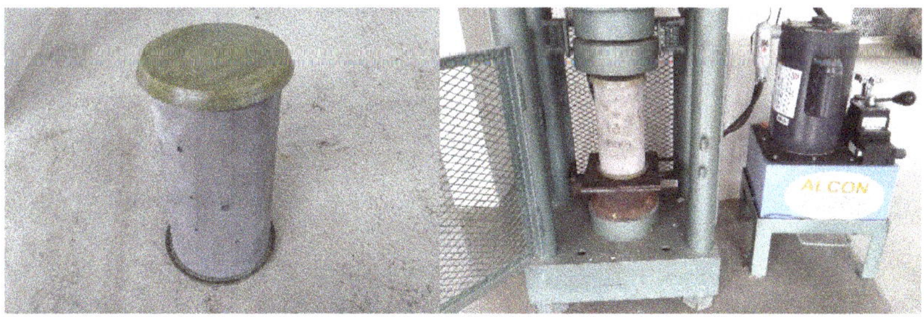

Figure 2. Compressive Strength Test.

2.4.2. Rebound Number

To determine the rebound number of specimens of the six concrete mixtures used in this study, tests were carried out according to the NMX-C-192-ONNCCE-2018 standard [54], in which a procedure is established to determine the rebound index for hardened concrete using a device known as a sclerometer or rebound hammer, to evaluate the compressive strength as well as the surface uniformity of the concrete. The results are considered relative rather than absolute values, but the test has the advantages of being non-destructive and widely used worldwide, and is used for evaluating the compressive strength of in-situ concrete [55] and in conjunction with the UPV test to predict the compressive strength of concrete in studies according to the findings of Amine et al. [56]. A rebound hammer, abrasive stone, spatula, flannel and brush were used to carry out the test. The test surface preparation was at least 150 mm in diameter and 100 mm thick. The surface was free of any layer other than concrete (see Figure 3).

Figure 3. Rebound Number test.

2.4.3. Electrical Resistivity

Electrical resistivity tests were carried out on the six concrete mixtures MC, ETC-10, ETC-20, ETC-30, ETC-40 and ETC-50. Electrical resistivity is considered a very important physical property to determine the quality and durability of concrete [57,58]. Several investigations have shown that the level of corrosion or resistance to corrosion of reinforcing steel in concrete exposed to aggressive media can be determined by electrical resistivity [59,60].

The electrical resistivity test was carried out according to the ASTM G57-07 standard [61], according to the specified equipment requirements and procedures for the measurement of resistivity in the laboratory and on site. The DURAR Network manual [62] indicates the criteria for interpretation of the resistivity results obtained and their relationship with the risk of corrosion of the reinforced concrete, which are presented in Table 4. The tests were carried out at 7, 14, 28, 90 and 180 days. Figure 4 shows the arrangement to carry out the electrical resistivity test.

Table 4. Electrical resistivity in concrete and risk of corrosion [20].

Electrical Resistivity	Risk of Corrosion in Reinforced Concrete
$\rho > 200$ kΩ-cm	Low Corrosion Risk
$200 > \rho > 10$ kΩ-cm	Moderate Corrosion Risk
$\rho < 10$ kΩ-cm	High Corrosion Risk

Figure 4. Characteristics of the electrical resistivity test.

3. Results and Discussion

3.1. Slump

Figure 5 shows the slumps in cm of the six study mixes, the control mix (MC) and the five Eco-friendly Ternary Concrete mixtures (ETC-10, ETC-20, ETC-30, ETC-40, ETC-50ETC).

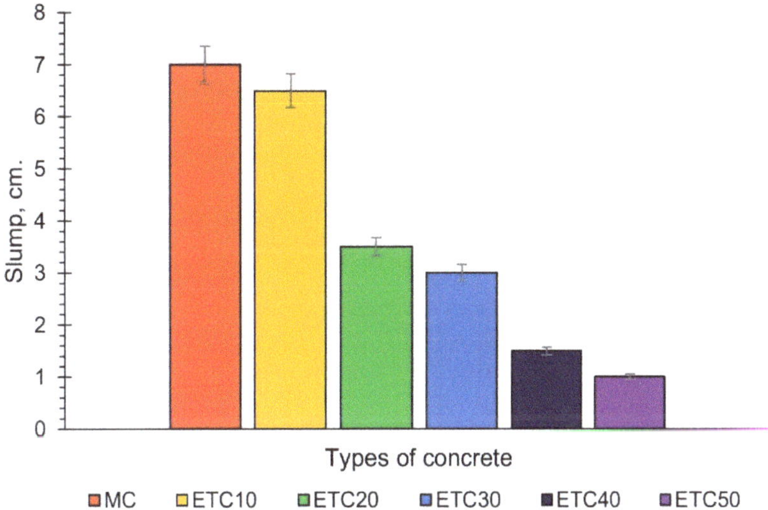

Figure 5. Slump of study mixtures (cm).

A decrease in workability or slump was observed in the five ETC mixtures; however, the ETC-10 mixture presented only a 7% decrease (0.5 cm) with respect to the control mixture (MC), with a value of 6.5 cm, which is considered an acceptable workability slump. With an increase to 20% in the percentage of substitution of CPC 30R with the combination of SCBA-SF, the slump showed a decrease of 50% (3.5 cm) with respect to the control mix (MC); this decrease in workability is attributed to the demand or absorption in excess of water due to pozzolanic materials [63,64], as is the case for SCBA and SF. For the ETC-30 mixture the slump was similar to that of the ETC-20 mixture, reaching a slump of 3 cm, which indicates a decrease of about 60% compared with the control mixture. In the case of the ETC-40 and ETC-50 mixtures, the effect of substituting CPC 30R by 40% and 50% respectively had a decisive effect in reducing the workability of these mixtures compared to the control mixture, with a decrease in slump of 80% for the ETC-40 mixture and 85% for the ETC-50. This behavior is due to excess water absorption by the supplementary materials used; therefore, in several investigations where concretes with large volumes of pozzolanic materials such as blast furnace slag or fly ash were used, water-reducing or super fluidizers additives were used to obtain slumps greater than 10 cm, which allowed adequate workability of the concrete mixtures [65,66].

3.2. Temperature

Figure 6 presents the behavior of the temperatures of the six studied concretes. It is observed that five mixtures presented a temperature of 25 °C and the ETC-50 mixture presented a temperature of 26 °C. The reported temperature values are within the specifications of the ASTM C 1064/C1064M-08 standard.

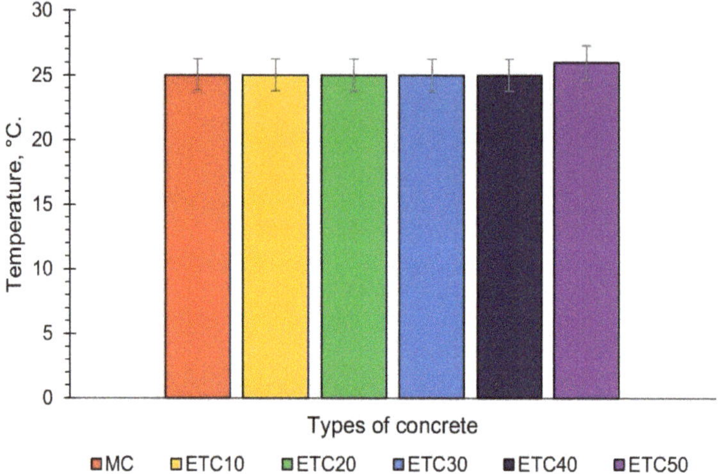

Figure 6. Temperatures of the studied concretes in fresh states.

3.3. Unit Weight

Figure 7 presents the unit weight results of the six study mixtures. There was minimal variation between the MC mixture and the ETC mixtures, and all of the unit weight values were within the specifications for the use of concrete in structural elements of civil works according to the NMX-C-155-ONNCCE-2014 standard, which indicates that hydraulic concretes for structural use must have a normal unit weight in fresh condition between 1900 kg/m^3 and 2400 kg/m^3 [67]. The lowest unit weight obtained was that of the ETC-50 mixture with 2149 kg/m^3, with a decrease of 5% compared to the unit mass of the MC mixture; the highest unit weight was presented by the ETC-30 mixture with a value of 2288 kg/m^3, 1.5% higher than the control mix. Khawaja et al. who evaluated concrete with Portland cement substitution in 5, 10, 15, 20, and 25% by SCBA, recorded an increase in unit weight of 3.13%, associated with the adhesive property of particles which reduced the concentration of induced air bubbles and consequently generated a stiffer matrix [68].

3.4. Mechanical and Durability Properties

3.4.1. Compressive Strength

Figure 8 shows the compressive strength results of each of the mixtures, which were tested at the ages of 7, 14, 28, 90 and 180 days. After 7 days, the concrete ETC hadlower compressive strength values than the control mixture, of 11.32, 7.66, 30.92, 44.55 and 75.31% respectively for the ETC-10, ETC-20, ETC-30, ETC-40 and ETC-50 mixtures; this negative effect was due to the presence of alternative pozzolanic materials to cement SCBA and SF, and is in agreement with Wu et al., who showed that FA had a negative effect on strength at early ages, but significantly enhanced the later-age strength [69]. In other studies a similar behavior has been shown even when the specimens of concrete were exposed to an aggressive medium such as sulfates [70]. At 14 days, increases in the resistance of the ETC concretes were observed, and this increase in compressive strength over time continued to 28 days, when the ETC-10 and ETC-20 concretes had 90% of the compressive strength values of the MC, with values of 28 and 29 MPa respectively, while for the ETC-30 and ETC-40 concretes the values were 22 and 23 MPa, and the ETC-50 mixture presenting the lowest compressive strength value with 13 MPa. These compressive strength values in the first 28 days coincide with the findings of various studies, where it has been shown that at 28 days sustainable or ecological concretes that substitute 20% of the CPC with supplementary materials obtain the best performance in compressive strength testing, as demonstrated by Mohamed [71], who found that a ternary concrete mixture made with the substitution of 10% FA + 10% silica fume for Portland Cement presented the highest resistance to compression in a study that covered substitutions from 10% to 50% of fly ash and silica fume for the fabrication of ternary and binary concretes exposed to different types of curing. Arif et al. found that sugar cane bagasse ash used as filler in concretes provided substantial improvements to compressive strength at substitution percentages of up to ≈20% [72]. In other studies, it has been shown that concretes with high FA contents—30%, 40% or higher—presented higher compressive strength values than the control mix, but this was due to the use of superfluidifiers and concretes with a low w/c ratio, equal to or less than 0.40 [73,74].

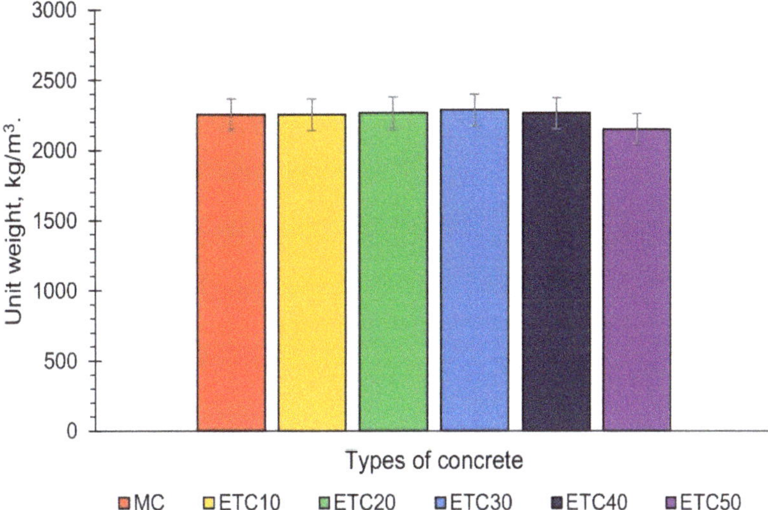

Figure 7. Unit weight of the studied concretes.

At 90 days the differences between the MC and the ETC-10, ETC-20 and ETC-30 concretes were minimal; however, lower values were observed for the specimens of the ETC-40 and ETC-50 mixtures. In percentages, the difference in compressive strength compared to the MC at 90 days was 7.22, 3.24, 1.07, 21.42 and 38.41% for the ETC-10, ETC-20, ETC-30, ETC-40, and ETC-50 mixtures respectively, with the ETC-30 mixture presenting the best performance. This result matches the findings of Le et al. [75], who concluded in their study that the compressive strength of a sample substituting OPC with 30% SCBA and 30% BFS was comparable to that of the control after 91 days [75]. At 180 days, the ETC-30, ETC-20 and ETC-40 specimens had a higher compressive strength than the specimen made with the MC control mixture; these results coincide with the literature, which indicates that at late ages the high amorphous silica content in the SCBA reacts with the calcium hydroxide product of the cement hydration process, giving rise to the formation of additional hydrated calcium hydroxide (C-S-H), which contributes to the increase in compressive strength over time [76]. In another investigation it was found that a concrete mix made with 25% of cement replaced with processed slag, which presented the highest SiO_2 content, obtained a superior compressive strength performance, reaching a value greater than 70 MPa at 90 days, which confirms the contribution to the increase in compressive strength due to pozzolanic material. A high content of SiO_2 presents a high capacity to yield tobermorite (calcium hydrosilicates (C–S–H)) by reacting with portlandite (a product of concrete mineral hydration) [77]. With the results of compressive strength at 180 days, it can be concluded that the optimal percentage of substitution of CPC with a combination of SCBA-SF is 30%, followed by 20%, with increases in compressive strength of 7.13 and 5.58% respectively compared to the MC, and in third place the ETC-40 mixture, which presented a compressive strength equal to the MC. Only the mixture of Ecofriendly Ternary Concrete with 50% substitution of SCBA-SF (ETC-50) failed to develop a mechanical resistance close to that of the control mix, reaching a resistance of 20.09 MPa at 90 days.

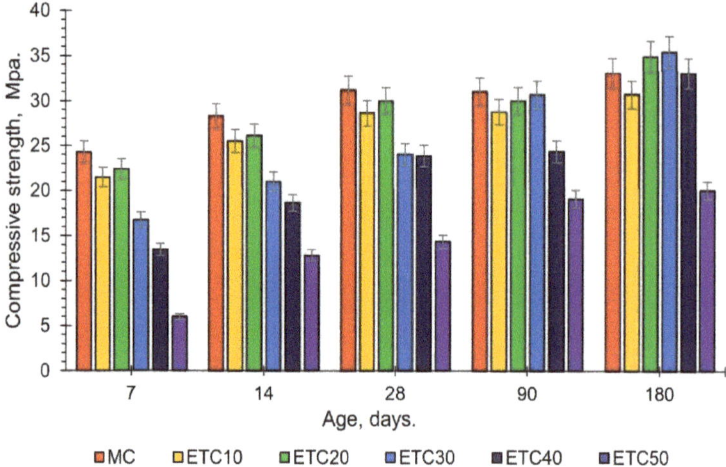

Figure 8. Compressive strength of the studied concretes.

3.4.2. Rebound Number

Figure 9 shows the results obtained from the rebound number tests to determine the compressive strength. It can be seen that the values obtained for the MC are similar to those presented in Figure 8: at the age of 28 days the control had an approximate value of 33 MPa for both simple compressive strength and rebound number tests. In the 7 day test, the ETC concretes reported compressive strength values lower than the MC by higher percentages than those reported in Figure 8, while over time these values increased in the five ETC mixtures, with this behavior likely being due to the effect of using materials with pozzolanic characteristics such as SCBA and SF. Unlike the compressive strength test using the cylinder, the values reported using the sclerometer for the five ETC concrete mixtures at the ages of 28, 90 and 180 days were always lower for the five ETC mixtures than for the MC; however, the mixtures with the best performances were still ETC20 and ETC30, which confirms the behavior in the compression test reported in Figure 8, where it was shown ETC20 and ETC30 were the best ETC mixtures at 90 and 180 days. The results obtained with the non-destructive rebound number test coincide with those reported in the literature on the use of said test to approximate the mechanical resistance of concrete elements in situ or in the laboratory, as reference values of resistance which must be supported by compressive strength tests of the evaluated concretes [78,79].

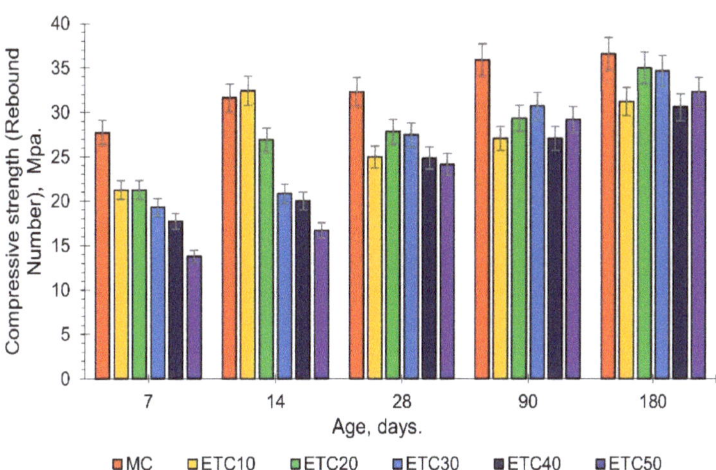

Figure 9. Compressive strength according to the rebound number.

3.4.3. Electrical Resistivity of Concrete Mixtures

Figure 10 shows the results obtained from the Electrical Resistivity test of the six concrete mixtures in this study. The tests were carried out at the ages of 7, 14, 28, 90 and 180 days. In Table 4 it can be seen that at the age of 7 days, almost all mixtures were found to have high corrosion risk, with values less than 10 kΩ-cm for the specimens of the MC, ETC-10, ETC-20, ETC-30 and ETC-40, mixtures. These values coincide with findings from the literature, where at early ages resistivity values are lower [80]. The only specimen that presented a higher value was the ETC50 mixture, reporting a resistivity of 13.45 kΩ-cm after 7 days, which indicates a moderate risk of corrosion. At the age of 14 days, all concretes showed a minimal increase in Electrical Resistivity, but both the ETC-40 and ETC-50 mixtures presented electrical resistivity values greater than 10 kΩ-cm, representing a moderate corrosion risk. At the age of 28 days, the benefit of the combination of the SCBA and the SF as pozzolanic materials was observed, as the durability of the ETC-20, ETC-30, ETC-40 and ETC-50 mixtures increased, with increases in the electrical resistivity values correlating with the percentage of substitution, with values that placed all of them in the moderate corrosion risk zone. ETC-20 and ETC-30 possessed electrical resistivities of 13.43 and 20.03 kΩ-cm while ETC-40 and ETC-50 presented the best performance with values of 47.5 and 51.4 kΩ-cm respectively. This is in agreement with the results of Bagheri et al., who evaluated concretes with different percentages of substitution of Portland cement with FA and SF, and found that the concretes with 20% and 30% FA and SF possessed electrical resistivity values at 28 days two times greater than that of the control mix [81]. At the age of 90 days, all Ecological Ternary Concretes (ETC-10, ETC-20, ETC-30, ETC-40, ETC-50ETC) reached the stage of moderate corrosion risk. The ETC-40 and ETC-50 concretes continued to present the best performances, with electrical resistivity values of 179.56 and 170.24 kΩ-cm respectively.

Finally, at the age of 180 days, the concretes that presented low electrical resistivity were the control mixture MC with a value of 10.88 kΩ-cm, followed by the concretes ETC-10 and ETC-20 with values of 12.74 and 54.39 kΩ-cm respectively. The concretes that presented the best performances were the ETC-30, ETC-40 and ETC-50 specimens, with values of 143.53, 191.44 and 156.20 kΩ-cm respectively. As can be seen, the Ecofriendly Ternary Concrete with 40% substitution of the SCBA-SF combination for Portland cement, mixture ETC-40, showed the best performance; this increase in electrical resistivity agrees with the results of Sadrmomtazi et al. [82], showing that including silica fume has positive effects on the fiber–matrix transition zone structure while increasing mechanical strength and specific electrical resistivity by up to 20 times compared to controls, due to the production of pozzolanic reactions and decreased concentration of portlandite, which increases uniformity and density as well as bond quality. It is observed that all ETC mixtures performed better in the electrical resistance test compared to the compression resistance test, and this behavior coincides with a report in the literature and is associated with the fact that the total volume of concrete pores is not reduced by pozzolanic reactions, but the pore structure becomes more discrete [83].

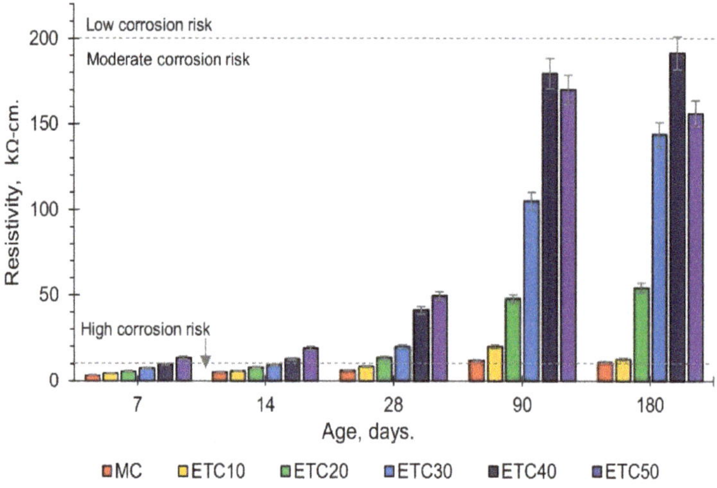

Figure 10. Electrical resistivity of the study concretes.

4. Conclusions

In all ETC mixtures there was a decrease in workability, which is attributed to the demand or absorption in excess of water due to pozzolanic materials. However, there were no significant variations in temperature or unit weight in the fresh state compared to the control mixture. The tested slump, temperature and unit weight of the ETC mixtures in their fresh state met requirements for the construction of civil works, such as bridges, pavements, buildings, dams, etc.

The results of compressive strength at 180 days indicated that the optimal percentage of substitution of CPC by combination of SCBA-SF was 30% followed by 20%, due to their increases in compressive strength of 7.13 and 5.58% respectively compared with the MC. The ETC-40 mixture also presented a compression resistance equal to that of the MC.

The rebound number test is a non-destructive test that can be used to evaluate the compressive strength of ETC concretes in the laboratory and on site, with the reservation that they are not considered as definitive values but rather as approximations, and it is always recommended to supplement rebound number tests with simple compression tests on cylinders and cubes.

All ETC mixtures presented better results in the electrical resistance test compared to the compression resistance test, suggesting that the ETC concretes were more durable and had a higher resistance to corrosion compared to the control mixture.

The Eco-friendly Ternary Concrete with 50% substitution of SCBA-SF (ETC-50) displayed a resistance of 20.09 MPa at 180 days, sufficient for the construction of minor works.

The use of ETC concretes has a very significant sustainability impact by contributing to the reduction of CO_2 emissions caused by Portland cement, replacing up to 50% of it with SCBA and SF waste and generating a culture of recycling in countries such as Mexico for the use of waste that, like SCBA, has lacked a defined use and previously been discarded as garbage.

Author Contributions: Conceptualization, M.A.B.-Z., R.C., G.S.-H. and V.M.M.-L.; Methodology, L.L.-R., A.L.-G., J.M.M.-R., H.A.-F., A.L.-S., C.T.M.-R., M.A.B.-Z. and R.C.; Data Curation, L.L.-R., J.M.M.-R., M.A.B.-Z., R.C., G.S.-H. and V.M.M.-L.; Writing—Review and Editing, L.L.-R., M.A.B.-Z., R.C., G.S.-H. and V.M.M.-L.; Visualization: M.A.B.-Z., R.C., G.S.-H. and V.M.M.-L.; Supervision: M.A.B.-Z., R.C., G.S.-H. and V.M.M.-L.; Funding acquisition: M.A.B.-Z. and R.C. All authors have read and agreed to the published version of the manuscript.

Funding: M.A. Baltazar-Zamora, et al., thank PRODEP for the support granted by the SEP to the Academic Body UV-CA-458 "Sustainability and Durability of Materials for Civil Infrastructure", within the framework of the 2018 Call for the Strengthening of Academic Bodies with IDCA 28593.

Institutional Review Board Statement: Not applicable.

Informed Consent Statement: Not applicable.

Data Availability Statement: The results are original of the research work.

Acknowledgments: M.A. Baltazar-Zamora, et al., thank PRODEP for the support granted by the SEP to the Academic Body UV-CA-458 "Sustainability and Durability of Materials for Civil Infrastructure", within the framework of the 2018 Call for the Strengthening of Academic Bodies with IDCA 28593. The authors thank Brenda Paola Baltazar García for technical support.

Conflicts of Interest: The authors declare no conflict of interest.

References

1. Raczkiewicz, W.; Wójcicki, A. Temperature impact on the assessment of reinforcement corrosion risk in concrete by galvanostatic pulse method. *Appl. Sci.* **2020**, *10*, 1089. [CrossRef]
2. Baltazar, M.A.; Márquez, S.; Landa, L.; Croche, R.; López, O. Effect of the type of curing on the corrosion behavior of concrete exposed to urban and marine environment. *Eur. J. Eng. Technol. Res.* **2020**, *5*, 91–95. [CrossRef]
3. Cramer, S.D.; Covino, B.S., Jr.; Bullard, S.J.; Holcomb, G.R.; Russell, J.H.; Nelson, F.J.; Laylor, H.M.; Soltesz, S.M. Corrosion prevention and remediation strategies for reinforced concrete coastal bridge. *Cem. Concr. Compos.* **2002**, *24*, 101–117. [CrossRef]
4. Troconis de Rincón, O.; Montenegro, J.C.; Vera, R.; Carvajal, A.M.; De Gutiérrez, R.M.; Del Vasto, S.; Saborio, E.; Torres-Acosta, A.; Pérez-Quiroz, J.; Martínez-Madrid, M.; et al. Reinforced Concrete Durability in Marine Environments DURACON Project: Long-Term Exposure. *Corrosion* **2016**, *72*, 824–833. [CrossRef]
5. Liang, M.T.; Lan, J.-J. Reliability analysis for the existing reinforced concrete pile corrosion of bridge substructure. *Cem. Concr. Res.* **2005**, *35*, 540–550. [CrossRef]
6. Baltazar, M.A.; Maldonado, M.; Tello, M.; Santiago, G.; Coca, F.; Cedano, A.; Barrios, C.P.; Nuñez, R.; Zambrano, P.; Gaona, C.; et al. Efficiency of galvanized steel embedded in concrete previously contaminated with 2, 3 and 4% of NaCl. *Int. J. Electrochem. Sci.* **2012**, *7*, 2997–3007.
7. Landa, L.; Croche, R.; Santiago, G.; Moreno, V.; Cuevas, J.; Méndez, C.; Jara, M.; Baltazar, M.A. Evaluation of the Influence of the Level of Corrosion of the Reinforcing Steel in the Moment-Curvature Diagrams of Rectangular Concrete Columns. *Eur. J. Eng. Technol. Res.* **2021**, *6*, 74–80. [CrossRef]
8. Santiago, G.; Baltazar, M.A.; Galván, R.; López, L.; Zapata, F.; Zambrano, P.; Gaona, C.; Almeraya, F. Electrochemical evaluation of reinforcement concrete exposed to soil Type SP contaminated with sulphates. *Int. J. Electrochem. Sci.* **2016**, *11*, 4850–4864. [CrossRef]
9. Saricimen, H.; Mohammad, M.; Quddus, A.; Shameem, M.; Barry, M.S. Effectiveness of concrete inhibitors in retarding rebar corrosion. *Cem. Concr. Compos.* **2002**, *24*, 89–100. [CrossRef]
10. Baltazar, M.; Almeraya, F.; Nieves, D.; Borunda, A.; Maldonado, E.; Ortiz, A. Corrosión del acero inoxidable 304 como refuerzo en concreto expuesto a cloruros y sulfatos. *Sci. Tech.* **2007**, *13*, 353–357.
11. Shaheen, F.; Pradhan, B. Influence of sulfate ion and associated cation type on steel reinforcement corrosion in concrete powder aqueous solution in the presence of chloride ions. *Cem. Concr. Res.* **2017**, *91*, 73–86. [CrossRef]
12. Baltazar, M.A.; Santiago, G.; Gaona, C.; Maldonado, M.; Barrios, C.P.; Nunez, R.; Perez, T.; Zambrano, P.; Almeraya, F. Evaluation of the corrosion at early age in reinforced concrete exposed to sulfates. *Int. J. Electrochem. Sci.* **2012**, *7*, 588–600.
13. Roventi, G.; Bellezze, T.; Giuliani, G.; Conti, C. Corrosion resistance of galvanized steel reinforcements in carbonated concrete: Effect of wet–dry cycles in tap water and in chloride solution on the passivating layer. *Cem. Concr. Res.* **2014**, *65*, 76–84. [CrossRef]
14. Baltazar, M.A.; Mendoza, J.M.; Croche, R.; Gaona, C.; Hernández, C.; López, L.; Olguín, F.; Almeraya, F. Corrosion behavior of galvanized steel embedded in concrete exposed to soil type MH contaminated with chlorides. *Front. Mater.* **2019**, *6*, 257. [CrossRef]
15. Mahasenan, N.; Smith, S.; Humphreys, K. The cement industry and global climate change current and potential future cement Industry CO_2 emissions. In *Greenhouse Gas Control Technologies—6th International Conference, Kyoto, Japan, 1–4 October 2002*; Elsevier: Amsterdam, The Netherlands, 2003; pp. 995–1000.

16. Worrell, E.; Price, L.; Martin, N.; Hendriks, C.; Ozawa, L.M. Carbon dioxide emissions from the global cement industry. *Annu. Rev. Energy Environ.* **2001**, *26*, 303–329. [CrossRef]
17. Monteiro, P.; Miller, S.; Horvath, A. Towards sustainable concrete. *Nat. Mater.* **2017**, *16*, 698–699. [CrossRef] [PubMed]
18. Mehdi, M.; Hamidreza, S.; Farhangi, V.; Karakouzian, M. Predicting the Effect of Fly Ash on Concrete's Mechanical Properties by ANN. *Sustainability* **2021**, *13*, 1469.
19. Shinga, D.; Sil, A. Effect of Partial Replacement of Cement by Silica Fume on Hardened Concrete. *Int. J. Emerg. Technol. Adv. Eng.* **2012**, *2*, 472–475.
20. Zhang, X.; An Yan, K.W. Carbonation property of hardened binder pastes containing super-pulverized blast-furnace slag. *Cem. Concr. Compos.* **2004**, *26*, 371–374. [CrossRef]
21. Safiuddin, M.; West, J.S.; Soudki, K.A. Hardened properties of self-consolidating high performance concrete including rice husk ash. *Cem. Concr. Compos.* **2010**, *32*, 708–717. [CrossRef]
22. Baltazar, M.A.; Ariza, H.; Landa, L.; Croche, R. Electrochemical evaluation of AISI 304 SS and galvanized steel in ternary ecological concrete based on sugar cane bagasse ash and silica fume (SCBA-SF) exposed to Na2SO4. *Eur. J. Eng. Res. Sci.* **2020**, *5*, 353–357. [CrossRef]
23. Jagadesha, P.; Ramachandramurthy, A.; Murugesan, R. Evaluation of mechanical properties of Sugar Cane Bagasse Ash concrete. *Constr. Build. Mater.* **2018**, *176*, 608–617. [CrossRef]
24. Baltazar, M.A.; Landa, A.; Landa, L.; Ariza, H.; Gallego, P.; Ramírez, A.; Croche, R.; Márquez, S. Corrosion of AISI 316 stainless steel embedded in sustainable concrete made with sugar cane bagasse ash (SCBA) exposed to marine environment. *Eur. J. Eng. Res. Sci.* **2020**, *5*, 127–131. [CrossRef]
25. Luo, M.; Jing, K.; Bai, J.; Ding, Z.; Yang, D.; Huang, H.; Go, Y. Effects of Curing Conditions and Supplementary Cementitious Materials on Autogenous Self-Healing of Early Age Cracks in Cement Mortar. *Crystals* **2021**, *11*, 752. [CrossRef]
26. Anandan, S.; Manoharan, S.V. Strength Properties of Processed Fly Ash Concrete. *J. Eng. Technol. Sci.* **2015**, *47*, 320–334. [CrossRef]
27. Baltazar, M.A.; Bastidas, D.M.; Santiago, G.; Mendoza, J.M.; Gaona, C.; Bastidas, J.M.; Almeraya, F. Effect of silica fume and fly ash admixtures on the corrosion behavior of AISI 304 embedded in concrete exposed in 3.5% NaCl solution. *Materials* **2019**, *12*, 4007. [CrossRef]
28. Atis, C.D.; Ozcan, F.; Kılıc, A.; Karahan, O.; Bilim, C.; Severcan, M.H. Influence of dry and wet curing conditions on compressive strength of silica fume concrete. *Build. Environ.* **2005**, *40*, 1678–1683. [CrossRef]
29. Bhanja, S.; Sengupta, B. Influence of silica fume on the tensile strength of concrete. *Cem. Concr. Res.* **2005**, *35*, 743–747. [CrossRef]
30. Ozcan, F.; Atis, C.D.; Karahan, O.; Uncuoglu, E.; Tanyildizi, H. Comparison of artificial neural network and fuzzy logic models for prediction of long-term compressive strength of silica fume concrete. *Adv. Eng. Softw.* **2009**, *40*, 856–863. [CrossRef]
31. Landa-Ruiz, L.; Baltazar-Zamora, M.A.; Bosch, J.; Ress, J.; Santiago-Hurtado, G.; Moreno-Landeros, V.M.; Márquez-Montero, S.; Méndez, C.; Borunda, A.; Juárez-Alvarado, C.A.; et al. Electrochemical Corrosion of Galvanized Steel in Binary Sustainable Concrete Made with Sugar Cane Bagasse Ash (SCBA) and Silica Fume (SF) Exposed to Sulfates. *Appl. Sci.* **2021**, *11*, 2133. [CrossRef]
32. Srinivasan, R. Experimental Study on Bagasse Ash in Concrete. *Int. J. Serv. Learn. Eng.* **2010**, *5*, 60–66. [CrossRef]
33. Landa-Sánchez, A.; Bosch, J.; Baltazar-Zamora, M.A.; Croche, R.; Landa-Ruiz, L.; Santiago-Hurtado, G.; Moreno-Landeros, V.M.; Olguín-Coca, J.; López-Léon, L.; Bastidas, J.M.; et al. Corrosion behavior of steel-reinforced green concrete containing recycled coarse aggregate additions in sulfate media. *Materials* **2020**, *13*, 4345. [CrossRef] [PubMed]
34. Kawade, U.R.; Rathi, V.R.; Girge, V.D. Effect of use of Bagasse Ash on Strength of Concrete. *Int. J. Innov. Res. Sci. Eng. Technol.* **2013**, *2*, 2997–3000.
35. Castaldelli, V.; Akasaki, J.; Melges, J.; Tashima, M.; Soriano, L.; Borrachero, M.; Monzó, J.; Payá, J. Use of Slag/Sugar Cane Bagasse Ash (SCBA) Blends in the Production of Alkali-Activated Materials. *Materials* **2013**, *6*, 3108–3127. [CrossRef] [PubMed]
36. Ariza-Figueroa, H.A.; Bosch, J.; Baltazar-Zamora, M.A.; Croche, R.; Santiago-Hurtado, G.; Landa-Ruiz, L.; Mendoza-Rangel, J.M.; Bastidas, J.M.; Almeraya-Calderón, F.A.; Bastidas, D.M. Corrosion behavior of AISI 304 stainless steel reinforcements in SCBA-SF ternary ecological concrete exposed to $MgSO_4$. *Materials* **2020**, *13*, 2412. [CrossRef]
37. Ganesan, K.; Rajagopal, K.; Thangavel, K. Evaluation of bagasse ash as corrosion resisting admixture for carbon steel in concrete. *Anti-Corros. Methods Mater.* **2007**, *54*, 230–236. [CrossRef]
38. Landa-Gómez, A.E.; Croche, R.; Márquez-Montero, S.; Villegas Apaez, R.; Ariza-Figueroa, H.A.; Estupiñan López, F.; Gaona Tiburcio, G.; Almeraya Calderón, F.; Baltazar-Zamora, M.A. Corrosion behavior 304 and 316 stainless steel as reinforcement in sustainable concrete based on sugar cane bagasse ash exposed to Na_2SO_4. *ECS Trans.* **2018**, *84*, 179–188. [CrossRef]
39. Ramakrishnan, K.; Ganesh, V.; Vignesh, G.; Vignesh, M.; Shriram, V.; Suryaprakash, R. Mechanical and durability properties of concrete with partial replacement of fine aggregate by sugarcane bagasse ash (SCBA). *Mater. Today Proc.* **2021**, *42*, 1070–1076. [CrossRef]
40. Ojeda, O.; Mendoza, J.M.; Baltazar, M.A. Influence of sugar cane bagasse ash inclusion on compacting, CBR and unconfined compressive strength of a subgrade granular material. *Rev. Alconpat* **2018**, *8*, 194–208.
41. Landa, L.; Márquez, S.; Santiago, G.; Moreno, V.; Mendoza, J.M.; Baltazar, M.A. Effect of the addition of sugar cane bagasse ash on the compaction properties of a granular material type hydraulic base. *Eur. J. Eng. Technol. Res.* **2021**, *6*, 76–79. [CrossRef]
42. Dong, P.; Van Tuan, N.; Thanh, L.; Thang, N.C.; Cu, V.; Mun, J.H. Compressive Strength Development of High-Volume Fly Ash Ultra-High-Performance Concrete under Heat Curing Condition with Time. *Appl. Sci.* **2020**, *10*, 7107. [CrossRef]

43. *NMX-C-414-ONNCCE-2014–Industria de la Construcción-Cementantes Hidráulicos-Especificaciones y Métodos de Ensayo*; ONNCCE, Cd.: México City, México, 2014.
44. ASTM C33/C33m-16e1. *Standard Specification for Concrete Aggregates*; ASTM International: West Conshohocken, PA, USA, 2016.
45. ASTM C29/C29M-07. *Standard Test Method for Bulk Density ("Unit Weight") and Voids in Aggregate*; ASTM International: West Conshohocken, PA, USA, 2007.
46. ASTM C127-15. *Standard Test Method for Relative Density (Specific Gravity) and Absorption of Coarse Aggregate*; ASTM International: West Conshohocken, PA, USA, 2015.
47. ASTM C128-15. *Standard Test Method for Relative Density (Specific Gravity) and Absorption of Fine Aggregate*; ASTM International: West Conshohocken, PA, USA, 2015.
48. ACI 211.1-2004 Standard. *Standard Practice for Selecting Proportions for Normal, Heavyweight, and Mass Concrete*; American Concrete Institute: Indianapolis, IN, USA, 2004.
49. NMX-C-156-ONNCCE-2010. *Determinación de Revenimiento en Concreto Fresco*; ONNCCE S.C.: México City, México, 2010.
50. ASTM C 1064/C1064M-08. *Standard Test Method for Temperature of Freshly Mixed Hydraulic-426 Cement Concrete*; ASTM International: West Conshohocken, PA, USA, 2008.
51. NMX-C-162-ONNCCE-2014. *Determinación de la Masa Unitaria, Cálculo del Rendimiento y Contenido de Aire del Concreto Fresco por el Método Gravimétrico*; ONNCCE S.C.: México City, México, 2014.
52. NMX-C-083-ONNCCE-2014. *Determinación de la Resistencia a la Compresión de Especímenes–Método de Prueba*; ONNCCE S.C.: México City, México, 2014.
53. NMX-C-159-ONNCCE-2004. Industria de la Construcción-Concreto-Elaboración y Curado de Especímenes en el Laboratorio. ONNCCE, Cd.: México City, México, 2004.
54. NMX-C-192-ONNCCE-2018. *Industria de la Construcción-Concreto-Determinación del Número de Rebote Utilizando el Dispositivo Conocido como Esclerómetro-Método de Ensayo*; ONNCCE, Cd.: México City, México, 2018.
55. Deng, P.; Sun, Y.; Liu, Y.; Song, X. Revised Rebound Hammer and Pull-Out Test Strength Curves for Fiber-Reinforced Concrete. *Adv. Civ. Eng.* **2020**, *2020*, 8263745. [CrossRef]
56. Amini, K.; Jalalpour, M.; Delatte, N. Advancing concrete strength prediction using non-destructive testing: Developmentand verification of a generalizable model. *Constr. Build. Mater.* **2016**, *102*, 762–768. [CrossRef]
57. Sengul, O. Use of electrical resistivity as an indicator for durability. *Constr. Build. Mater.* **2014**, *73*, 434–441. [CrossRef]
58. Lubeck, A.; Gastaldini, A.L.G.; Barin, D.S.; Siqueira, H.C. Compressive strength and electrical properties of concrete with white Portland cement and blast-furnace slag. *Cem. Concr. Comp.* **2012**, *34*, 392–399. [CrossRef]
59. Hornbostel, K.; Larsen, C.K.; Geiker, M.R. Relationship between concrete resistivity and corrosion rate—A literature review. *Cem. Concr. Comp.* **2013**, *39*, 60–72. [CrossRef]
60. Morris, W.; Vico, A.; Vazquez, M. Chloride induced corrosion of reinforcing steel evaluated by concrete resistivity measurements. *Electrochim. Acta.* **2004**, *49*, 4447–4453. [CrossRef]
61. ASTM C1760-12. *Standard Test Method for Bulk Electrical Conductivity of Hardened Concrete (Withdrawn 2021)*; ASTM International: West Conshohocken, PA, USA, 2012.
62. Troconis De Rincón, O.; Carruyo, A.R.; Andrade, C.; Helene, P.R.L. ; *Díaz. I. Manual de Inspección, Evaluación y Diagnóstico de Corrosión en Estructuras de Hormigón Armado. Red DURAR*; CYTED: Caracas, Venezuela, 1997.
63. Vouk, D.; Nakic, D.; Stirmer, N.; Cheeseman, C.R. Use of sewage sludge ash in cementitious materials. *Rev. Adv. Mater. Sci.* **2017**, *49*, 158–170.
64. Minnu, S.N.; Bahurudeen, A.; Athira, G. Comparison of sugarcane bagasse ash with fly ash and slag: An approach towards industrial acceptance of sugar industry waste in cleaner production of cement. *J. Clean. Prod.* **2021**, *285*, 124836. [CrossRef]
65. Patra, R.K.; Mukharjee, B.B. Influence of incorporation of granulated blast furnace slag as replacement of fine aggregate on properties of concrete. *J. Clean. Prod.* **2017**, *165*, 468–476. [CrossRef]
66. Choi, S.; Kim, Y.; Oh, T.; Cho, B. Compressive Strength, Chloride Ion Penetrability, and Carbonation Characteristic of Concrete with Mixed Slag Aggregate. *Materials* **2020**, *13*, 940. [CrossRef] [PubMed]
67. NMX-C-155-ONNCCE-2014. *Industria de la Construcción–Concreto Hidráulico–Dosificado en Masa–Especificaciones y Métodos de Ensayo*; ONNCCE, Cd.: México City, México, 2014.
68. Khawaja, S.A.; Javed, U.; Zafar, T.; Riaz, M.; Zafar, M.S.; Khan, M. Eco-friendly incorporation of sugarcane bagasse ash as partial replacement of sand in foam concrete. *Clean. Eng. Technol.* **2021**, *4*, 100164. [CrossRef]
69. Wu, W.; Wang, R.; Zhu, C.; Meng, Q. The effect of fly ash and silica fume on mechanical properties and durability of coral aggregate concrete. *Constr. Build. Mater.* **2018**, *185*, 69–78. [CrossRef]
70. Landa, L.; Ariza, H., Santiago, G.; Moreno, V.; López, R.; Villegas, R.; Márquez, S.; Croche, R.; Baltazar, M. Evaluation of the behavior of the physical and mechanical properties of green concrete exposed to magnesium sulfate. *Eur. J. Eng. Res. Sci.* **2020**, *5*, 1353–1356. [CrossRef]
71. Mohamed, H.A. Effect of fly ash and silica fume on compressive strength of self-compacting concrete under different curing conditions. *Ain Shams Eng. J.* **2011**, *2*, 79–86. [CrossRef]
72. Arif, E.; Clark, M.W.; Lake, N. Sugar cane bagasse ash from a high-efficiency co-generation boiler as filler in concrete. *Constr. Build. Mater.* **2017**, *151*, 692–703. [CrossRef]

73. Nath, P.; Sarker, P. Effect of Fly Ash on the Durability Properties of High Strength Concrete. *Procedia Eng.* **2011**, *14*, 1149–1156. [CrossRef]
74. Yu, J.; Lu, C.; Leung, K.Y.; Li, G. Mechanical properties of green structural concrete with ultrahigh-volume fly ash. *Constr. Build. Mater.* **2017**, *147*, 510–518. [CrossRef]
75. Le, D.; Sheen, Y.; Lam, M. Fresh and hardened properties of self-compacting concrete with sugarcane bagasse ash–slag blended cement. *Constr. Build. Mater.* **2018**, *185*, 138–147. [CrossRef]
76. Xu, Q.; Ji, T.; Gao, S.; Yang, Z.; Wu, N. Characteristics and Applications of Sugar Cane Bagasse Ash Waste in Cementitious Materials. *Materials* **2018**, *12*, 39. [CrossRef]
77. Parron-Rubio, M.E.; Perez-García, F.; Gonzalez-Herrera, A.; Rubio-Cintas, M.D. Concrete Properties Comparison When Substituting a 25% Cement with Slag from Different Provenances. *Materials* **2018**, *11*, 1029. [CrossRef] [PubMed]
78. Poorarbabi, A.; Ghasemi, M.; Azhdary Moghaddam, M.A. Concrete compressive strength prediction using non-destructive tests through response surface methodology. *Ain Shams Eng. J.* **2020**, *11*, 939–949. [CrossRef]
79. Qasrawy, H.Y. Concrete strength by combined nondestructives methods Simply and reliably predicted. *Cem. Concr. Res.* **2000**, *30*, 739–746. [CrossRef]
80. Landa, A.; Croche, R.; Márquez-Montero, S.; Galván-Martínez, R.; Tiburcio, C.G.; Almeraya Calderón, F.; Baltazar, M. Correlation of compression resistance and rupture module of a concrete of ratio w/c = 0.50 with the corrosion potential, electrical resistivity and ultrasonic pulse speed. *ECS Trans.* **2018**, *84*, 217–227. [CrossRef]
81. Bagueri, A.; Zanganeh, H.; Alizadeh, H.; Shakerinia, M.; Marian, M.A. Comparing the performance of fine fly ash and silica fume in enhancing the properties of concretes containing fly ash. *Constr. Build. Mater.* **2013**, *47*, 321–333.
82. Sadrmomtazi, A.; Tahmouresi, B.; Saradar, A. Effects of silica fume on mechanical strength and microstructure of basalt fiber reinforced cementitious composites (BFRCC). *Constr. Build. Mater.* **2018**, *162*, 321–333. [CrossRef]
83. Buenfeld, N.R.; Glass, G.K.; Hassanein, A.M.; Zhang, J.Z. Chloride transport in concrete subjected to electrical field. *J. Mater. Civ. Eng.* **1998**, *10*, 220–228. [CrossRef]

 crystals

MDPI

Article

The Influence of Water/Binder Ratio on the Mechanical Properties of Lime-Based Mortars with White Portland Cement

Dejan Vasovic *, Jefto Terzovic, Ana Kontic, Ruza Okrajnov-Bajic and Nenad Sekularac

Faculty of Architecture, University of Belgrade, 11000 Belgrade, Serbia; jefto@arh.bg.ac.rs (J.T.); an.ko@arh.bg.ac.rs (A.K.); ruza@arh.bg.ac.rs (R.O.-B.); nenad.sekularac@arh.bg.ac.rs (N.S.)
* Correspondence: d.vasovic@arh.bg.ac.rs; Tel.: +381-631-119-838

Abstract: Protecting the built cultural heritage is one of the most important tasks in architectural practice. The process of repair is time-consuming, weather-dependent, and sensitive to materials applied. Introducing new materials in historic building repair in order to decrease the time needed for repair, brings some risk in the preservation process. The most common material for masonry repair is lime mortar. Adding cement to lime mortar can improve the mechanical properties of mortar and speed up the repair process. The high amount of cement may increase the strength, but decrease ductility and permeability of mortar, causing damages to protected buildings. An increase in strength with the smallest amounts of cement demands optimization of water content in the mixture. Tests were performed to investigate the influence of the water/binder (w/b = water/(lime + cement) ratio on mortar strength and water permeability. An air-entraining agent (AEG) was introduced to improve permeability. Results confirmed that adding small amounts of cement to lime (20% by weight) and decreasing of w/b ratio, improves the strength, with almost negligible influence on water permeability. The addition of very small amounts of AEG did not decrease the strength, nor the permeability.

Keywords: lime-cement mortar; compressive strength; air-entrained agent; heritage conservation; reconstruction and restoration of historical buildings

Citation: Vasovic, D.; Terzovic, J.; Kontic, A.; Okrajnov-Bajic, R.; Sekularac, N. The Influence of Water/Binder Ratio on the Mechanical Properties of Lime-Based Mortars with White Portland Cement. *Crystals* **2021**, *11*, 958. https://doi.org/10.3390/cryst11080958

Academic Editors: Antonella Sola, Sumit Chakraborty, Valentina Volpini and Cesare Signorini

Received: 18 July 2021
Accepted: 14 August 2021
Published: 16 August 2021

1. Introduction

Historical buildings have cultural significance, witness to architectural history, but also preserve information about structural systems, materials applied and building techniques used [1]. Until the end of the 18th century, the use of lime mortar was considered as an element of continuity of architectural heritage, whether it was applied for rendering or as a structural component. The oldest data, from 6000 BC (before Christ) regarding the application of lime in Serbia, is located in Lepenski Vir, an archaeological site from the Neolithic era on the banks of the Danube River, where the lime mortar floor, made with quicklime, was found [2]. The process of lime mortars hardening, i.e. transformation of $Ca(OH)_2$, into $CaCO_3$ as a carrier of strength, occurs in the presence of carbon dioxide. According to Van Balen [3], diffusion of CO_2 is roughly 10,000 times lower in water than in air [4,5]. Literature review [6–10], has shown that after 180 days from the production day, lime mortar reaches approximately the final values of its compressive strength. Elongated setting and requirements for dry environmental conditions, slow and undetermined carbonation, low mechanical properties, and internal cohesion, were the main reasons for the introduction of Portland Cement (PC) into conservation practice. Lime and alumina silicates in cement develop crystalline substances during hydration which improves the strength of mortar and its adhesion to the brick and stone. [11]. Nevertheless, as conservation of historic structures is quite complex and its accomplishment lies in the adequate interaction between the inherited structure and a new one, the rapid diffusion of PC was stopped because of its incompatibility with the original structure. A high concentration of soluble salts found in PC could cause damage to the original materials as salt crystallization occurs. Moreover, higher mechanical strengths of the PC than of the original mortar, as well as low deformability, appeared to be the source of degradation of historical heritage. International organizations, such as International Council on Monuments and Sites (ICOMOS) or International Centre for the Study of the Preservation and Restoration of Cultural Property (ICCROM), prefer practitioners to use the same or materials similar to originals, both in composition and characteristics, for the restoration works [7,12].

Research problems in the area of mechanical characteristics of lime-based mortars are various, moving from deeper exploration and understanding of carbonation process [3,8,13,14], different factors affecting it [15–17], to studies focusing on mechanical characteristics of the final product with different variables [7,18–28]. However, the addition of cement to lime-based mortars is still a common practice in most conservation works in order to increase the hardening reaction by cement hydration [29].

Long-term exposure to environmental loads damage and decay the walls of the historical buildings. Mortar for the restoration of the walls of historical monuments should bind the masonry elements (stone, brick) into a sound structure providing load-bearing capacity in the shortest possible time, while its compressive and tensile strengths must not be higher than those of the stone or brick, so that future environmental loads should damage restored mortar first, instead of the stone or brick that we want to protect. The adhesion of mortar should be lower than the tensile strength of the base material, to ensure the separation of the mortar from stone and brick instead of the crushing and falling stone or brick pieces together with the binder.

Figure 1 shows the retaining wall of a 14th-century fortification in the south of Serbia. Figure 1a was taken before the reconstruction in 2012 and shows the condition of the damaged wall: after several hundred years, the stones show no visible signs of damage; the white lime mortar that it had been built with collapsed because it had weaker mechanical properties. The wall was repaired with mortar containing a large amount of ordinary grey Portland cement (OPC). After hardening, this mortar had an inappropriate colour (as the white colour was requested by Heritage Protection Authorities), became significantly stronger and more impermeable than stone. The idea that stronger mortar will better strengthen the wall proved to be completely wrong. Figure 1b, taken in 2021, less than nine years after the reconstruction, shows the consequences of the application of the inappropriate mortar: the stone is damaged and the surface layer, few centimetres thick, fell off. Significantly reduced permeability of the mortar caused the transport of water vapour and moisture through the stone material instead of through the mortar and resulted in accelerated deterioration of the stone wall.

Figure 1. Malpractice of introducing cement-lime mortar in Serbia (**a**) before July 2012; (**b**) nine years after repair of the wall—July 2021).

Many researchers are trying to define an adequate proportion of cement, to improve lime mortar properties, keeping its compatibility with historic structures. Nevertheless, it has been found that by blending cement with the lime, it is possible to reduce cracking, to quicken the application and hardening of the mortar, thus providing protection from the rain before carbonation has been completed, and to ensure reliability and predictability of its properties [30]. Literature review regarding the issue of adding PC to lime mortars shows that there are no clear recommendations. Some studies promote the use of PC to a significant extent of a minimum of 20% of lime weight, justifying it by high initial strengths [11,19]. Other studies promote the use of PC in smaller amounts, referencing unchanged water absorption with regard to lime mortars with no additives [31], significant reduction of porosity due to the increase of cement content [29], or loss of elastoplastic behaviour or mortar mixtures, which makes them able to adapt to movements and deformation under critical stress in the masonry [32].

It is notable, that the information available regarding this topic is diversified, with no unique attitude concerning the optimum amount of Portland cement in lime-based mortars. Literature review showed that criteria of water/binder ratio in lime mortars with the addition of cement, is quite neglected, therefore, this paper aims at discussing early mechanical properties (up to 28 days) of the abovementioned mixtures. It is well known that the water/binder ratio in cement mortars has a strong influence on the mechanical characteristics of the hardened material. However, this rule is not applicable to lime mortars, due to smaller particle size. Research made by Lawrence and Walker [33], demonstrates that the water/binder ratio has almost no influence at all on the mechanical properties of lime mortars. On the other hand, according to Lanas and Alvarez [7], a higher water/binder ratio results in greater porosity, which enables better diffusion of CO_2, a key component in the process of carbonation. Hence, the main hypothesis of this paper is that with an adequate determination of water/binder ratio in lime mortars with the addition of White Portland Cement (WPC), higher initial mechanical strengths could be acquired with amounts of cement, up to 20% of the lime weight. Use of the WPC is mandatory, as heritage protection requirements demand the white colour of the mortar.

The literature reviews also showed different studies of lime mortars with incomparable mixtures. Some mixture proportions are determined by weight, others by volume; some use hydrated lime in powder, others, lime putty. However, the water content of the mixtures is determined arbitrarily: "on the basis of masonry experience", "adequate for common workability", by volume or by weight of binder (regardless of the lime/cement ratio), etc. Such determination of water content is especially inadequate when cement is added. Excessive water may prevent the expected cement hardening. The mechanical characteristics of cement-based mortars and concrete crucially depend on the W/C factor. It is, therefore, necessary to precisely define the amount of water in lime mortar with the addition of cement. This study shows that it is necessary to keep the same consistency of the lime-cement mortars mixtures, in order to adequately compare their mechanical properties.

2. Materials and Methods

2.1. Composition of Mortar Mixtures

Mortars were made by using the following components:

- Hydrated lime CL90-S (EN 459-1, produced by Ingram, Bosnia, and Herzegovina).
- Calcium carbonate aggregate produced by OMYA (Venčac, Serbia); it was used in two grades (0.0–0.8 mm and 0.8–2.0 mm), equally dosed.
- Lime/Aggregate ratio was defined among the commonest dosages described in the literature, as 1:3.
- WPC CEM I 52.5 N was used (EN 197-1, produced by CRH Slovakia). Different amounts of WPC were added, as shown in Table 1.

Table 1. Mortar mixtures and properties of fresh mortars.

Series	Specimen Code	w/b Ratio	WPC	Lime/WPC (by Weight)	AEG/WPC (by Weight)	Slump Flow (mm)	Slump (mm)
1	E	0.875	-	-	-	68.0	5
1	C5	0.875	WPC	5%	-	80.0	10
1	C10	0.875	WPC	10%	-	87.5	16
1	C20	0.875	WPC	20%	-	91.0	17
2	KC5	0.845	WPC	5%	-	71.5	9
2	KC10	0.8125	WPC	10%	-	70.0	9
2	KC20	0.755	WPC	20%	-	68.0	9
3	AC5	0.845	WPC	5%	0.4%	75.5	10
3	AC10	0.8125	WPC	10%	0.4%	75.0	10
3	AC20	0.755	WPC	20%	0.4%	78.5	11

Table 1 shows mixture proportions and properties of one reference (E) and nine different mixtures in three series (C, KC, and AC) that were made. Six specimens of each mixture were casted (three for testing after 7 and three for testing after 28 days).

- Series 1, with w/b ratio at fixed value (0.875);
- Series 2, with w/b ratio at three different values, but same workability (slump);
- Series 3, with the same workability (slump), but with the added air-entraining agent (AEG-*Chryso Air G 100*, 0.4% by weight of cement).

The amount of water added (w/l = 0.875) for the reference lime mortar (E) was based on experience. The first series of samples were casted with the same water/binder ratio, which increased the slump. In order to obtain the same slump, in series 2 and 3, the water/binder ratio was decreased.

Mixtures were prepared following European standard EN 196-1. Binders were mixed with water for 1 min, the aggregate was added, and mixed for 2 minutes. Afterwards, the prepared paste was casted in moulds, dimensions 40 × 40 × 160 mm, and covered with absorbent paper in order to decrease evaporation.

The specimens were cured under dry environment conditions (T = 20 ± 5 °C, RH = 65 ± 5%), removed from moulds after 5 days (as reference samples did not gain sufficient strength to be removed earlier), and tested on the 7th and 28th day.

2.2. Methods for Measurement of Properties of Hardened Mortars

2.2.1. Compressive Strength

Compressive strength was measured in accordance with European standard EN 1015-11 using CONTROLS Pilot Pro Multipurpose 500/15 kN Cement Compression and Flexural Machine, with a loading rate of 40 N/s.

2.2.2. Flexural Strength

A three-point flexural test was performed using CONTROLS Pilot Pro Multipurpose 500/15 kN Cement Compression and Flexural Machine, with the loading rate of 1 N/s.

2.2.3. Carbonation Depth

A carbonation depth test was performed using phenolphthalein on the broken surfaces of the specimens, immediately after the flexural test. Phenolphthalein indicated the development of carbonation, as it changed the colour from colourless to purple at pH > 9, in the presence of OH^- ions. This method, although not very precise, was chosen as the fastest and easily visible. Other more precise methods for measuring the carbonation depth (X-ray Diffraction Analysis, Fourier Transform Infrared Spectroscopy, Thermo-gravimetric Analysis and Loss on Ignition, Dust digestion [34] required special equipment and trained professionals for the application, to which the authors had no access.

2.2.4. Capillary Absorption

The capillary absorption tests were conducted in accordance with European standard EN 1015-18, by placing the specimens in shallow water, around 3 mm depth. The weight of specimens was taken in different time intervals, at 1, 3, 5, 10, 15, 20, 30, 45, 60, 90, 120, and 150 min, until the absorption reached an asymptotic value.

3. Results

Results of compressive and flexural strength testing after 7 and 28 days are shown in Table 2. Values in Table 2 are average values of three specimens for flexure, and six specimens for compressive strength. It shows that the addition of WPC in the range of 20% (specimens C20, KC20, and AC20) improves compressive strength in comparison to the reference mortar (E).

Table 2. Average compressive (f_c) and flexural strength (f_f), at 7 and 28 days, with their corresponding standard deviation (σ).

Series	Specimen Code	Compressive Strength, f_c (MPa)				Flexural Strength, f_f (MPa)			
		7 Days		28 Days		7 Days		28 Days	
		$f_{c,7}$	σ	$f_{c,28}$	σ	$f_{f,7}$	σ	$f_{f,28}$	σ
1	E	0.580	0.056	1.515	0.131	0.145	0.022	0.300	0.029
1	C5	0.430	0.032	1.125	0.051	0.160	0.024	0.335	0.060
1	C10	0.540	0.079	1.315	0.065	0.185	0.029	0.340	0.062
1	C20	0.940	0.028	1.705	0.109	0.325	0.051	0.525	0.041
2	KC5	0.910	0.050	1.425	0.094	0.130	0.017	0.460	0.037
2	KC10	0.940	0.038	1.100	0.131	0.130	0.021	0.505	0.029
2	KC20	1.430	0.071	2.020	0.112	0.145	0.021	0.535	0.037
3	AC5	0.795	0.051	1.415	0.061	0.150	0.140	0.465	0.027
3	AC10	0.820	0.056	1.620	0.140	0.185	0.190	0.510	0.037
3	AC20	1.345	0.063	2.135	0.093	0.455	0.310	0.520	0.036

3.1. Compressive Strength

The testing of specimens from Series 1 (Figure 2), with a constant water/binder ratio, shows that the addition of 5% of WPC reduces compressive strength by 27%, whilst the addition of 10% of WPC reduces it by 36%. Significant change occurs with the addition of WPC of 20%, where the increase is more than 120% for the initial, 7-day compressive strength, comparing to the reference specimen (E) without WPC. This trend of insignificant change of mechanical strengths with smaller additions of WPC (5% and 10%), can be noted throughout results from all series. This observation is in accordance with results obtained from previous studies [11,19].

When it comes to water/binder ratio, by comparing results obtained from Series 1 (Figure 2) and Series 2 (Figure 3), it can be concluded that decreasing of water content in lime mortars with the addition of WPC, can have a major impact on the initial mechanical strengths. In order for the mixtures with the addition of cement to have the same consistency as the reference mixture (mixture E), it was necessary to reduce the total amount of water with the increase in the amount of added cement. This is notable for both 7th day and 28th-day strengths. In relation to results from Series 1, results from Series 2 show higher strengths, increasing as the amount of WPC increases, which was expected. It can be concluded that the addition of 20% of WPC, with a water binder ratio of 0.775, increases compressive strength by 90% in the first 7 days, and by 60% on the 28th day. The test results of KC10 samples with 10% WPC, showed an unexpected decrease in strength, which may be due to internal defects.

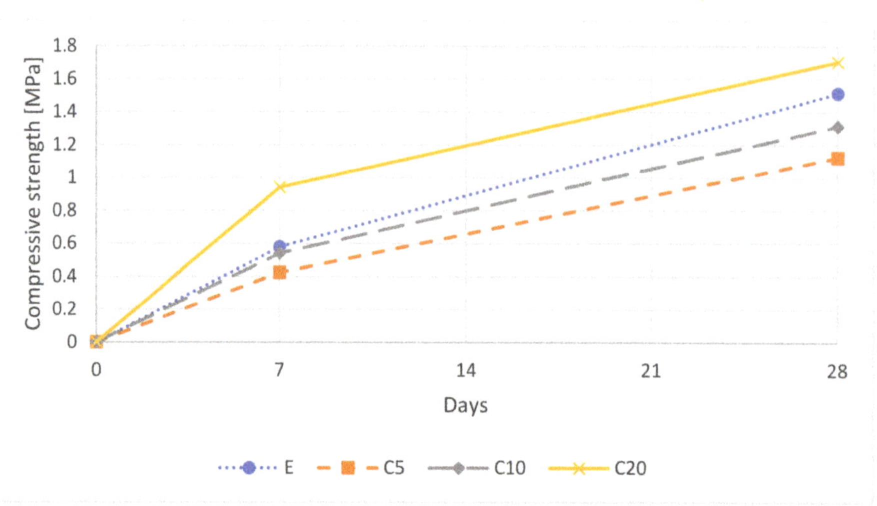

Figure 2. Development of compressive strength from 7–28 days in Series 1 (same w/b ratio, different percentage of WPC).

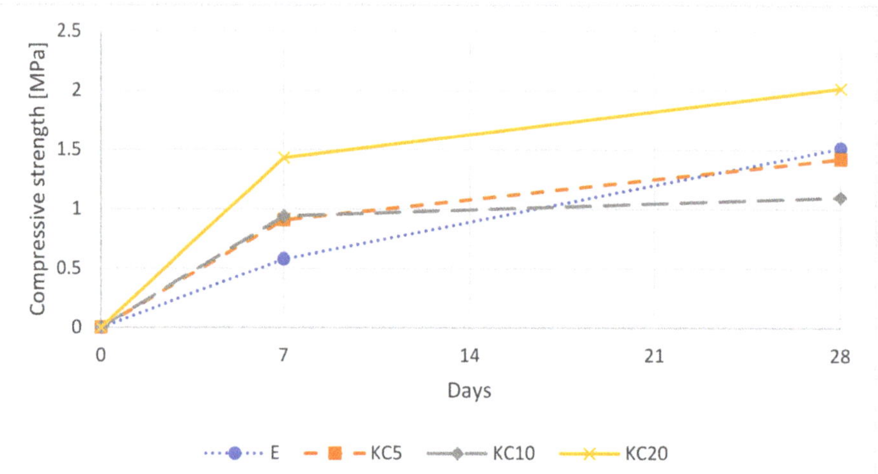

Figure 3. Development of compressive strength from 7–28 days in Series 2 (varied w/b ratio and percentage of WPC).

The main idea behind the addition of air-entraining agents (AEG) in the amount of 0.4% of WPC amount, was an increase of porosity as the addition of WPC decreases it. As the addition of AEG was the only variable between Series 2 and Series 3 specimens, it was expected that this addition would reduce mechanical strengths, but the result showed that this mixture gave almost the same results as the previous ones (Figure 4). The highest 28th-day compressive strength (2.135 MPa) was for a specimen named AC20, with 20% WPC and 0.4% of AEG by weight of WPC. Concluding from the experiment, the addition of AEG does not reduce the compressive strength of lime mortars with the addition of WPC.

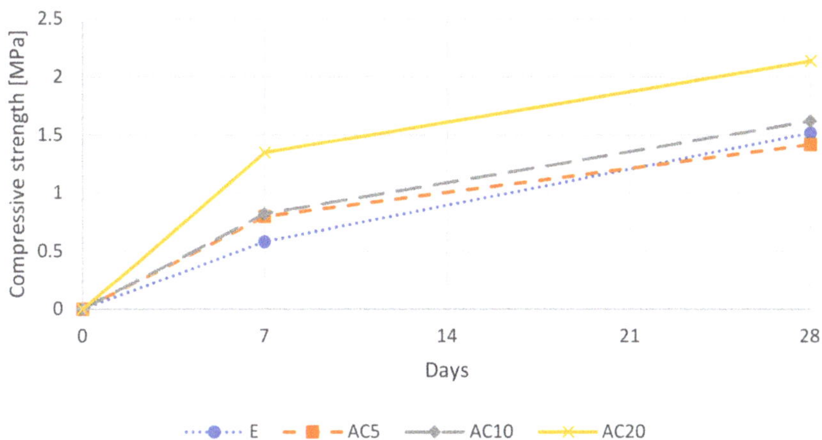

Figure 4. Development of compressive strength from 7–28 days in Series 3 (varied w/b ratio and percentage of WPC, 0.4% AEG agents added).

3.2. Flexural Strength

It is noticeable that the addition of 5–10% of WPC to lime mortar slightly improves the flexural strength compared to the reference mortar (E) (results in Table 2). Regarding the specimens from Series 1 (Figure 5), with a constant water/binder ratio, it can be observed that the addition of 5% of WPC increases flexural strength by 28%, whilst the addition of 10% of WPC, increases it by 48%. Significant change occurs with the addition of WPC of 20%, where the increase is more than 250% for the initial 7-day compressive strength, regarding the reference specimen with no WPC added. The abovementioned trend, of a slight change of mechanical strengths with the addition of WPC between 5–10%, is diagnosed for flexural strength, as well.

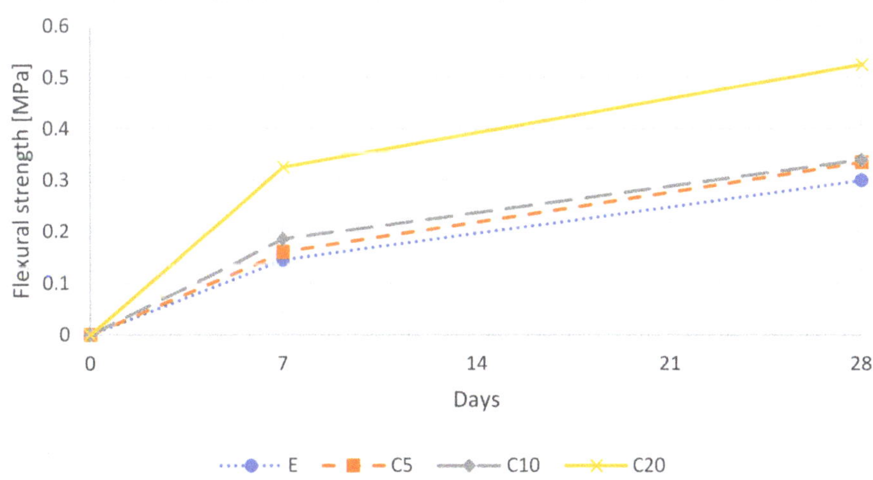

Figure 5. Development of flexural strength from 7–28 days in Series 1 (same w/b ratio, different percentage of WPC).

With regard to water/binder ratio, comparing results obtained from Series 1 and Series 2 (Figure 6), it can be concluded that decreasing of water content in lime mortars with the addition of WPC, had a small influence only on 28th-day strengths, but did not improve the 7th-day strengths. It can be noted that an increase in the amount of WPC, slightly increases the flexural strengths of specimens.

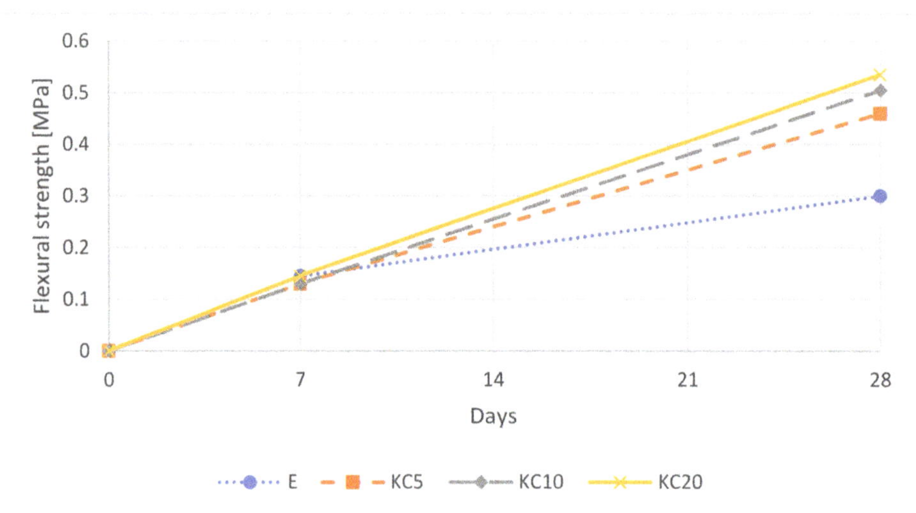

Figure 6. Development of flexural strength from 7–28 days in Series 2 (varied w/b ratio and percentage of WPC).

As for the specimens from Series 3 (Figure 7), the same trend can be noticed. The addition of AEG actually had no impact on flexural strengths, in comparison to the mixtures with no AEG added (Series 2).

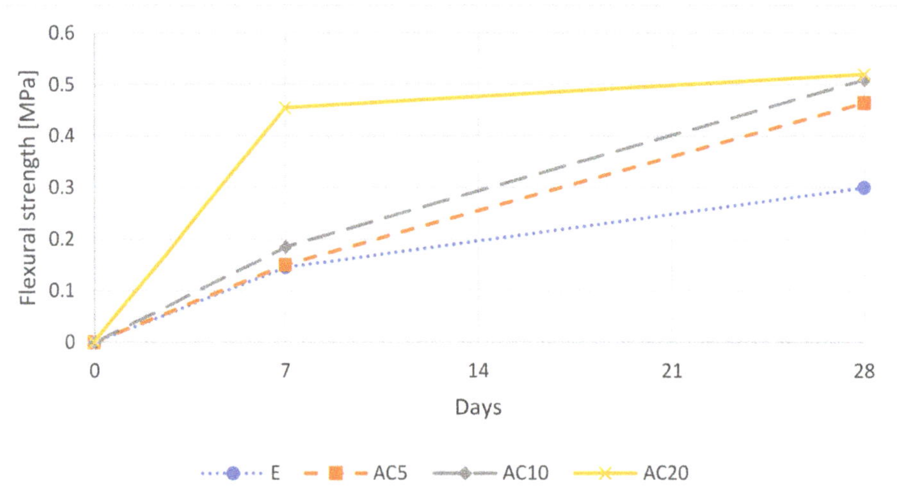

Figure 7. Development of flexural strength from 7–28 days in Series 3 (varied w/b ratio and percentage of WPC, 0.4% AEG agents added).

3.3. Carbonation Depth

Carbonation depth, determined by the phenolphthalein test after 28 days, showed that carbonation progressed from exterior to interior, but the measurement showed negligible differences between different specimens. For specimens with the addition of WPC, hardening is a result of two processes, carbonation of lime and hydration of cement. The result of these two processes can be seen by comparing specimens KC5, KC10, KC20 (Figure 8), where the increase of WPC amount decreases the carbonation depth.

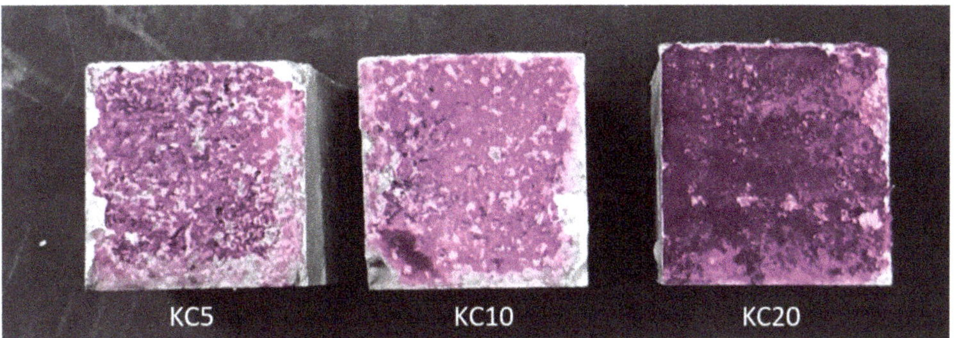

Figure 8. Phenolphthalein test specimens in Series 2 (varied w/b ratio and percentage of WPC).

With the decrease of WPC, the effect of both the carbonation and hydration process is apparent. It is expected that carbonation will proceed gradually until the 180th day of the porosity. Results from specimens with the addition of AEG show different patterns, with gradual carbonation in the core, as well as by the edges (Figure 9).

Figure 9. Phenolphthalein test, AC20 specimen (varied w/b ratio and percentage of WPC, 0.4% AEG agent added).

3.4. Capillary Absorption

Measured values of capillary absorption are shown in Table 3. It is commonly observed that the addition of WPC decreases the capillary absorption coefficient as a result of the changed pore structure. Results obtained from this research shows an insignificant change in capillary water absorption between different mixtures.

Although a test of porosity was not a part of this research, capillary water absorption results indicated that small amounts of WPC did not change capillary pore structure, as the absorption coefficient did not change significantly. For specimens with a constant water/binder ratio, the percentage of water absorbed was slightly increased. This is probably due to the fact that excess water in these mixtures resulted in greater porosity than for mixtures with constant consistency, where less water was used, thus, lower porosity was obtained. Results are shown in Figures 10 and 11.

Table 3. Capillary water absorption: mass of the specimens (m), mass of the absorbed water (Δm), and its corresponding standard deviation (σ).

Series	Specimen Code	m (g)	Δm (g)	σ	t (min)	A (kg/m^2 × $\sqrt{h}$)	U (%)
1	E	222.01	35.24	1.16	150	13.93	15.87
1	C5	230.18	35.92	3.40	120	15.87	15.60
1	C10	226.17	37.50	1.37	150	14.82	16.58
1	C20	223.09	38.30	0.30	150	15.14	17.17
2	KC5	220.14	32.96	0.91	150	13.03	14.97
2	KC10	225.40	37.51	1.79	120	16.58	16.64
2	KC20	224.85	37.57	0.88	050	14.85	16.71
3	AC5	226.71	31.61	0.12	90	16.13	13.94
3	AC10	224.35	32.98	1.36	90	16.83	14.70
3	AC20	236.96	34.64	0.68	120	15.31	14.62

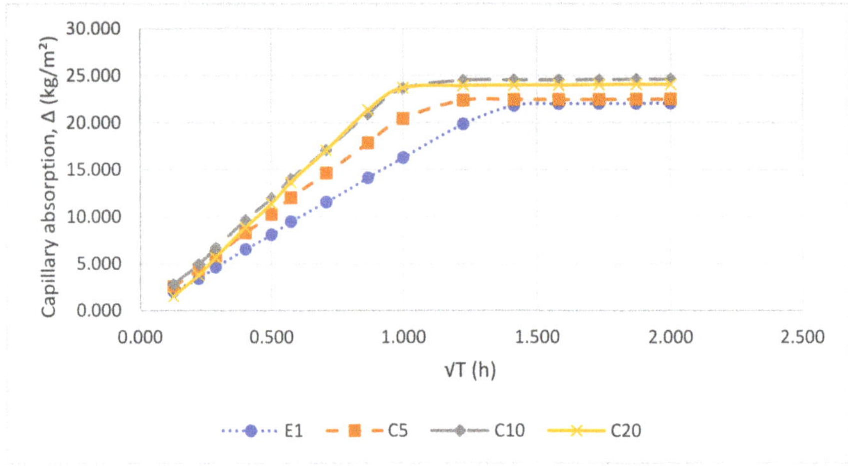

Figure 10. Capillary water absorption in Series 1 (same w/b ratio, different percentage of WPC).

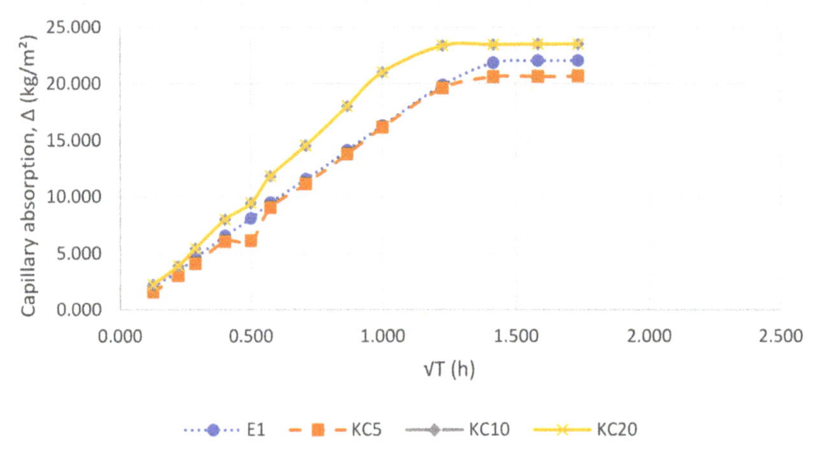

Figure 11. Capillary water absorption in Series 2 (varied w/b ratio and percentage of WPC).

Specimens from Series 3 (Figure 12), with the addition of AEG, did not show any significant difference. It was expected that due to increased porosity, the capillary water absorption coefficient would become higher. Hence, results obtained show different situations, referring authors to the assumption from compressive strength tests, as well. It is assumed that AEG actually did not change porosity, i.e., did not react with WPC, but with lime, resulting in greater calcification of specimens.

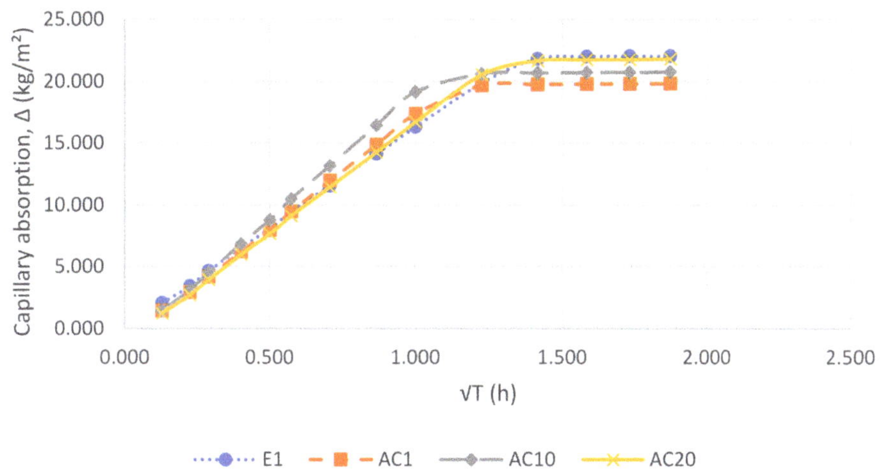

Figure 12. Capillary water absorption in Series 3 (varied w/b ratio and percentage of WPC, 0.4% AEG agents added).

4. Discussion

The addition of WPC makes some difference in regard to the mechanical characteristics of specimens tested. Comparison of the results obtained from the specimens in Series 1 and 2, where the only variable is water/binder ratio, shows that adequate dosing of water can have a major impact on the initial mechanical strengths of lime mortars with the addition of WPC. In order to maintain the same slump, the w/b ratio was reduced as the amount of WPC was increased. This is due to the well-known fact that the strength of cement mortar is inversely proportional to the water/cement ratio [8], which is not applicable to the lime mortars due to their fine particles content. Analysing the w/b ratio used, it could be numerically presented that this ratio is a composition of a water/lime ratio as of the reference specimen (0.875), and a standard water/cement ratio of 0.5. Concerning this and the amount of WPC added to each specimen, the final w/b ratio was as presented in Table 2. Results from Series 2, in regard to results from Series 1, show higher strengths as the amount of WPC increases, which was expected. It can be concluded that the addition of 20% of WPC, with a water binder ratio of 0.775, increases compressive strength by 90% in the first 7 days, and 60% on the 28th day.

As for the specimens from Series 3, with a constant w/b ratio and the addition of AEG of 0.4% to the weight of WPC added, unexpected observations occurred. AEG, as small bubbles of oxygen, was added to cement in order to improve its freeze-thaw behaviour by reducing initial stresses in the structure of the material. On the other hand, by adding AEG, the porosity is reduced. The initial idea behind Series 3 composition was to increase the porosity for specimens with WPC added, as its addition significantly reduces porosity. However, the obtained results point out different assumptions. This situation may be due to the fact that air lime mortar, as it is commonly known, hardened as a result of the transition from hydroxide $Ca(OH)_2$, to carbonate $CaCO_3$, in the presence of CO_2. The obtained results indicate the possibility of reaction between AEG and lime mortar, as AEG are CO_2 carriers. As the amount of AEG depended on the amount of WPC, AEG/lime ratio was increasing from specimen AC5 to AC20. The assumption obtained from this part of the research is that regardless of the decrease of porosity with adding AEG, its reaction with $Ca(OH)_2$ actually promotes carbonation inside the specimen.

When it comes to flexural strength results, it is shown that adding WPC has a slightly positive impact as the final value is increasing with the addition of WPC. That increase is almost insignificant, so the obtained results in this research are in accordance with Arandigoyen and Alvarez [32], who concluded that 40% (by volume, approximately 26% by weight) of cement was the lower limit for improvement of flexural mechanical characteristics of lime-cement mortars. Values of flexural strength, moving from 0.125–0.455 MPa for 7-day tests and from 0.3–0.535 MPa for 28-day tests, are in accordance with the results of previous studies [29,35]. It can be noted that absolute values of flexural strengths are low, these values increase faster than those of compressive strength, which is in accordance with Ramesh, Azenha and Lourenco [36].

When it comes to capillary absorption, it can be observed that the addition of 5, 10, and 20% of WPC, does not significantly change capillary absorption capacity. It is evident from the obtained results that these results are almost the same for all mixtures, moving from 13.94% to 17.17%, which is believed not to be a significant factor of different water and steam permeability.

5. Conclusions

In this paper, different amounts of WPC were used in order to understand how and if this addition has an influence on the mechanical and physical characteristics of lime mortars. Throughout this research, some assumptions were confirmed, whilst some findings motivated new research questions. The main findings of this research are:

- The addition of WPC to air lime mortars, of 5–10% of lime, has almost no influence on the mechanical characteristics of these mixtures. For compressive strength, results show that these small amounts actually degrade the material. Significant change happens with the addition of 20% of WPC, where both compressive and flexural strength increase. Further research should examine the mixtures with higher cement content in order to determine the optimal quantity of cement, to achieve higher strengths while maintaining the vapour permeability.
- The w/b ratio has an important influence on the initial mechanical strengths of lime mortars with the addition of WPC. To achieve the same workability, the w/b ratio decreased with the increasing amounts of additional cement. The excess of water prevents cement from gaining expected mechanical strengths, thus decreasing the overall strength of blended lime. Future research should examine the possibility of further reduction of the w/b ratio while maintaining the required consistency using superplasticizers.
- The addition of AEG in lime mixtures, as a function of the weight of WPC, was expected to decrease mechanical strengths as a result of increased porosity. Nevertheless, this addition increased mechanical strengths, while, as authors assume, changing porosity slightly. This conclusion was reached from capillary water absorption results, as the absorption coefficient remained almost the same as of the reference mixture. The main assumption is that AEG while introducing the CO_2 in the structure of the material, improves calcification and consequently increases mechanical properties of mortars, does not change porosity significantly. Different types of air-entraining agents, possible increase of the amount of AEG, and the consequence of such increase to the early strengths of the cement-blended lime mortars should be studied.
- Future research should test the adhesion of the cement-blended lime mortars to masonry units, as well as the porosity of the samples using mercury intrusion porosimetry.

Author Contributions: Conceptualization, methodology and resources, D.V. and N.S.; validation, J.T. and R.O.-B.; investigation, all authors; writing—original draft preparation, A.K.; writing—review and editing, D.V. All authors have read and agreed to the published version of the manuscript.

Funding: This research received no external funding.

Institutional Review Board Statement: Not applicable.

Informed Consent Statement: Not applicable.

Data Availability Statement: The data presented in this study are available on request from the corresponding author.

Acknowledgments: The authors extend their gratitude to Vojislava Vasovic for her contribution in preparing tables and graphs for this text.

Conflicts of Interest: The authors declare no conflict of interest.

References

1. Friedman, D. *Historical Building Construction: Design, Materials, and Technology*, 2nd ed.; W. W. Norton & Company: New York, NY, USA, 2010.
2. Stanojević, D. Kreč kao istorijski material. In *Zbornik radova Seminara i radionice "Kreč kao istorijski materijal", Proceedings of the Sopoćani, Serbia, 25–27 August 2014*; Republički zavod za zaštitu spomenika kulture: Beograd, Serbia, 2014.
3. Van Balen, K. Carbonation reaction of lime, kinetics at ambient temperature. *Cem. Concr. Res.* **2005**, *35*, 647–657. [CrossRef]
4. Richardson, M. *Carbonation of Reinforced Concrete. Its Causes and Management*; Citis: Dublin, Ireland, 1988.
5. Welty, J.; Wicks, C.; Wilson, R.; Rorrer, G. *Fundamental of Momentum, Heat and Mass Transfer*; Wiley: New York, NY, USA, 1969.
6. Bromblet, P. Evaluation of the durability and compatibility of traditional repair lime based mortars in three limestones. *Int. J. Restor. Build. Monum.* **2000**, *6*, 513–528.
7. Lanas, J.; Alvarez, J.I. Masonry repair lime-based mortars: Factors affecting the mechanical behavior. *Cem. Concr. Res.* **2003**, *33*, 1867–1876. [CrossRef]
8. Lawrence, M. A Study of Carbonation in Non-Hydraulic Lime Mortars. Ph.D. Thesis, University of Bath, Bath, UK, 2006.
9. Papayianni, I. Design and manufacture of repair mortars for interventions on monuments and historical buildings. In Proceedings of the RILEM Technical Committee International Workshop on Repair Mortars for Historic Masonry, Delft, The Netherlands, 26–28 January 2005.
10. Valek, J.; Bartos, P. Influences affecting compressive strength of modern non-hydraulic lime mortars used in masonry conservation. In *Structural Studies, Repair and Maintenance of Historical Buildings*; Advances in Architecture Series VII:13; WIT Press: Southampton, UK, 2001.
11. Abdel-Mooty, M.; Khedr, S.; Mahfouz, T. Evaluation of lime mortars for the repair of historic buildings. In *Structural Studies, Repairs and Maintenance of Heritage Architecture XI*; WIT Press: Southampton, UK, 2009.
12. Venice Charter, International Charter for the Conservation and Restoration of Monuments and Sites. 1964. Available online: http://www.icomos.org/charters/venice_e.pdf (accessed on 8 June 2021).
13. Moorehead, D. Cementation by the carbonation of hydrated lime. *Cem. Concr. Res.* **1986**, *16*, 700–708. [CrossRef]
14. Van Balen, K.; Van Gemert, D. Modelling lime mortar carbonation. *Mater. Struct.* **1994**, *27*, 393–398. [CrossRef]
15. Oliveira, M. A Multi–Physics Approach Applied to Masonry Structures with Non–Hydraulic Lime Mortars. Ph.D. Thesis, Universidade do Minho, Escola de Engenharia, Guimarães, Portugal, 2015.
16. Oliveira, M.A.; Azenha, M.; Lourenco, P.; Meneghini, A.; Guimarães, E.T.; Castro, F.; Soares, D. Experimental analysis of the carbonation and humidity diffusion processes in aerial lime mortar. *Constr. Build. Mater.* **2017**, *148*, 38–48. [CrossRef]
17. Cultrone, G.; Sebastian, E.; Huertas, M.O. Forced and natural carbonation of lime-based mortars with and without additives: Mineralogical and textural ch. *Cem. Concr. Res.* **2005**, *35*, 2278–2289. [CrossRef]
18. Peroni, S.; Tersigni, G.; Torraca, S.; Cerea, S.; Forti, M.; Guidobaldi, F.; Rossi-Doria, P.; de Rege, A.; Picchi, D.; Pietrafitta, F.; et al. Lime based mortars for the repair of ancient masonry and possible substitutes. In *Mortars, Cements and Griuts Used in the Conservation of Historic Buildings*; ICCROM: Rome, Italy, 1981.
19. Teutonico, J.M.; McCaig, I.; Burns, C.; Ashurst, J. The Smeaton Project: Factors Affecting the Properties of Lime-Based Mortars. *APT Bull.* **1993**, *25*, 32. [CrossRef]
20. Henriques, F.; Charola, A. Comparative study of standard test procedures for mortars. In Proceedings of the 8th International Congress in Deterioration and Conservation of Stone, Berlin, Germany, 30 September–4 October 1996.
21. Baronio, G.; Binda, L.; Saisi, A. Mechanical and physical behaviour of lime mortars after the cgaracterisation of historic mortar. In Proceedings of the International RILEM Workshop on Historic Mortars: Characteristics and Tests, Cachan, France, 31 August–2 September 2000.
22. Moropoulou, A.; Bakolas, A.; Moundoulas, P.; Aggelakopoulou, E.; Anagnostopoulou, S. Strength development and lime reaction in mortars for repairing historic masonries. *Cem. Concr. Compos.* **2005**, *27*, 289–294. [CrossRef]

23. Lanas, J.; Sirera, R.; Alvarez, J. Compositional changes in lime-based mortars exposed to different environments. *Thermochim. Acta* **2005**, *429*, 219–226. [CrossRef]
24. Válek, J.; Matas, T. Experimental Study of Hot Mixed Mortars in Comparison with Lime Putty and Hydrate Mortars. In Proceedings of the 2nd Historic Mortars Conference HMC2010 and RILEM 203-RHM Final Workshop, Prague, Czech Republic, 22–24 September 2010.
25. Faria, P.; Martins, A. Influence of Air Lime type and Curing Conditions on Lime and Lime-Metakaolin Mortars. In *Case Studies of Building Pathology in Cultural Heritage*; Springer: Berlin/Heidelberg, Germany, 2013.
26. Papayianni, I.; Pachta, V. Experimental study on the performance of lime-based grouts used in consolidating historic masonries. *Mater. Struct.* **2014**, *48*, 2111–2121. [CrossRef]
27. Vejmelková, E.; Keppert, M.; Kersner, Z.; Rovnanikova, P.; Černý, R. Mechanical, fracture-mechanical, hydric, thermal, and durability properties of lime–metakaolin plasters for renovation of historical buildings. *Constr. Build. Mater.* **2012**, *31*, 22–28. [CrossRef]
28. Duran, A.; Navarro-Blasco, I.; Fernández, J.; Alvarez, J. Long-term mechanical resistance and durability of air lime mortars. *Constr. Build. Mater.* **2014**, *58*, 147–158. [CrossRef]
29. Cizer, O.; van Balen, K.; van Gemert, D.; Elsen, J. Blended lime-cement mortars for conservation purposes: Microstructure and strenght development. In *Structural Analysis of Historic Construction: Preserving Safety and Significance*; Taylor & Francis Group: London, UK, 2008; pp. 965–972.
30. Arizzi, A.; Cultrone, G. Negative Effects of the use of White Portlan Cement as additive to aerial lime mortars set at atmospheric conditions: A chemical, mineralogical and physical-mechanical investigation. In *Brick and Mortar Research*; Manuel Rivera, S., Pena Diaz, A., Eds.; Nove Science Publisher, Inc.: New York, NY, USA, 2012; pp. 231–243.
31. Gulbe, L.; Vitina, I.; Setina, J. The Influence of Cement on Properties of Lime Mortars. *Procedia Eng.* **2017**, *172*, 325–332. [CrossRef]
32. Arandigoyen, M.; Alvarez, J. Pore structure and mechanical properties of cement–lime mortars. *Cem. Concr. Res.* **2007**, *37*, 767–775. [CrossRef]
33. Walker, P.; Lawrence, R. The impact of the water/lime ratio on the structural characteristics of air lime mortars. In *Structural Analysis of Historic Construction: Preserving Safety and Significance*; Fodde, D., Ed.; Taylor & Francis Group: London, UK, 2008; pp. 885–889.
34. Duggan, A.R.; Goggins, J.; Clifford, E.; McCabe, B.A. The Use of Carbonation Depth Techniques on Stabilized Peat. *Geotech. Test. J.* **2017**, *40*, 20160223. [CrossRef]
35. Pozo-Antonio, J. Evolution of mechanical properties and drying shrinkage in lime-based and lime cement-based mortars with pure limestone aggregate. *Constr. Build. Mater.* **2015**, *77*, 472–478. [CrossRef]
36. Ramesh, M.; Azenha, M.; Lourenço, P.B. Mechanical properties of lime–cement masonry mortars in their early ages. *Mater. Struct.* **2019**, *52*, 13. [CrossRef]

Article

Effects of Curing Conditions and Supplementary Cementitious Materials on Autogenous Self-Healing of Early Age Cracks in Cement Mortar

Mian Luo [1,2,3,*], Kang Jing [1,2], Jingquan Bai [1,2], Ziqi Ding [1,2], Dingyi Yang [1,2], Haoliang Huang [4,*] and Yongfan Gong [1,2]

1 College of Civil Science and Engineering, Yangzhou University, Yangzhou 225127, China; www.2910252115@outlook.com (K.J.); baijingquan2021@163.com (J.B.); dzq960825@163.com (Z.D.); ydy1991@163.com (D.Y.); yfgong@yzu.edu.cn (Y.G.)
2 Research Institute of Green Building Materials, Yangzhou University, Yangzhou 225127, China
3 State Key Laboratory of Green Building Materials, Beijing 100024, China
4 School of Materials Science and Engineering, South China University of Technology, Guangzhou 510641, China
* Correspondence: luomian@yzu.edu.cn (M.L.); huanghaoliang@scut.edu.cn (H.H.)

Citation: Luo, M.; Jing, K.; Bai, J.; Ding, Z.; Yang, D.; Huang, H.; Gong, Y. Effects of Curing Conditions and Supplementary Cementitious Materials on Autogenous Self-Healing of Early Age Cracks in Cement Mortar. *Crystals* **2021**, *11*, 752. https://doi.org/10.3390/cryst11070752

Academic Editors: Cesare Signorini, Antonella Sola, Sumit Chakraborty and Valentina Volpini

Received: 9 June 2021
Accepted: 24 June 2021
Published: 27 June 2021

Abstract: The autogenous healing potential of cement-based materials is affected by multiple factors, such as mix composition, crack width, pre-cracking age and external environmental conditions. In this study, the effects of curing conditions and supplementary cementitious materials (SCMs) on autogenous self-healing of early age cracks in cement mortar were investigated. Three curing conditions, i.e., standard curing, wet–dry cycles and incubated in water, and two SCMs, i.e., fly ash (FA) and blast furnace slag (BFS) with various contents (cement replacement ratio at 0%, 20%, and 40%) were examined. A single early age crack (pre-cracking age of 3 days) with a width of 200~300 μm was generated in cylindrical mortar specimens. Autogenous crack self-healing efficiency of mortar specimens was evaluated by performing a visual observation and a water permeability test. Moreover, microstructure analysis (XRD, SEM and TG/DTG) was utilized to characterize the healing products. The results indicated that the presence of water was essential for the autogenous self-healing of early age cracks in cement mortar. The efficiency of self-healing cracks was highest in specimens incubated in water. However, no significant self-healing occurred in specimens exposed to standard curing. For wet–dry cycles, a longer healing time was needed to obtain good self-healing compared to samples incubated in water. SCMs type and content significantly affected the autogenous self-healing ability of early age cracks. The self-healing efficiency of early age cracks decreased with increases in FA and BFS content. BFS mortars exhibited greater recovery in relation to water penetration resistance compared to the reference and FA mortars. Almost the same regain of water tightness and a lower crack-healing ratio after healing of 28 days in FA mortars were observed compared to the reference. The major healing product in the surface cracks of specimens with and without SCMs was micron-sized calcite crystals with a typical rhombohedral morphology.

Keywords: crack; autogenous self-healing; cement mortar; curing conditions; supplementary cementitious materials (SCMs)

1. Introduction

The presence of cracks is inevitable in concrete structures and may reduce concrete's durability due to the easy ingress of aggressive substances [1–4]. Fortunately, an autogenous crack-healing phenomenon has been found under the right conditions in concrete, which is mainly due to the ongoing hydration of unhydrated cement particles and the formation of calcium carbonate precipitation [5–7]. However, the autogenous healing capacity of concrete itself is limited in most instances [8,9]. Therefore, in recent years, enhancing the self-healing ability of concrete has received increasing attention.

In reported studies, various methods have been proposed to promote the self-healing efficiency of concrete [10]. These methods can be divided broadly into two categories [9]. The first method, defined as autonomic healing, promotes self-healing efficiency by adding additional self-healing agents into a concrete matrix, such as microcapsules containing adhesives [11], bacteria [12–15] and active mineral additives [16,17]. The second method, defined as autogenic or autogenous healing, relies on the self-healing ability of a concrete matrix itself by optimally designing the mix composition [18], creating suitable environmental conditions [19] and using fibers to constrain the width of potential cracks [20,21]. Despite lower healing efficiency compared to autonomic healing, autogenous healing exhibits advantages in terms of cost and compatibility, as no other components that usually are not found in concrete are introduced. Therefore, it is important to develop cement-based materials with robust autogenous healing.

Fly ash (FA) and blast furnace slag (BFS) are widely utilized as supplementary cementitious materials (SCMs) in modern concrete [22,23]. The partial replacement of cement with SCMs can save costs, decrease CO_2 emissions and improve the performance of concrete. Recently, several studies [18,24–30] have reported that incorporating SCMs could improve the autogenous crack healing of concrete. Van Tittelboom et al. [18] concluded that cement replacement by blast furnace slag or fly ash improved autogenous healing by enhancing further hydration, and blast furnace slag showed a better effect than fly ash. Sahmaran et al. [24] also found that the type of supplementary cementitious materials greatly affected the self-healing capability of cementitious composites. Huang et al. [25] reported that slag cement paste exhibited higher autogenous healing potential compared to Portland cement paste. Qiu et al. [26] investigated the effect of slag content (0%, 30% and 60% cement replacement) on the autogenous healing behavior of engineered cementitious composite (ECC) at a pre-cracking age of 40 days. The results indicated that replacing cement with blast furnace slag at 30% exhibited a better crack width reduction compared to the 0% and 60% replacement. Şahmaran et al. [28] and Termkhajornkit et al. [29] pointed out that the self-healing ability of cement-based materials increased with fly ash content.

It is notable that, in the above reported studies, most of specimens used to evaluate autogenous healing potential were pre-cracked at an age of 28 days or later. Considering that cement-based materials easily crack at an early age, more information is needed regarding the autogenous self-healing of early age cracks in cement-based materials containing supplementary cementitious materials. Therefore, in this study, so as to better understand the self-healing behavior of cement-based materials containing supplementary cementitious materials, the effects of various supplementary cementitious materials (FA and BFS) on the autogenous self-healing of early age cracks (pre-cracking age of 3 days) in cement mortar were investigated. Moreover, autogenous healing is also greatly influenced by exposure to environmental conditions [31–33]. The influence of curing conditions on autogenous self-healing was also investigated. Three curing conditions (standard curing, wet–dry cycles and incubated in water) that are common exposure environments in the field of engineering were considered. The best one among the three curing conditions for autogenous healing was used for subsequent research in cement mortar containing supplementary cementitious material. Autogenous crack self-healing efficiency was evaluated by performing a visual observation and a water permeability test. Moreover, X-ray diffraction (XRD), scanning electron microscopy (SEM) and thermogravimetry (TG/DTG) were utilized to characterize the healing products. The effect mechanisms of curing conditions, as well as the influence of supplementary cementitious material, on autogenous self-healing were further addressed.

2. Materials and Methods

2.1. Materials and Specimen Preparation

Portland cement, river sand (fineness modulus of 2.1) and tap water were mixed to prepare mortar mixtures. The water-to-cement ratio (W/C) was 0.5 and the sand-to-cement mass ratio was 3. Two cement types (P·O 42.5 and P·I 42.5) were used. For mortar specimens used to investigate the effect of curing conditions on autogenous crack self-healing, P·O 42.5 cement was used. For mortar specimens used to investigate the effect of supplementary cementitious material on autogenous crack self-healing, P·I 42.5 cement, which only contains Portland clinker, was used to avoid the influence of supplementary cementitious materials in the cement itself. Although some research results [18] show that the types of cement may affect the degree of hydration and self-healing ability, they do not affect the discussion of test results in the paper because curing conditions and supplementary cementitious materials were the main factors considered in this study. Portland cement was partially replaced by fly ash (FA) or blast furnace slag (BFS), and two SCMs (FA and BFS) with various contents (cement replacement ratio at 0%, 20% and 40%) were considered. The mixing proportions of the specimens are shown in Table 1. Chemical compositions of Portland cement, fly ash (FA) and blast furnace slag (BFS) are shown in Table 2. The polypropylene fibers with a diameter of 31 μm and a length of 12 mm were utilized to keep the specimens intact after being pre-cracked. The mixtures were poured into cylindrical molds with a diameter of 10cm and a height of 2.5 cm. The specimens were unmolded after 24 h and subsequently stored in a standard curing room (20 ± 2 °C and >95% RH) for further curing.

Table 1. Mixing proportions of cement mortars (FA and BFS are short for fly ash and blast furnace slag, respectively).

Mix.	Cement (g)	FA (g)	BFS (g)	Sand (g)	Water (g)
Reference	450	/	/	1350	225
20FA	360	90	/	1350	225
40FA	270	180	/	1350	225
20BFS	360	/	90	1350	225
40BFS	270	/	180	1350	225

Table 2. Chemical compositions of Portland cement, fly ash (FA) and blast furnace slag (BFS).

Components (wt.%)	P·O 42.5	P·I 42.5	FA	BFS
SiO_2	22.94	17.56	55.81	29.51
CaO	59.77	68.80	4.48	42.55
Al_2O_3	6.95	3.41	27.31	12.21
SO_3	3.24	2.69	1.07	2.24
Fe_2O_3	3.98	3.64	5.81	0.30
MgO	1.89	2.46	0.91	11.23
Others	1.23	1.44	3.93	1.91

2.2. Creation of Crack

Before evaluating the autogenous crack self-healing efficiency of the mortar specimen, it was necessary to make cracks that met the test requirements. In this study, a new crack production method was developed; with this method, it is easy to control crack width and shape (Figure 1). At 3 days of age, the cylindrical mortar specimen was taken out from the curing room and tightly wrapped along the side of the specimen by the steel hose clamp. Then, the splitting tensile strength test (Figure 1) was used to generated cracks in the specimens. Crack width was controlled at 200~300 μm by adjusting the tightness of the hose clamp.

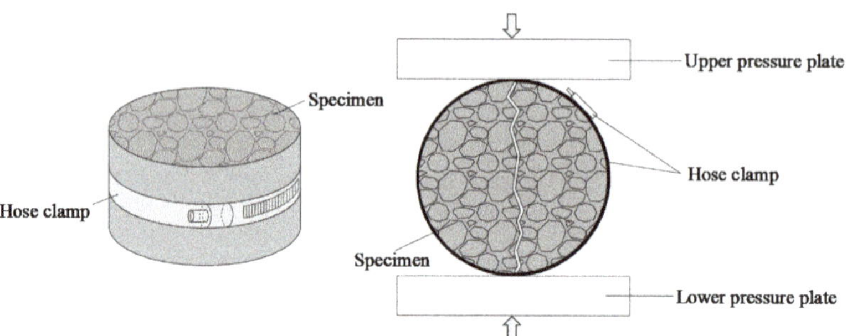

Figure 1. Creation of crack in mortar specimen.

2.3. *Self-Healing Curing Conditions*

To investigate the effect of curing conditions on autogenous crack self-healing in the cement mortar specimens, pre-cracked specimens were placed in three curing conditions: (a) standard curing (>95% RH); (b) wet–dry cycles (for each cycle, the cracked specimen was incubated in water for 12 h and then taken out to air dry for 12 h); (c) submerged in water. After healing for 0, 3, 14, 28 and 56 days, crack-healing quantification of specimens was conducted by a stereo optical microscope and a water permeability test. To investigate the effect of supplementary cementitious material on autogenous crack self-healing, all pre-cracked specimens were submerged in water for crack-healing quantification.

2.4. *Crack-Healing Quantification*

2.4.1. The Crack-Healing Ratio

The closure of surface cracks on the specimen during the self-healing process was observed by a stereo optical microscope (LIOO, Attendorn, Germany). Crack self-healing efficiency was evaluated based on the change in crack area before and after healing. Crack area was measured based on the number of pixels in the recorded images through Image-J software. The crack-healing ratio is defined as the Equation (1), where A_0 and A_t represent the crack area before and after healing for time t, respectively. For more details, see our earlier papers [14,34].

$$\text{The crack healing ratio} = \frac{A_0 - A_t}{A_0} \times 100\% \tag{1}$$

2.4.2. Water Permeability Test

A constant water head permeability test (Figure 2) was carried out to obtain the initial water permeability coefficient of the specimens k_0 (cm/s), as well as the water permeability coefficient of the specimens after healing for a certain time k_t (cm/s). For the detailed calculation of the water permeability coefficient, please refer to our previous research [34]. In addition, the relative permeability coefficient was calculated to evaluate the self-healing efficiency of mortar specimens according to Equation (2).

$$\text{The relative permeability coefficient} = \frac{k_t}{k_0} \tag{2}$$

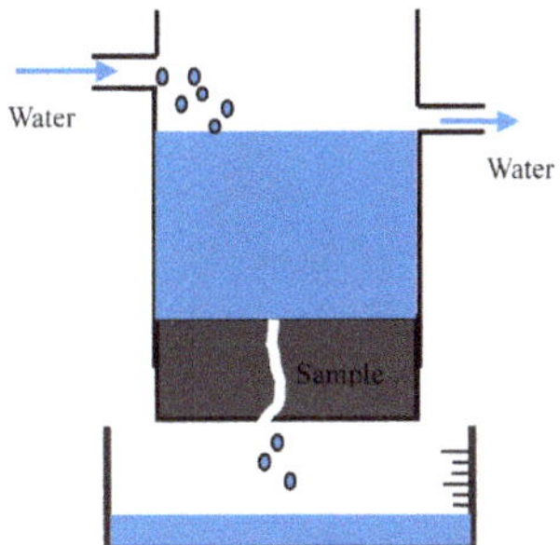

Figure 2. Setup of water permeability test.

2.5. Characterization of Reaction Products of Self-Healing Formed in Early Age Cracks

After crack-healing quantification, the healing products that formed at the surface cracks of the mortar specimens were collected and analyzed with XRD, SEM and TG/DTG, respectively. XRD analyses adopted the D8-Advance X-ray diffractometer produced by the Bruker-AXS company (Karlsruhe, Germany). Powder samples were scanned at diffraction angle 2θ from 5° to 90°. For TG/DTG analysis, powder samples were heated from room temperature to 800 °C at a rate of 10 °C/min in an N_2 atmosphere.

3. Results and Discussion

3.1. Crack-Healing Quantification under Different Curing Conditions

3.1.1. The Crack-Healing Ratio

Figure 3 shows that the closure of surface cracks in mortar specimens under different curing conditions and after different healing times. It can be seen that no visual healing was observed in specimens exposed to standard curing, even after a healing period of 56 days. However, complete crack sealing was found in specimens incubated in water for 14 days. For specimens under wet–dry cycles, only partial crack filling was showed.

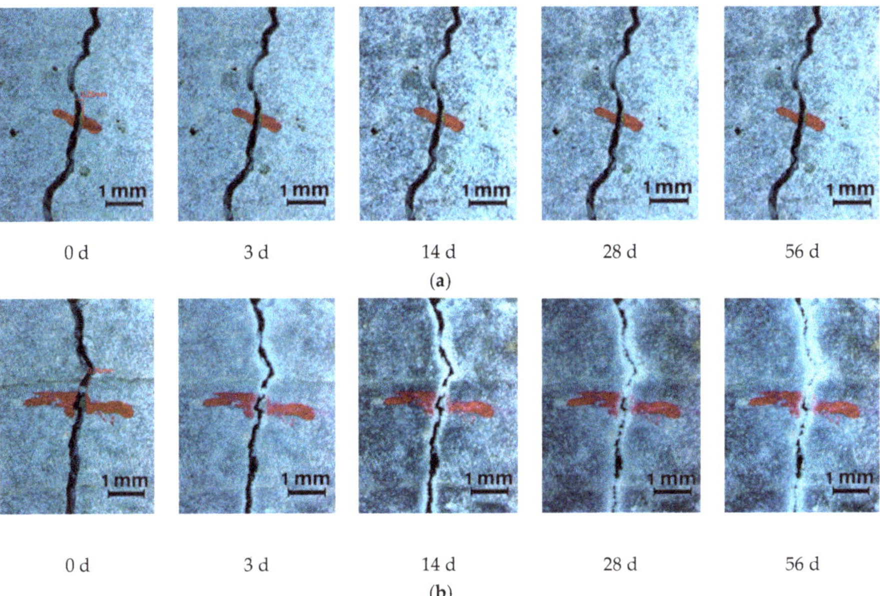

Figure 3. *Cont.*

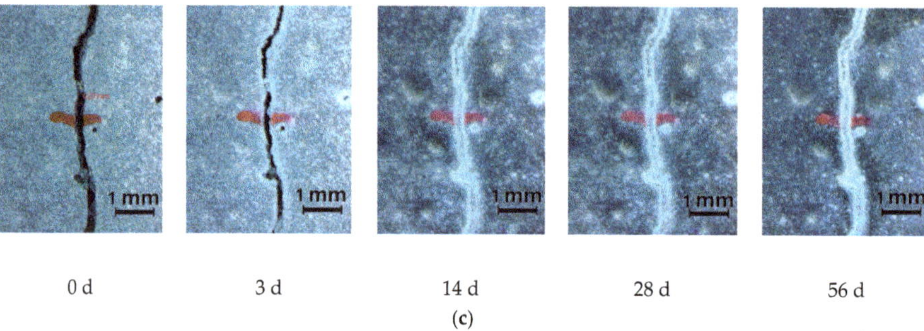

Figure 3. The closure of surface cracks in specimens under different curing conditions after different healing times: (**a**) standard curing (>95% RH); (**b**) wet–dry cycles; (**c**) submerged in water (*d* denotes days).

To quantitatively assess self-healing efficiency, the images of surface cracks in mortar specimens monitored by a stereo optical microscope were processed using Image-J software (National Institutes of Health, Bethesda, MD, America) (Figure 4); the crack-healing ratio was obtained according to Equation (1), as shown in Figure 5. The crack-healing ratio was highest in specimens incubated in water, followed by wet–dry cycles; standard curing was the lowest ratio. It was noted that the crack-healing ratio of specimens under wet–dry cycles gradually increased with the extension of healing time, although these exhibited a lower crack-healing ratio than those specimens incubated in water. The autogenous self-healing of cracks is mainly attributable to the hydration of unhydrated cement particles and the precipitation of calcium carbonate. In the opening of the crack, the main mechanism of self-healing may be attributable to the precipitation of calcium carbonate. Inside the crack, the ongoing hydration of unreacted cement may be dominant because the carbonization process is limited; this is because carbon dioxide struggles to enter deep into the crack, especially in the specimens incubated in water. For specimens incubated in water, adequate water supply is beneficial to the further hydration of unhydrated cement particles and the precipitation of calcium carbonate in the surface crack due to more Ca^{2+} dissolution emanating from said crack. Carbon dioxide from the air can dissolve in water to form carbonates and produce calcium carbonate precipitations when in contact with calcium ions in the mouth of the crack. For specimens under wet–dry cycles, in each cycle, the cracked specimen was incubated in water for 12 h and then taken out to air dry for 12 h. More carbonate ions can be formed because carbon dioxide is more likely to reach the crack opening while being air dried. However, less hydration of cement particles occurred and there was less Ca^{2+} supply at the same healing time. The crack-healing ratio for wet–dry cycle specimens depends on the combined result of the above two aspects. Therefore, although curing in water may inhibit carbon dioxide supply, specimens submerged in water exhibited a higher crack-healing ratio than that of specimens under wet–dry cycles (Figure 5). No significant self-healing (less than 10% of crack-healing ratio) occurred in specimens exposed to standard curing due to lack of water. Calcium carbonate precipitation struggles to form, and the healing of the cracks can only be attributed to the ongoing hydration of partial unreacted cement particles in high humidity conditions.

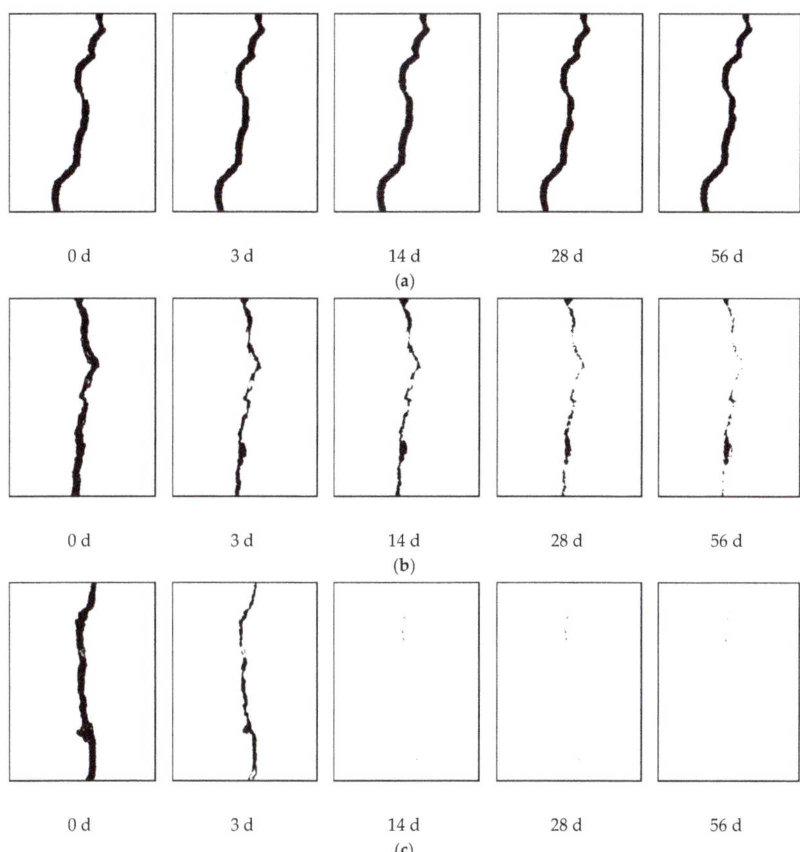

Figure 4. Binarization images of surface cracks in specimens under different curing conditions after different healing times: (**a**) standard curing (>95% RH); (**b**) wet–dry cycles; (**c**) submerged in water (*d* denotes days).

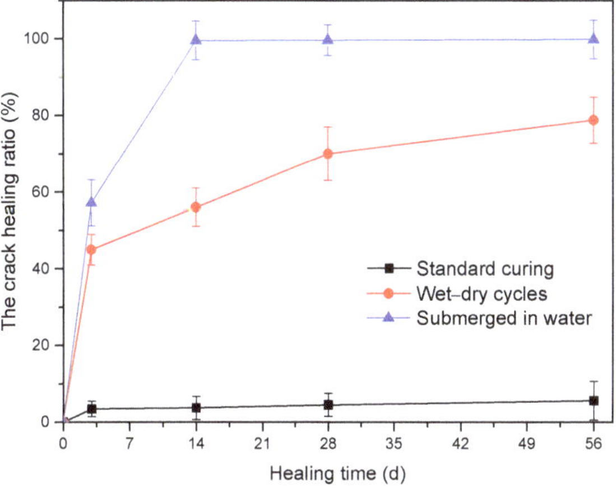

Figure 5. Crack-healing ratio of specimens under different curing conditions after different healing times (*d* denotes days).

3.1.2. Water Permeability Test

The measurements of crack width, crack area and water permeability were widely used to characterize the healing effect [9,14,31,32]. In Section 3.1.1, the crack-healing ratio based on a reduction in crack area was used to visualize the self-healing of surface cracks, but it could not reflect the internal healing effect. Therefore, the water permeability test, which is not only related to the sealing of surface cracks but also to the healing of internal cracks, was conducted to further evaluate the self-healing efficiency of the cracks. Figure 6 shows the change in relative permeability of specimens under different curing conditions and healing times. The result is consistent with that of the crack-healing ratio. The specimens incubated in water and under wet–dry cycles exhibited a better recovery in water penetration resistance compared to specimens exposed to standard curing. An obvious decline in relative permeability was observed in specimens incubated in water and under wet–dry cycles, which indicated that the presence of water is essential for the autogenous self-healing of early age cracks in cement mortar. In addition, it is worth noting that, although the crack-healing ratio for wet–dry cycles specimens was obviously inferior to that of specimens incubated in water, the change in relative permeability coefficient for wet–dry cycle specimens was similar to that of specimens incubated in water. This may be attributed to more healing products being formed in internal cracks for specimens under wet–dry cycles.

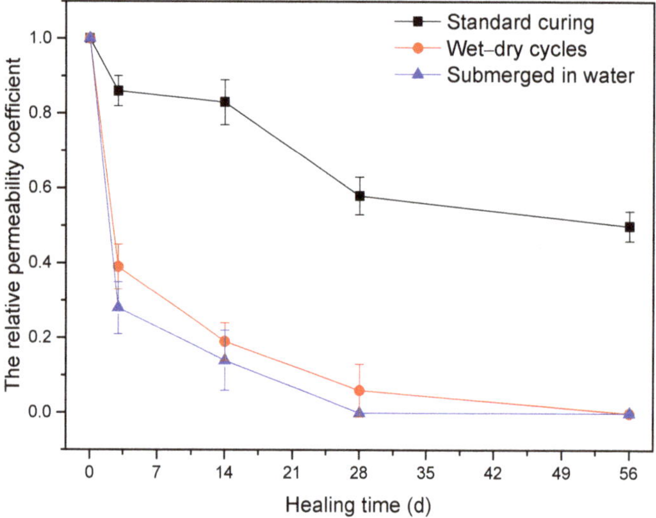

Figure 6. Relative permeability coefficient changes with healing time under different curing conditions (*d* denotes days).

3.2. *Crack-Healing Quantification under Different Supplementary Cementitious Materials*

3.2.1. The Crack-Healing Ratio

The crack-healing ratios of mortar specimens containing different SCMs with various content are shown in Figure 7. As can be seen, the addition of SCMs changed the autogenous self-healing behavior of surface cracks in mortar specimens. The influence of SCMs on the crack-healing ratio is related to their type and content. For BFS specimens, the crack-healing ratio decreased with the increase in BFS content. Moreover, 20BFS exhibited a higher crack-healing ratio compared to the reference; however, 40BFS showed a lower crack-healing ratio compared to the reference. For FA specimens, the crack-healing ratio also decreased with the increase in FA content. However, both 20FA and 40FA showed a lower crack-healing ratio compared to the reference after a healing period of 28 days. In general, it is not beneficial to replace cement with FA or BFS for the self-healing of early age surface cracks (pre-cracking age of 3 days) in mortar specimens (except 20BFS specimen). Comparing FA with BFS, at a given content of 20%, BFS specimens exhibited a higher crack-healing ratio compared to FA specimens. However, FA specimens showed a higher crack-healing ratio compared to BFS specimens when the content was increased to 40%.

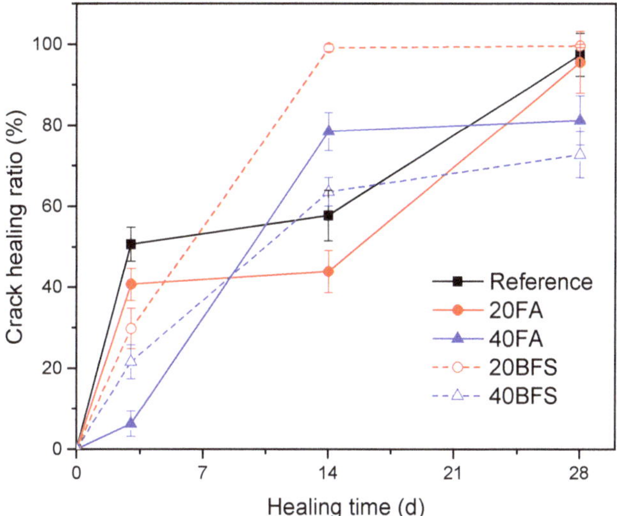

Figure 7. Crack-healing ratio of specimens containing different SCMs with various content (*d* denotes days).

The autogenous self-healing of surface cracks is mainly attributable to two mechanisms: (1) ongoing hydration of unreacted cement or SCMs particles; (2) formation of calcium carbonate precipitation. The crack-healing ratio of specimens with SCMs depends on the coupled effect of the abovementioned two aspects. In this study, early age cracks (pre-cracking age of 3 days) were generated. After cracking, specimens with SCMs had more unreacted binder particles due to slower hydration of FA or BFS. The ongoing hydration of unreacted binder particles is beneficial to promote self-healing of the crack. However, the ongoing hydration of FA or BFS needs to be activated by $Ca(OH)_2$, which is mainly attributed to cement hydration. In specimens with SCMs, the portlandite content is lower than that of the referenced specimens due to less cement composition and the partial consumption of portlandite by the hydration reaction of FA or BFS [18,27]. The reduction in $Ca(OH)_2$ greatly influenced the formation of calcium carbonate precipitations and ongoing hydration of FA or BFS at the surface cracks. When the formation of calcium carbonate precipitation is the main mechanism, FA or BFS may not be good for the autogenous self-healing of surface cracks. It was noted that 20BFS exhibited a higher crack-healing ratio compared to the reference. This is because that there are still large amounts of unhydrated cement particles in early cracking specimens when the BFS content is low; further hydration of unreacted cement particles provides additional $Ca(OH)_2$ to promote the hydration of BFS and the formation of calcium carbonate precipitation at the crack's surface. The 40FA specimen exhibited a higher crack-healing ratio compared to the 40BFS specimens, which may be attributed to the lower hydration activity of FA, resulting in more $Ca(OH)_2$ for calcium carbonate formation.

3.2.2. Water Permeability Test

Figure 8 shows the change in relative permeability of specimens containing different SCMs with various content over different healing times. The changing trend of regaining water tightness is different from that of the crack-healing ratio. This is because regaining the water tightness of the pre-cracked mortar specimens is not only related to the sealing of surface crack but also to the healing of internal cracks. For BFS mortars, both 20BFS and 40BFS showed better regaining abilities of water tightness compared to the reference specimens. Although a lower crack-healing ratio after healing for 28 days was observed for 40BFS mortar compared to the reference, the higher regaining capacity for water tightness may be attributed to more healing products being formed in internal cracks due to the ongoing hydration of more unreacted BFS particles. As for FA mortars, the regaining of water tightness declined with the increasing FA content. This is because that the hydration activity of FA is low and not enough calcium hydroxide can be utilized to activate the hydration of unreacted FA particles in internal cracks for mortars with a high content of FA. Moreover, it was noted that 20FA and 40FA exhibited a lower decline in the relative permeability coefficient after healing for 3 days compared to the reference, but almost the same decline in the relative permeability coefficient after healing for 28 days was observed compared to the reference. This is because unreacted FA particles in the crack cannot be activated in the early healing process but can be activated as the healing time increases. It also can be found that BFS mortars exhibited a better regaining of water tightness compared to FA mortars; this is due to their higher hydration activity and CaO content for slag.

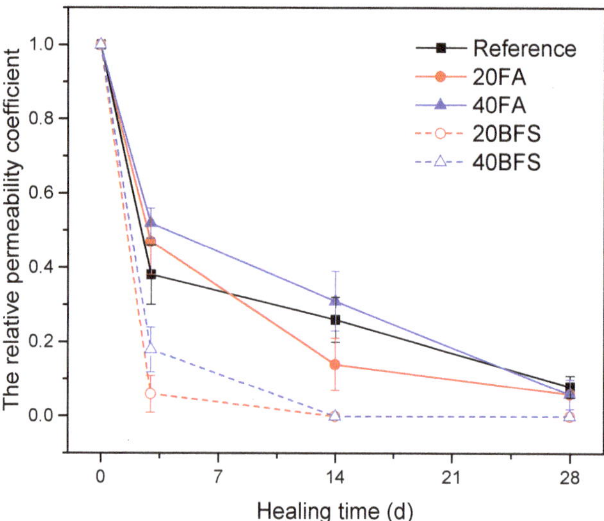

Figure 8. Relative permeability coefficient of specimens containing different SCMs with various content (*d* denotes days).

3.3. Mineralogy of Reaction Products of Self-Healing in Early Age Cracks

After crack-healing quantification, XRD, SEM and TG/DTG were conducted on the healing products scraped from the healed surface cracks of the referenced specimens, as well as the specimens that contained different supplementary cementitious materials (SCMs). Figure 9 shows the XRD patterns of the healing products formed in mortar specimens containing different supplementary cementitious materials (SCMs) compared to the reference. No significant differences were found for the diffraction peaks of all samples. Calcite was detected as the major crystal healing product for all samples. SEM observations indicated that micron-sized calcite crystals with a typical rhombohedral morphology were closely packed together, as shown in Figure 10. In addition, the TG/DTG curves of the healing products in the crack mouth are shown in Figure 11. An obvious weight loss in the range of 600–800 °C was found, and the corresponding peaks related to the decomposition of calcium carbonate at about 750 °C were observed in all samples. The weight losses between 600 and 800 °C of reference, 20FA and 20BFS were 41.12%, 40.93% and 40.45%, respectively. According to the thermal decomposition equation of calcium carbonate, the percentages of calcium carbonate in the healing products from reference, 20FA and 20BFS were calculated as 93.46%, 93.03%, and 91.94%. The results of TG analysis were consistent with XRD and SEM, indicating that the major healing product from the healed surface cracks of the referenced specimens and the specimens containing different supplementary cementitious materials (SCMs) was calcium carbonate.

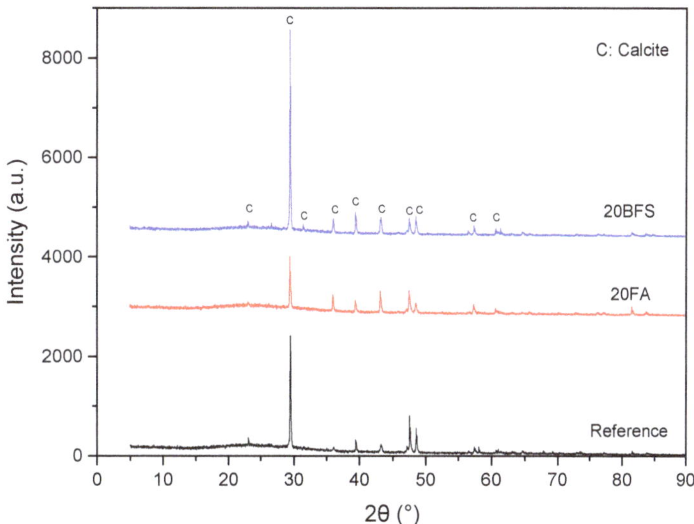

Figure 9. X-ray diffraction (XRD) patterns of the healing products in the mouths of the cracks.

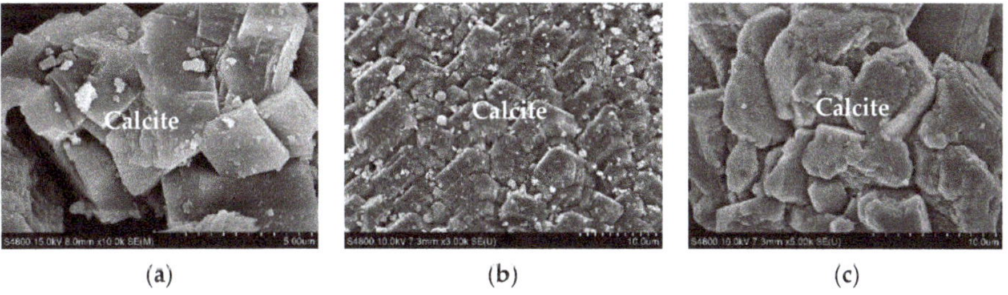

Figure 10. Scanning electron microscopic (SEM) observations of the healing products in the mouths of the cracks: (**a**) Reference; (**b**) 20FA; (**c**) 20BFS.

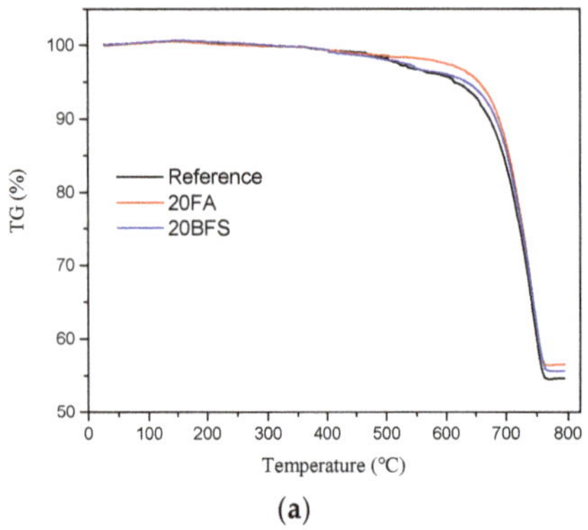

(a)

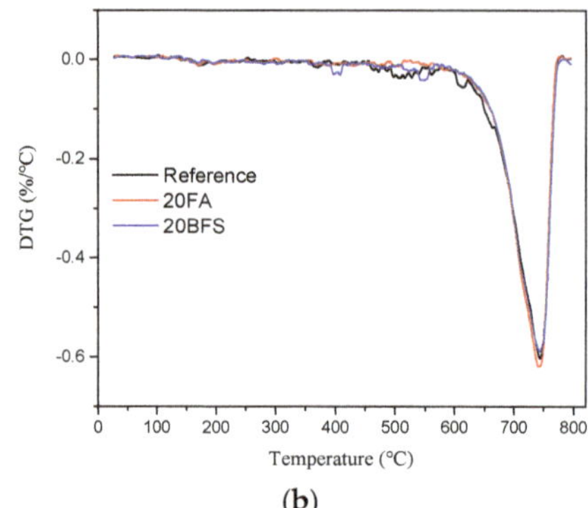

(b)

Figure 11. Thermogravimetric (TG/DTG) curves of the healing products in the mouths of the cracks: (**a**) TG curves; (**b**) DTG curves.

4. Conclusions

The effects of curing conditions and supplementary cementitious materials on the autogenous self-healing of early age cracks in cement mortar were investigated. Three curing conditions (standard curing, wet–dry cycles and incubated in water) and two SCMs (FA and BFS) with various contents (cement replacement ratio at 0%, 20%, and 40%) were considered. Autogenous crack self-healing efficiency of mortar specimens was evaluated by performing a visual observation and a water permeability test. Moreover, XRD, SEM and TG/DTG were conducted to characterize the healing products. Based on the experimental results, the following conclusions can be drawn:

(1) The presence of water is essential for autogenous self-healing of early age cracks in cement mortar. Crack self-healing efficiency was highest in specimens incubated in water. However, no significant self-healing occurred in specimens exposed to standard curing. For wet–dry cycles, a longer healing time was needed to obtain good self-healing compared to those samples incubated in water.

(2) SCMs type and content significantly affected the autogenous self-healing ability of early age cracks. Crack self-healing efficiency for early age cracks decreased with the increase in FA and BFS content. BFS mortars exhibited better recovery of water penetration resistance compared to the reference and FA mortars. Almost the same regaining of water tightness and a lower crack-healing ratio after healing for 28 days in FA mortars were observed in comparison to the reference.

(3) The major healing product in surface cracks of specimens with and without SCMs was micron-sized calcite crystals with a typical rhombohedral morphology.

Author Contributions: Conceptualization, M.L. and H.H.; methodology, M.L., K.J. and J.B.; investigation, M.L., K.J. and J.B.; data curation, K.J. and Z.D.; writing—original draft preparation, M.L.; writing—review and editing, M.L. and H.H.; visualization, J.B. and Z.D.; supervision, M.L., D.Y., Y.G. and H.H. All authors have read and agreed to the published version of the manuscript.

Funding: This research was funded by the National Natural Science Foundation of China (No. 51808483), the Natural Science Foundation of Jiangsu Province (No. BK20180930), the Opening Project of State Key Laboratory of Green Building Materials, and the Opening Fund of Jiangsu Key Laboratory of Construction Materials (No. CM2018-10).

Institutional Review Board Statement: Not applicable.

Informed Consent Statement: Not applicable.

Data Availability Statement: The data presented in this study are available on request from the corresponding authors.

Conflicts of Interest: The authors declare no conflict of interest.

References

1. Wang, K.; Jansen, D.C.; Shah, S.P. Permeability study of cracked concrete. *Cem. Concr. Res.* **1997**, *27*, 381–393. [CrossRef]
2. Zhang, P.; Wittmann, F.H.; Vogel, M.; Müller, H.S.; Zhao, T. Influence of freeze-thaw cycles on capillary absorption and chloride penetration into concrete. *Cem. Concr. Res.* **2017**, *100*, 60–67. [CrossRef]
3. Huseien, G.F.; Shah, K.W.; Sam, A.R.M. Sustainability of nanomaterials based self-healing concrete: An all-inclusive insight. *J. Build. Eng.* **2019**, *23*, 155–171. [CrossRef]
4. Wang, Y.R.; Cao, Y.; Zhang, P.; Ma, Y.; Zhao, T.; Wang, H.; Zhang, Z. Water absorption and chloride diffusivity of concrete under the coupling effect of uniaxial compressive load and freeze-thaw cycles. *Constr. Build. Mater.* **2019**, *209*, 566–576. [CrossRef]
5. Lauer, K.R.; Slate, F.O. Autogenous healing of cement paste. *ACI Mater. J.* **1956**, *52*, 1083–1098. [CrossRef]
6. Ter Heide, N. Crack Healing in Hydrating Concrete. Master's Thesis, Delft University of Technology, Delft, The Netherlands, 2005.
7. Huang, H.; Ye, G.; Damidot, D. Characterization and quantification of self-healing behaviors of microcracks due to further hydration in cement paste. *Cem. Concr. Res.* **2013**, *52*, 71–81. [CrossRef]
8. Reinhardt, H.W.; Jooss, M. Permeability and self-healing of cracked concrete as a function of temperature and crack width. *Cem. Concr. Res.* **2003**, *33*, 981–985. [CrossRef]
9. Rooij, M.; van Tittelboom, K.; Belie, N.; Schlangen, E. (Eds.) *Self-Healing Phenomena in Cement-Based Materials: State-of-the-Art Report of RILEM Technical Committee 221-SHC. Self-Healing Phenomena in Cement-Based Materials*; Springer: Berlin/Heidelberg, Germany, 2013.
10. Li, W.; Dong, B.; Yang, Z.; Xu, J.; Chen, Q.; Li, H.; Xing, F.; Jiang, Z. Recent Advances in Intrinsic Self-Healing Cementitious Materials. *Adv. Mater.* **2018**, *30*, 1705679. [CrossRef]
11. Van Tittelboom, K.; De Belie, N.; Van Loo, D.; Jacobs, P. Self-healing efficiency of cementitious materials containing tubular capsules filled with healing agent. *Cem. Concr. Compos.* **2011**, *33*, 497–505. [CrossRef]
12. Wiktor, V.; Jonkers, H.M. Quantification of crack-healing in novel bacteria-based self-healing concrete. *Cem. Concr. Compos.* **2011**, *33*, 763–770. [CrossRef]
13. Wang, J.Y.; Soens, H.; Verstraete, W.; De Belie, N. Self-healing concrete by use of microencapsulated bacterial spores. *Cem. Concr. Res.* **2014**, *56*, 139–152. [CrossRef]
14. Luo, M.; Qian, C.-X.; Li, R.-Y. Factors affecting crack repairing capacity of bacteria-based self-healing concrete. *Constr. Build. Mater.* **2015**, *87*, 1–7. [CrossRef]
15. Xu, J.; Wang, X.Z. Self-healing of concrete cracks by use of bacteria-containing low alkali cementitious material. *Constr. Build. Mater.* **2018**, *167*, 1–14. [CrossRef]
16. Sisomphon, K.; Copuroglu, O.; Koenders, E.A.B. Self-healing of surface cracks in mortars with expansive additive and crystalline additive. *Cem. Concr. Compos.* **2012**, *34*, 566–574. [CrossRef]
17. Alghamri, R.; Kanellopoulos, A.; Litina, C.; Al-Tabbaa, A. Preparation and polymeric encapsulation of powder mineral pellets for self-healing cement based materials. *Constr. Build. Mater.* **2018**, *186*, 247–262. [CrossRef]
18. Van Tittelboom, K.; Gruyaert, E.; Rahier, H.; De Belie, N. Influence of mix composition on the extent of autogenous crack healing by continued hydration or calcium carbonate formation. *Constr. Build. Mater.* **2012**, *37*, 349–359. [CrossRef]
19. Jiang, Z.; Li, W.; Yuan, Z. Influence of mineral additives and environmental conditions on the self-healing capabilities of cementitious materials. *Cem. Concr. Compos.* **2015**, *57*, 116–127. [CrossRef]
20. Yang, Y.G.; Yang, E.H.; Li, V.C. Autogenous healing of engineered cementitious composites at early age. *Cem. Concr. Res.* **2011**, *41*, 176–183. [CrossRef]
21. Zhang, Z.; Qian, S.; Ma, H. Investigating mechanical properties and self-healing behavior of micro-cracked ECC with different volume of fly ash. *Constr. Build. Mater.* **2014**, *52*, 17–23. [CrossRef]
22. Saillio, M.; Baroghel-Bouny, V.; Pradelle, S.; Bertin, M.; Vincent, J.; d'Espinose de Lacaillerie, J.B. Effect of supplementary cementitious materials on carbonation of cement pastes. *Cement Concr. Res.* **2021**, *142*, 106358. [CrossRef]
23. Haridharan, M.K.; Matheswaran, S.; Murali, G.; Abid, S.R.; Fediuk, R.; Mugahed Amran, Y.H.; Abdelgader, H.S. Impact response of two-layered grouted aggregate fibrous concrete composite under falling mass impact. *Constr. Build. Mater.* **2020**, *263*, 120628. [CrossRef]
24. Sahmaran, M.; Yildirim, G.; Erdem, T.K. Self-healing capability of cementitious composites incorporating different supplementary cementitious materials. *Cem. Concr. Compos.* **2012**, *35*, 89–101. [CrossRef]
25. Huang, H.; Ye, G.; Damidot, D. Effect of blast furnace slag on self-healing of microcracks in cementitious materials. *Cem. Concr. Res.* **2014**, *60*, 68–82. [CrossRef]
26. Qiu, J.; Tan, H.S.; Yang, E.-H. Coupled effects of crack width, slag content, and conditioning alkalinity on autogenous healing of engineered cementitious composites. *Cem. Concr. Compos.* **2016**, *73*, 203–212. [CrossRef]
27. Darquennes, A.; Olivier, K.; Benboudjema, F.; Gagné, R. Early-age self-healing of cementitious materials containing ground granulated blast-furnace slag under water curing. *J. Adv. Concr. Technol.* **2016**, *14*, 717–727.
28. Şahmaran, M.; Keskin, S.B.; Ozerkan, G.; Yaman, I.O. Self-healing of mechanically-loaded self consolidating concretes with high volumes of fly ash. *Cem. Concr. Compos.* **2008**, *30*, 872–879. [CrossRef]

29. Termkhajornkit, P.; Nawa, T.; Yamashiro, Y.; Saito, T. Self-healing ability of fly ash–cement systems. *Cem. Concr. Compos.* **2009**, *31*, 195–203. [CrossRef]
30. Parashar, A.; Bishnoi, S. A comparison of test methods to assess the strength potential of plain and blended supplementary cementitious materials. *Constr. Build. Mater.* **2020**, *256*, 119292. [CrossRef]
31. Liu, H.; Huang, H.; Wu, X.; Peng, H.; Li, Z.; He, J.; Yu, Q. Effects of external multi-ions and wet-dry cycles in a marine environment on autogenous self-healing of cracks in cement paste. *Cem. Concr. Res.* **2019**, *120*, 198–206. [CrossRef]
32. Suleiman, A.R.; Nehdi, M.L. Effect of environmental exposure on autogenous self-healing of cracked cement-based materials. *Cem. Concr. Res.* **2018**, *111*, 197–208. [CrossRef]
33. Zhang, W.; Zheng, Q.; Ashour, A.; Han, B. Self-healing cement concrete composites for resilient infrastructures: A review. *Compos. Part B Eng.* **2020**, *189*, 107892. [CrossRef]
34. Luo, M.; Bai, J.Q.; Jing, K.; Ding, Z.Q.; Yang, D.Y.; Qian, C.X. Self-healing of early-age cracks in cement mortars with artificial functional aggregates. *Constr. Build. Mater.* **2021**, *272*, 121846. [CrossRef]

MDPI

Article

Experimental Evaluation of Cement Mortars with End-of-Life Tyres Exposed to Different Surface Treatments

Eduardo García [1], Bárbara Villa [2], Mauricio Pradena [1,3,*], Bruno Urbano [2], Víctor H. Campos-Requena [2], Carlos Medina [4] and Paulo Flores [4]

1 Departamento de Ingeniería Civil, Facultad de Ingeniería Civil, Universidad de Concepción, Edmundo Larenas 219, Casilla 160-C Correo 3, Concepción 4030000, Chile; eduargarcia@udec.cl
2 Departamento de Polímeros, Facultad de Ciencias Químicas, Universidad de Concepción, Edmundo Larenas 129, Casilla 160-C Correo 3, Concepción 4030000, Chile; bvilla2016@udec.cl (B.V.); burbano@udec.cl (B.U.); vcamposr@udec.cl (V.H.C.-R.)
3 Unidad de Desarrollo Tecnológico, UDT, Universidad de Concepción, Casilla 4051, Concepción 4030000, Chile
4 Departamento de Ingeniería Mecánica, Facultad de Ingeniería, Universidad de Concepción, Edmundo Larenas 219, Casilla 160-C Correo 3, Concepción 4030000, Chile; cmedinam@udec.cl (C.M.); pfloresv@udec.cl (P.F.)
* Correspondence: mpradena@udec.cl

Abstract: An end-of-Life Tyre (ELT) is a type of waste that can generate negative social and environmental impacts due to its disposal. Considering that rubber can improve concrete properties and the massive use of concrete as construction material, the addition of ELT rubber in concrete mixes is attractive. However, concrete mechanical properties are negatively affected due to the rubber-cementitious matrix interaction. Although rubber treatments have been developed to minimise the negative effects, the geo-dependency of the mix makes necessary to find cost-effective and practical solutions that will allow a real use of the ELT waste. Therefore, the objective of the present study is to characterise the properties of cement mortars with the addition of ELT rubber under three surface treatments: hydration, oxidation-sulphonation, and hydrogen peroxide. The results show that hydration is the most favourable treatment from a technical, practical, and economical point of view. In fact, with this treatment, it is possible to add up to 5% ELT rubber, with respect to the aggregate weight, and still exceed the design strength without adding more cement or additives as other investigations. The use of Portland Pozzolana Cement, with local fly ash waste, contributes as well to the promissory results obtained.

Keywords: cement mortar; End-of-Life Tyre; waste; surface treatment; compressive strength; flexural strength; workability; fly ash

Citation: García, E.; Villa, B.; Pradena, M.; Urbano, B.; Campos-Requena, V.H.; Medina, C.; Flores, P. Experimental Evaluation of Cement Mortars with End-of-Life Tyres Exposed to Different Surface Treatments. *Crystals* **2021**, *11*, 552. https://doi.org/10.3390/cryst11050552

Academic Editors: Cesare Signorini, Antonella Sola, Sumit Chakraborty and Valentina Volpini

Received: 1 April 2021
Accepted: 3 May 2021
Published: 15 May 2021

1. Introduction

The end-of-life tyre (ELT) is a type of waste that has become an environmental and social problem. In effect, the accumulation of ELT produces concentration of rats, larvae, mice, insects, and it increases the risk of fires difficult to extinguish [1].

Moreover, globally, 1000 million tonnes of ELTs are generated annually, with more than 50% destined to landfills or left as untreated garbage [2]. In Chile, for instance, more than 145,000 tonnes of ELT were generated in 2019 and only 17% was recycled [3,4]. In this regard, efforts have been made to improve the management of this waste through Law N°20920, which establishes the framework for waste management, extends producer responsibility, and the promotion of recycling in Chile [5]. Although, in this law, tyres are defined as priority products, there is no special provision for the management of ELTs. Therefore, the problems associated with ELT disposal remain.

On the other hand, concrete is a relatively inexpensive material, with the ability to develop high strengths and different shapes, which makes it suitable for multiple applications. In fact, concrete is the most widely used material in the construction industry worldwide, producing approximately 25 billion tonnes annually [6].

However, the concrete manufacturing process requires large amounts of energy, raw materials, and has a large impact on the environment. Therefore, alternatives have been developed to reduce these impacts through the use of non-conventional materials, such as recycled waste. Among the alternatives developed, the behaviour of concrete with added rubber has been studied by different authors [7–12]. These studies report that the inclusion of rubber can improve properties of the concrete, such as energy absorption capacity, ductility, thermal insulation, resistance to lead cycles, post-cracking behaviour, and durability [7–10,13]. Considering these possibilities of improvement and the massive use of concrete, the study of the incorporation of ELT in the concrete material is very attractive as a solution for reusing this waste.

However, although the potential benefits of adding ELT rubber into the concrete, there are challenges related to the weak rubber-cementitious matrix interaction, which results in basic mechanical concrete properties negatively affected [9,11,12]. Furthermore, the dependence of the concrete properties on the individual components, which vary from region-to-region [14], makes local studies on the impact of incorporating ELT in concrete mixes necessary. For instance, evaluations of fresh concrete mixes indicate an increase in workability [15,16], while others report reductions in this property [17–19], and even cases of no appreciable change [20,21]. Moreover, studies show that it has not been possible to find a relationship between replacement percentage and grain size of ELT with the mix workability [22]. However, other studies report such relationships. For instance, the results of Su et al. [23] indicate that concretes with larger rubber show better workability than those with smaller rubber particles. Additionally, the case of concrete with continuous rubber granulometry offers better workability and strength to water permeability compared to concretes with single rubber size.

Additionally, different investigations conclude that rubber incorporation can have a negative impact on the basic mechanical concrete properties, i.e., the flexural and compressive strength [7–9,13,24–27]. Indeed, Liu et al. [28] studied the performance of rubber-based concretes at replacement percentages of 5%, 10%, and 15% with respect to fine aggregate volume. The results indicate that, compared to standard concrete specimens, the rubber-incorporated specimens decrease the compressive and flexural strength. Moreover, the reported decrease in compressive strength is twice the reduction in flexural strength. However, in general, the experimental results show a significant improvement in the cracking behaviour of the material, together with a higher resistance to load cycles and higher toughness.

Yu and Zhu [24] report that rubber content and size can affect the porosity structures of cement mortars. Actually, they state that the reasons behind the reduced strength are the combined changes in rubber content and porosity structures.

In order to minimise the reduction of the mechanical properties of cementitious materials, the application of rubber treatments before the addition to the mix has been evaluated [25–27]. For instance, Mohammad et al. [26] developed a treatment consisting of soaking ELT rubber in water for 24 h, before adding it to the rest of the concrete components. This treatment reduces the amount of air trapped on the surface of the material, which, after the hydration time, produces a decrease in the amount of air bubbles around the material. The result is a relatively minor concrete strength reduction, being more favourable the results of compressive strength than flexural strength. Furthermore, samples with replacement less than 20% showed improvements of the fatigue strength under load cycles [26].

Another treatment developed consists of an oxidation process using a solution of potassium permanganate (KMnO4) and sulphonated with sodium bisulphite (NaHSO3), at different concentrations and contact times. The results showed that the treatment modified the rubber surface, decreasing the contact angle of the rubber with water and significantly improving the interaction between the cementitious matrix and the rubber particles. Therefore, properties, such as compressive strength and impact resistance, were also positively affected. The strength improvements of rubber modified concrete as a function of the rubber percentage was evident and accentuated at around 4% rubber content, with an improvement close to 10 MPa [27].

However, the properties of concrete mixes with rubber are locally dependent due to differences in cement manufacture and components, aggregate characteristics, and ELT properties.

In fact, geo-dependency is one of the characteristics that help to explain the massive use of concrete. Geo-dependency as well is crucial when dealing with waste materials and concrete, especially because the alternative solutions must be practical and feasible to implement in order to effectively reuse the waste [29].

The work presented in this article is part of a wider investigation on the effects of ELT rubber in the concrete material. In particular, and considering factors as geo-dependency and the complexity of rubber treatments, this stage focuses on evaluating traditional properties of mortars with ELT rubber using Chilean cements. More specifically, the objective of the present study is to characterise the properties of cement mortars with the addition of ELT rubber under three surface treatments. The study considers different variables to define the substitution percentages and the appropriate rubber granulometries according to the performance of mortar properties in the fresh and hardened state. In this way, it is expected to define mortar mixes with the best mechanical performance considering technical, practical, and economic aspects.

2. Materials and Methods

2.1. Materials

2.1.1. Cement

Two commercial brands of cement available in the local market were used in this research: Bío Bío Especial (C1) and Polpaico Especial (C2). This choice was based in the necessity of proposing practical and sustainable alternatives for the effective use of ELT rubber.

According to the standard NCh 148, based on ASTM C150/C150M-20, C1 and C2 cements were classified as standard grade Portland pozzolanic cement [30]. The properties of the cements used in this study are presented in Table 1.

Table 1. Cement characterization.

Properties		Bío Bío Especial (C1)	Polpaico Especial (C2)	Requirements NCh 148
Specific gravity (g/cm^3)		2.8	2.7–3.2	-
Autoclave expansion (%)		0.1	0.1	<1.0
Initial setting (h:m)		2:40	1:30	>01:00
Final setting (h:m)		3:40	5:50	<12:00
Compressive	7 d	320	187	>180
Strength (kg/cm^2)	28 d	410	275	>250

One way to chemically characterize the cement is in terms of its oxide's composition, which is directly related to the final properties that a mortar or concrete can develop. Tapia [31] performed a chemical analysis in terms of the oxide components of the cements, including the two cements used in this research (Table 2). Calcium oxide (CaO) and silicon oxide (SiO_2) are the most important for this research, since, from them, dicalcium silicate (C2S) and tricalcium silicate (C3S) are formed, which are the main components of the clinker and they are responsible for the strength of the hydrated cement paste.

Table 2. Oxide components of cements, in percentage [31].

Oxides	Cement C1	Cement C2
CaO	35.0	32.7
SiO_2	24.2	19.3
Al_2O_3	12.6	9.0
Fe_2O_3	4.3	4.8
MgO	2.5	3.0
SO_3	2.4	2.8
K_2O	1.2	0.7
Na_2O	1.8	2.2
TiO_2	0.6	0.1

Furthermore, Tapia [31] made a SEM–EDX analysis where the presence of fly ash in the cement C1 was evidenced, the fly ash being an agent known for the increment of the concrete strength on time.

2.1.2. Sand

Sand consists of a stone material composed of hard particles with a stable shape and size that pass through the 4.75 mm aperture sieve and it is retained on the 0.075 mm sieve [32]. In the present study, Bío Bío sand, available at the Concrete Laboratory, Universidad de Concepción, was used. Bío Bío sand is the typical fine aggregate for making concrete in the local market. The sand properties according to the standards NCh 1239:2009 [33], based on ISO 7033:1987, and NCh 165:2009 [34] based on ISO/DIS 20290-1, are presented in Table 3.

Table 3. Sand characterization.

Properties	
Relative Density (kg/m^3)	2668.2
Relative Density SSD (kg/m^3)	2728.5
Apparent Relative Density (kg/m^3)	2839.3
Water Absorption (%)	2.3
Fineness Modulus	2.7

2.1.3. Water

The water used was from the public water supply that complied with NCh 409/1, and was not contaminated prior to use [35].

2.1.4. Rubber

The rubber was provided by the company Polambiente, which recycles ELT from truck and car tyres. Once received, it was sieved in order to obtain three different sizes, which are presented in Table 4.

Table 4. Rubber particle sizes.

Sizes	Range (mm)
T1	2.36–4.75
T2	1.18–2.36
T3	0.3–1.18

Each of these sizes was characterised using the Chilean regulations associated with fine aggregate [32,33], which is shown in Table 5.

Table 5. Rubber properties.

Properties	T1	T2	T3
Relative Density (kg/m^3)	1099.4	1020.9	1004.1
Relative Density SSD (kg/m^3)	1127.1	1047.1	1029.9
Apparent Relative Density (kg/m^3)	1130.8	1048.4	1030.7
Water Absorption (%)	2.5	2.6	2.6
Fineness Modulus	5.0	4.0	2.8

2.2. *ELT Modification*

In the search for improving the adhesion and interaction between the cementitious matrix and the ELT rubber aggregate, three methods were applied to the surface of the ELT grains before the incorporation in the cement mortar mix.

2.2.1. Treatment 1: Hydration (R1)

Treatment 1 is the simplest of the three treatments applied. The method is based on the work by Mohammadi et al. [26], and consists of soaking the rubber in water for 24 h before incorporating it into the mortar mix. This treatment reduces the amount of air trapped on the surface of the material after the hydration time, which improves the interaction with the cement paste.

2.2.2. Treatment 2: Oxidation–Sulphonation (R2)

Treatment 2 is based on the work by He et al. [27]. It consists of an immersion of the rubber in a 15% sodium hydroxide solution stirring at 150 rpm for 45 h at room temperature. After washing and filtering, the sample is put in contact with a 7% KMnO4 solution. This solution is adjusted to pH 2 and stirred at 150 rpm for 2 h at 70 °C. The product is then transferred to a saturated NaHSO3 solution, which is reacted at 150 rpm for another 2 h at 70 °C. After washing and filtering, the sample is dried for 12 h at 50 °C. By this means, polar groups, such as carbonyl, hydroxyl, and sulfonic groups, are introduced on the surface of the rubber, which increases the hydrophilicity.

2.2.3. Treatment 3: Contact with Hydrogen Peroxide (R3)

This method is based on the study by Shatanawi et al. [25] originally developed to incorporate rubber into asphalt mixes. The treatment consists of contacting the rubber with a hydrogen peroxide solution at a concentration of 50% at a temperature of 60 °C, for a time of 60 min, where a Fenton reaction is subsequently provoked, using a concentration of Fe2+ 12 mM and H2O2 600 mM at pH 3–4. These reactions produce homolytic cleavage of single C–C bonds to produce terminal hydroxyls.

2.3. *Experimental Program*

Due to the large number of combinations, and in order to reduce the uncertainty of the results, the evaluation was divided into three stages. The variables considered in each phase were the type of cement, the size of the rubber grains, the surface treatment applied to the rubber, and the test age to which the samples were subjected. Moreover, the percentage of rubber replacement varies from a range of 0% (control samples) to 12.5%, with respect to the weight of the fine aggregate.

The first stage aimed to make a fast characterization of the samples in order to take useful decisions for the next phases. For that, samples with the addition of untreated rubber, at early test ages of 7 and 14 days, with two cements, were included in the experimental analysis. Furthermore, two replacement percentages were considered, a minimum of 2.5% and a maximum of 12.5%, with respect to the weight of the fine aggregate and with three rubber grain sizes (T1, T2, and T3). Figure 1 presents the details of stage 1.

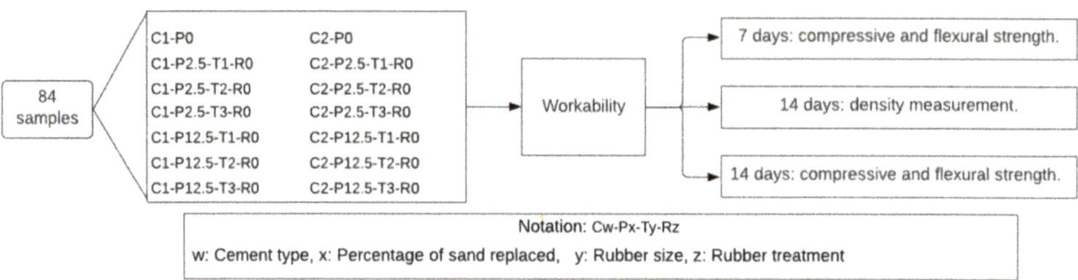

Figure 1. Experimental plan description of stage 1.

Based on the results obtained in the first stage, the second stage is developed (Figure 2). This phase contemplates substitution percentages with rubber of 5% and 7.5% with respect to the weight of the fine aggregate in conditions without treatment and with treatment R1, and test ages up to 28 days. This, in order to have more information on the properties of the samples studied, maintaining the two cements and the three sizes of rubber as variables.

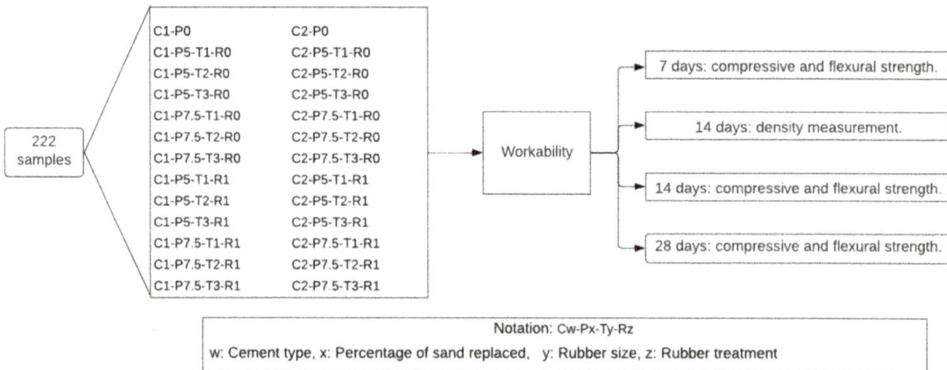

Figure 2. Experimental plan description of stage 2.

Due to the technical complexity of applying the R2 and R3 treatments, stage 3 is based on the previous results in order to optimize the number of samples evaluated. Therefore, in this stage, the best results obtained in the previous stages are considered in the analysis. Figure 3 presents the details of stage 3.

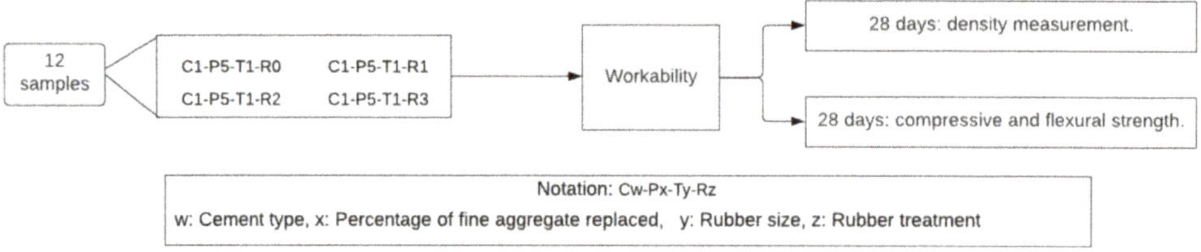

Figure 3. Experimental plan description of stage 3.

In the three stages, the tests were conducted on specimens with dimensions 40 × 40 × 160 mm according to the Réunion Internationale des Laboratoires et Experts des Maté-riaux, systèmes de construction et ouvrages (RILEM) [36]. The workability of each mixture was evaluated according to the NCh 2257/3 standard [37], based on ISO 1920-2: 2016, using the reduced cone method. Additionally, the flexural and compressive strengths were evaluated according to the standard NCh 158 [38], whose international equivalent corresponds to the UNE-EN 196-1: 2018 standard. In order to determine the density, the weight and size measurements of the samples were performed at the hardened state in accordance to NCh 158 [38]. In addition to the macro-characterisation of the samples, i.e., the fresh and hardened mechanical behaviour of cement mortars, a micro-level analysis of the rubber grains under the treatments was performed. The micro-characterisation consists of the measurement of the contact angle between the rubber samples and the water, in addition to the analysis of the SEM (Scanning Electron Microscopy) images. The aim of this approach is to have a better understanding of the macro behaviour of the cement mortar samples.

2.4. Cement Mortar Mix Dosage

For the mix dosage, the Mortar Manual developed by the Chilean Institute of Cement and Concrete was applied. This manual establishes generalities, properties and dosage methods for cement mortars [39]. The procedure based on compliance with workability and compressive strength, modified due to the partial replacement of the aggregate by ELT rubber, was applied.

The reference compressive strength used in this study corresponds to 300 kg/cm^2, which, according to the applied design method [39], implies a design strength of 350 [kg/cm^2], with a medium workability, which means a drop of between 3 and 8 cm measured by the reduced cone method. The mix dosage is presented in Table 6.

Table 6. Mix dosage to produce 1 m^3 of mortar.

Material	C1 (kg)	C2 (kg)
Cement	662.2	662.2
Fine aggregate	1206.5	1239.3
Water	291.3	291.3
Total	2160.0	2192.8

3. Results

3.1. Stage 1

3.1.1. Workability

The results of the reduced cone test for the samples with untreated rubber addition are presented in Figure 4. With respect to the control mortar, without rubber addition, there is an important variation in the workability of the mix with high dependence on the percentage of replacement. With larger amounts of rubber, although all the results are within the design range [38], the samples with smaller rubber size have a much lower workability compared to the others. This is explained due to the higher specific surface area of the smaller rubber particles, causing a greater friction with the rest of the mixture, resulting in a lower cone slump.

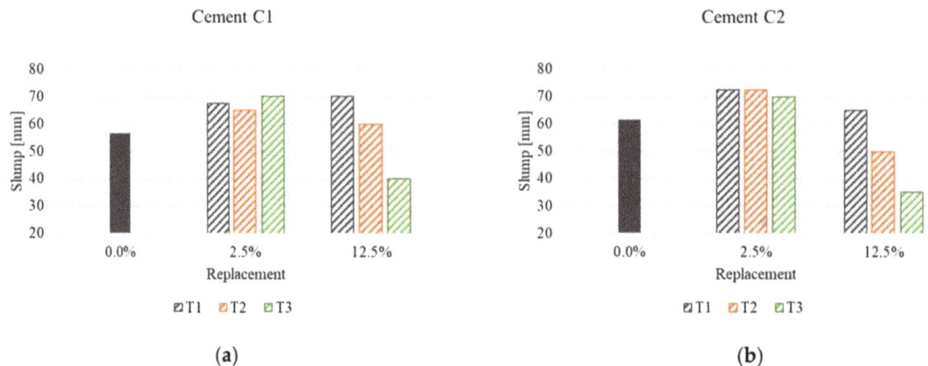

(a)

(b)

Figure 4. Mortar cone slump with addition of untreated rubber 2.5% and 12.5% replacement: (**a**) cement C1; (**b**) cement C2.

3.1.2. Density

The results of the densities for the samples with the addition of untreated rubber are presented in Figure 5. The error bars correspond to the standard deviation for the two cements used; the measured densities are similar, with minimal differences of around 2% in the case of the control mortar, being generally higher in samples with cement C1. Due to the lower density of the ELT rubber with respect to the aggregate, an inverse relationship is observed between the amount of rubber added and the mortar density. Regarding the size of the rubber aggregate, it is not possible to establish a relationship between this and the density since the results do not show a trend in this respect.

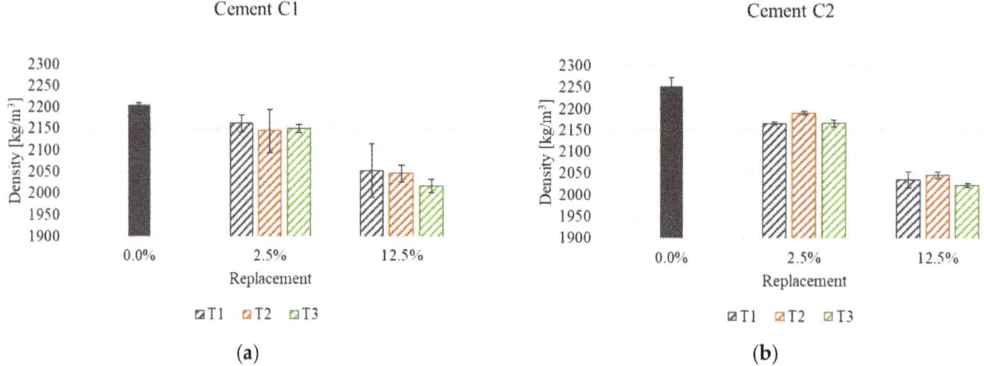

(a)

(b)

Figure 5. Density of mortar with addition of untreated rubber 2.5% and 12.5% replacement: (**a**) cement C1; (**b**) cement C2.

3.1.3. Flexural Strength

The results of the flexural strength for the samples with the addition of untreated rubber are presented in Figure 6. The control mortar has practically equal strengths with both cements at 7 days. This changes at 14 days, where the samples with cement C1 have higher strengths. It is observed that, as the percentage of ELT rubber increases, the strength of the samples decreases.

At a test age of 14 days, when the cement used is C1, the average strength reduction (considering the three rubber sizes) for replacement percentages of 2.5% and 12.5% is 20% and 39%, respectively, with respect to the control mortar. In the case of mortars made with C2 cement, the reduction in strength for both replacement percentages reaches 15% and 38%, respectively.

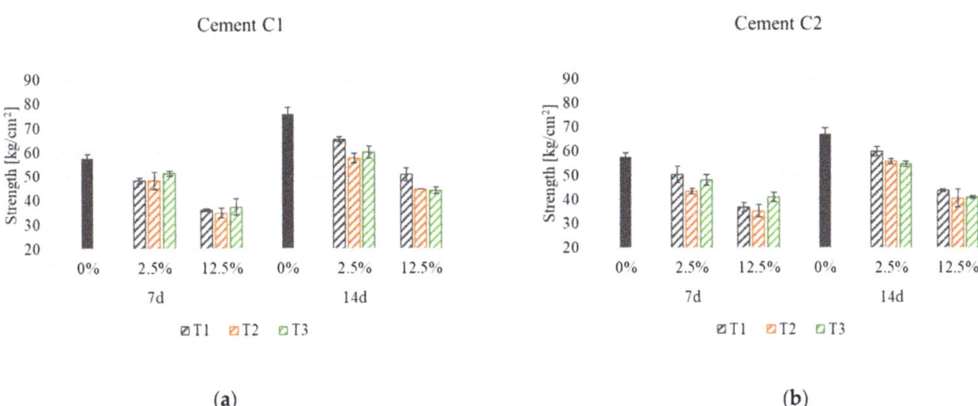

Figure 6. Flexural strength of mortar samples with addition of untreated rubber at 2.5% and 12.5% replacement, after 7 and 14 days: (**a**) cement C1; (**b**) cement C2.

3.1.4. Compressive Strength

Figure 7 presents the results of the compressive strength tests for the samples with untreated rubber addition. Similar to the results previously shown, a strength reduction is observed as the amount of rubber added increases. In general, the strength reduction is greater than in the case of flexural strength, which is in agreement with what is found in the literature [23]. At a test age of 14 days, when the cement used is C1, the average strength reduction (considering the three rubber sizes) for replacement percentages of 2.5% and 12.5% is 18% and 45%, respectively, with respect to the control mortar. In the case of mortars made with C2 cement, the reduction in strength for both replacement percentages reaches 19% and 48%, respectively.

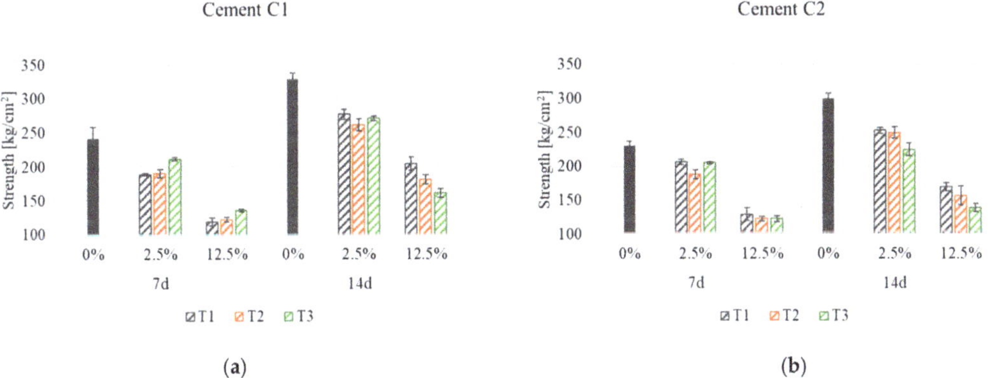

Figure 7. Compressive strength of mortar samples with untreated rubber addition at 2.5% and 12.5% replacement, after 7 days and 14 days: (**a**) cement C1; (**b**) cement C2.

From the results, is possible to conclude that the 12.5% replacement is not recommended. In the case of cement C2, for example, the reduction of compressive strength is approximately 50% with this replacement.

3.2. Stage 2

3.2.1. Workability

Figures 8 and 9 show the slump measurements of mortar samples with untreated rubber and under treatment R1 (hydration), respectively. When untreated rubber is added to the mix, the trend is similar with both cements, i.e., the workability tends to be maintained or increase with larger rubber sizes. For the T3 size, the workability decreases as more rubber is added. Again, it should be noted that, in all cases, the workability remains within the design range [39] and the general relationship is that the larger the rubber size, the higher the workability.

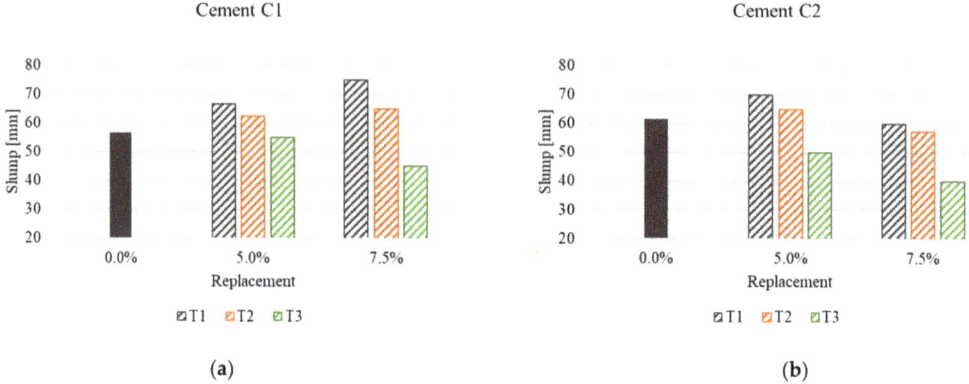

Figure 8. Mortar cone slump with addition of untreated rubber, at 5% and 7.5% replacement: (**a**) cement C1; (**b**) cement C2.

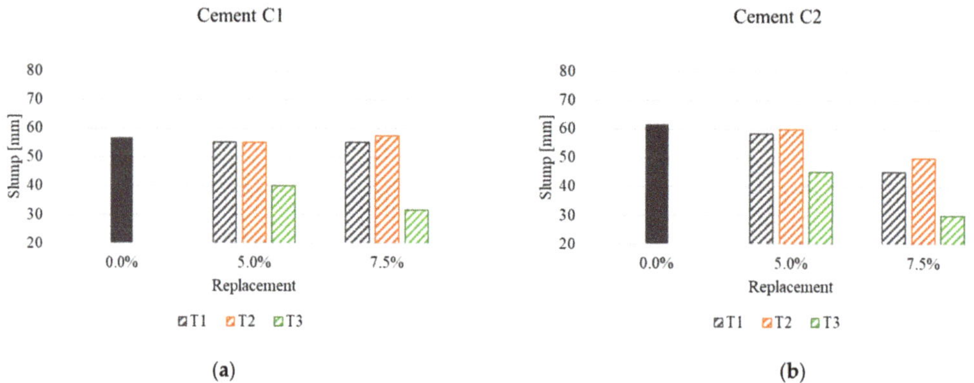

Figure 9. Rubber-added mortar cone slump under treatment R1, at 5% and 7.5% replacement: (**a**) cement C1; (**b**) cement C2.

When rubber is added to the mortar under hydration treatment, workability tends to decrease as the amount of ELT added increases. When the replacement percentage is 5%, considering both cements, workability tends to be maintained for samples with larger rubber grain sizes. For samples with rubber size T3 (0.3–1.18 mm), the workability decreases drastically. In the case of 7.5% replacement, this behaviour is repeated and the reduction of workability is accentuated with rubber size T3, reaching the minimum of the design range of 3 cm [39].

3.2.2. Density

The densities for the samples with untreated rubber are presented in Figure 10. The results for the samples with cement C1 show that with rubber incorporated at 5% and 7.5%, the density decreases by 2% and 8%, respectively. Similarly, in samples with cement C2 under the same conditions, the density decreases by 3% and 10%, respectively. This fact shows an inverse relationship between the amount of rubber added and the density of the mortar. Regarding the size of the incorporated rubber, in the case of cement C1 mortars, the densities are higher when the rubber used is larger (T1). In the case of cement C2 mortars, medium size rubber mortars (T2) have slightly higher densities.

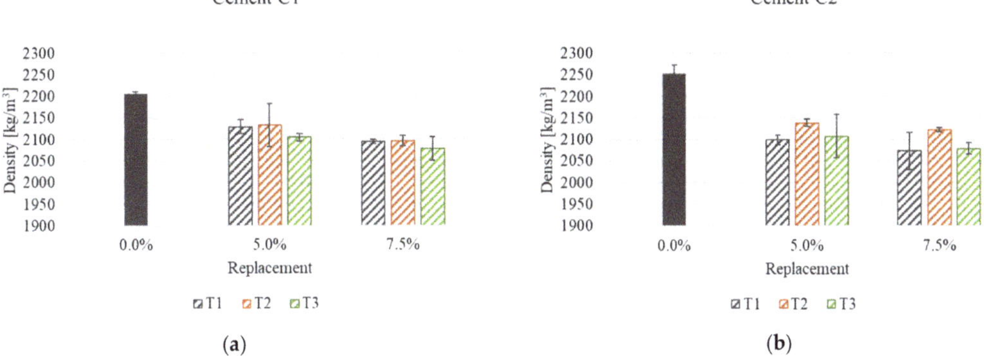

Figure 10. Density of mortar with addition of untreated rubber at 5% and 7.5% replacement, after 14 days: (**a**) cement C1; (**b**) cement C2.

In the case of mortars with rubber addition under treatment R1, it is observed that the smaller the rubber size produces lower densities (Figure 11). For the samples of cement C1 with percentages of 5% and 7.5% of rubber incorporated, the density reductions are on average 3% and 5%, respectively. For the same percentages of incorporated rubber, but with cement C2 samples, the density reductions are 5% and 7% respectively.

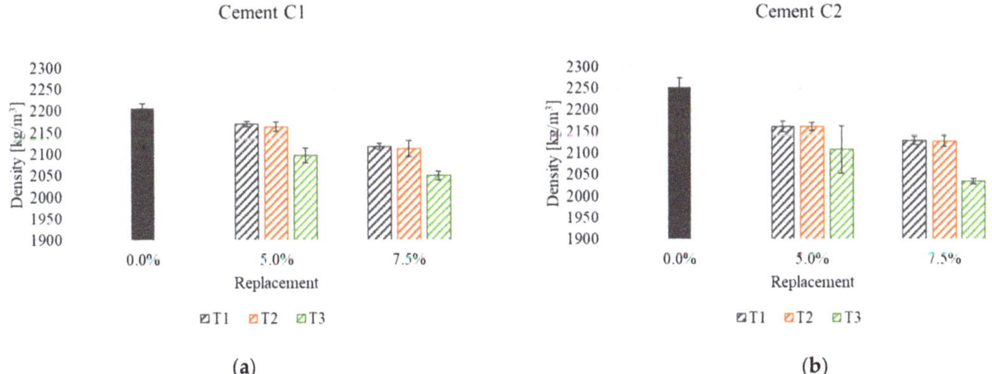

Figure 11. Density of rubber-added mortar under treatment R1 at 5% and 7.5% replacement, after 14 days: (**a**) cement C1; (**b**) cement C2.

3.2.3. Flexural Strength

The results of the flexural strength tests for the samples with untreated rubber addition and under treatment R1 are presented in Figures 12 and 13, respectively. When comparing the samples without rubber, at early ages, the strength is similar for both cements. However, as the test age advances, the difference increases, with the samples of cement C1 having higher strength.

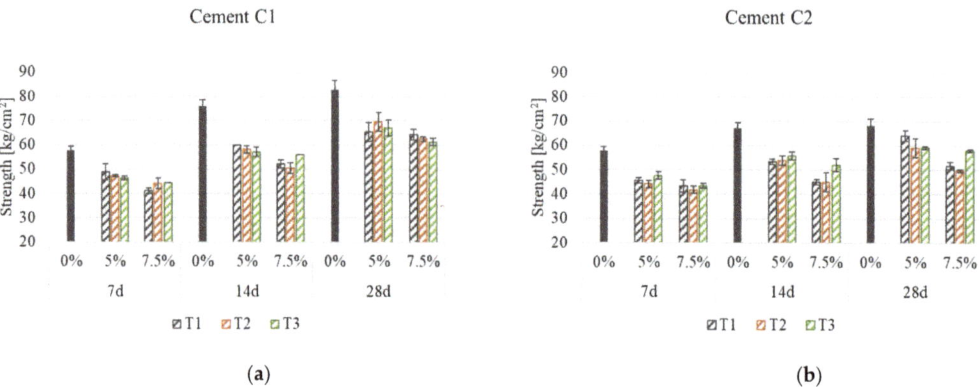

(a) (b)

Figure 12. Flexural strength of mortar samples with addition of untreated rubber at 5% and 7.5% replacement, after 7, 14, and 28 days: (**a**) cement C1; (**b**) cement C2.

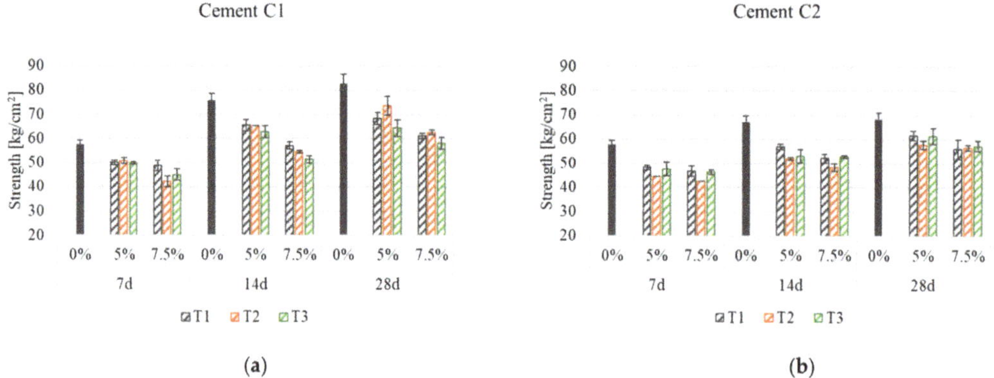

(a) (b)

Figure 13. Flexural strength of rubber-added mortar samples under treatment R1 at 5% and 7.5% replacement, after 7, 14, and 28 days: (**a**) cement C1; (**b**) cement C2.

In the samples with untreated rubber, there is no clear trend about the rubber size giving the best results. However, for samples with cement C1, the T1 size shows similar or slightly higher strengths than the other grain sizes. In the case of cement C2, the T3 and T1 sizes showed the highest strengths.

At a test age of 14 days, when the cement used is C1, the average strength reduction (considering the three rubber sizes) for replacement percentages of 5% and 7.5% is 23% and 30%, respectively, with respect to the control mortar. In the case of mortars made with C2 cement, the reduction in strength for both replacement percentages reaches 19% and 29%, respectively. Although the strength reduction of samples with cement C2 is smaller compared to the results of samples with C1, in absolute terms, the strength of these last samples is higher in all cases.

The flexural strength of the mortars with rubber under R1 treatment has a similar behaviour to the one observed in stage 1, i.e., a slight improvement in strength with respect to the samples with untreated rubber. This fact is observed to a greater extent in the samples with cement C1. In this case, at a test age of 14 days, when the cement C1 is used, the average strength reduction (considering the three rubber sizes) for replacement percentages of 5% and 7.5% is 13% and 21%, respectively, with respect to the control mortar. In the case of mortars made with C2 cement, the reduction in strength for both replacement percentages reaches 19% and 22%, respectively.

3.2.4. Compressive Strength

The results of the compressive strength tests for the samples with untreated rubber addition and under treatment R1 (hydration) are shown in Figures 14 and 15, respectively. At a test age of 14 days, when the cement used is C1, the average reduction in strength (considering the three rubber sizes) for replacement percentages of 5% and 7.5% is 18% and 31%, respectively, with respect to the control mortar. In the case of mortars made with C2 cement, the reduction in strength for both replacement percentages reaches 34% and 39%, respectively. There is a clear difference between the results of the two cements, being the samples made with cement C1, the ones with the highest strength at 28 days. In all cases, it is observed that the strength of the samples containing untreated rubber is higher when cement C1 is used. In addition, the only case of samples containing ELT and fulfilling the design strength of 350 (kg/cm^2) is with 5% replacement ELT and size T1.

The results of rubber samples made with cement C1, and under treatment R1, indicate a slight increase in strength compared to the untreated rubber samples. This does not occur in the samples made with cement C2. In this case, at a test age of 14 days, when the cement used is C1, the average strength reduction (considering the three rubber sizes) for replacement percentages of 5% and 7.5% is 9% and 23%, respectively, with respect to the control mortar. In the case of mortars made with C2 cement, the reduction in strength for both replacement percentages reaches 30% and 39%, respectively.

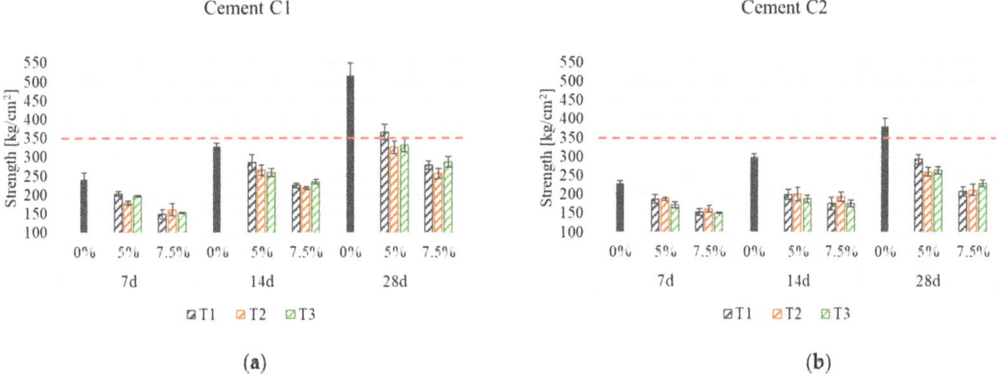

Figure 14. Compressive strength of rubber-added mortar samples with addition of untreated rubber at 5% and 7.5% replacement, aged 7, 14, and 28 days: (**a**) cement C1; (**b**) cement C2.

The results show very significant differences between the strengths obtained at 28 days with C1 and C2 when ELT rubber is added (untreated and with R1 treatment). In a similar way to the previous case, when untreated ELT is added, the design strength is only fulfil with 5% replacement, in this occasion with sizes T1 and T2. For 7.5% replacement, only the average result with size T1 reaches the design strength, but not all the individual samples, as it is shown by the standard deviation. The results evidence a trend of better technical performance associated with T1 size. However, other relevant factors allow considering the T1 size for the last stage of the study. In effect, considering the characteristics of the

ELT rubber, the largest size (T1) is easier to handle, less expensive to produce, and requires less energy to manufacture it.

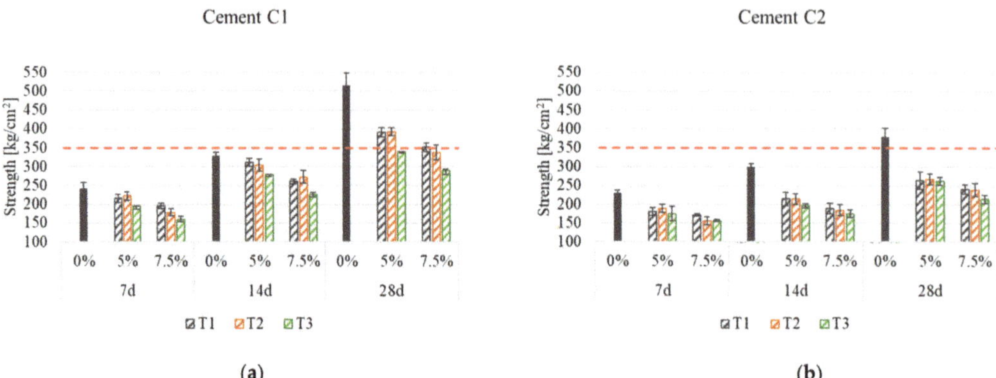

Figure 15. Compressive strength of rubber-added mortar samples under treatment R1 at 5% and 7.5% replacement, after 7, 14, and 28 days: (**a**) cement C1; (**b**) cement C2.

Tapia [31] analysed the composition of three brands of cement, including the two used in this work. Similar results were found in that study in terms of the difference in compressive strength when comparing control mortars made with both cements (C1 and C2). This difference can be attributed to the direct relationship between concrete strength and the presence of dicalcium silicate (C2S) and tricalcium silicate (C3S) in the cement composition, which in the case of cement C1 is greater than in cement C2. Another reason found is the presence of fly ash in cement C1, which is not present in cement C2 [39] and which, in the long term, causes concretes with this compound to continue acquiring strength.

The lower reduction in compressive strength observed in rubber-based mortars with cement C1 can be explained by the presence of fly ash. Indeed, this component has been used in other investigations to improve the interaction between the rubber and the cementitious matrix, showing adequate efficacy in mitigating the reduction of concrete strength [40–42].

For these reasons, C1 cement was chosen to continue with the last stage of the study.

3.3. Stage 3

3.3.1. Workability

Figure 16 shows the cone slump of the rubber-added mortar samples under the three treatments applied. The figure shows as well, the results with untreated rubber, under the same conditions, i.e., 5% substitution rate, grain size T1 (2.36–4.75 mm) and use of cement C1.

With respect to the control mortar, treatment 2 (oxidation–sulphonation) and treatment 3 (hydrogen peroxide), show slightly greater cone slumps, with the greatest corresponding to treatment R2. However, these slump values are always lower than the results of the mortars with untreated rubber. Furthermore, in all cases, the results are within the design range [39].

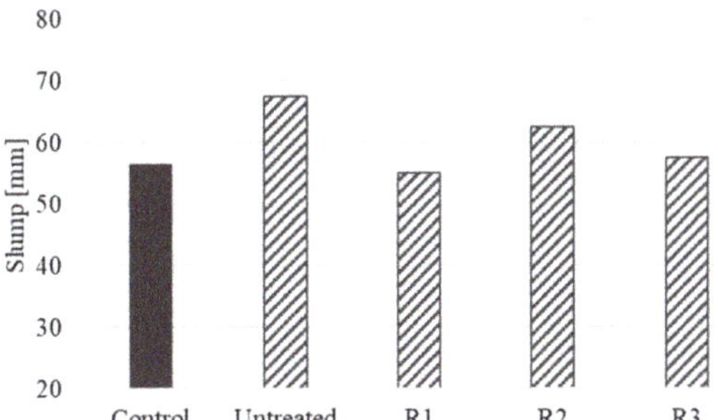

Figure 16. Cone slumps of cement mortars with 5% replacement rubber, grain size T1, using cement C1.

3.3.2. Density

Figure 17 shows the densities of mortars with rubber addition under the three treatments used, and furthermore, shows the results with untreated rubber, under the same conditions, i.e., 5% substitution rate, grain size T1 (2.36–4.75 mm) and use of cement C1.

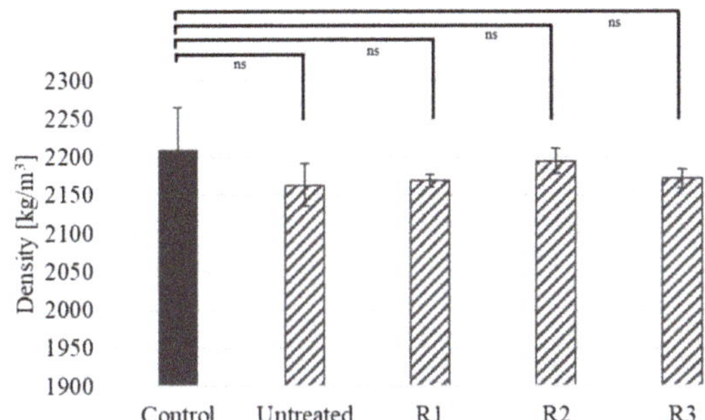

Figure 17. Density of mortars with addition of rubber at 5% replacement, grain size T1, using cement C1. ns = non-significant.

With respect to the control mortar, the samples present slightly lower densities due to the lower density of the rubber used, but when compared to the samples containing untreated rubber, they are higher, which is attributable to a lower presence of air in the mix, which in turn is due to the treatments that reduce the hydrophobicity of the rubber. This effect is mostly visible in the mortars under treatment R2 (oxidation–sulphonation), where the density decrease is insignificant compared to control mortar.

Using one-way analysis of variance (ANOVA) with Dunnett's post hoc, it is possible to observe that there is no statistically significant difference (ns) between the samples containing ELT, either untreated or with any of the three treatments, compared to the control mortar.

3.3.3. Flexural Strength

The results of the flexural strength tests for the rubber added samples under the three treatments are presented in Figure 18. This figure shows as well, the results with untreated rubber, under the same conditions, i.e., 5% substitution rate, grain size T1 (2.36–4.75 mm), and use of cement C1.

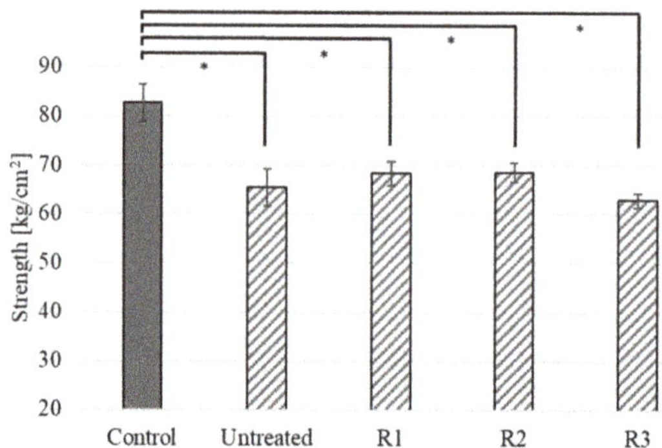

Figure 18. Flexural strength of mortars with addition of rubber at 5%, grain size T1, using cement C1, after 28 days. * ($p < 0.05$).

When comparing the treatments with untreated rubber, both treatments, R1 (hydration), and treatment R2 (oxidation–sulphonation), present mortars with better results. Only in the case of treatment R3 (contact with hydrogen peroxide) the strength of the mortars is lower. Compared to control mortar, in all cases the flexural strength is lower, decreasing by 21% in the case of mortar with untreated rubber, 17% for rubber with treatment R1, 17% for rubber with treatment R2, and 24% in the case of rubber with treatment R3.

Through one-way ANOVA, it was determined that there is a statistically significant difference between the control mortar and the samples containing ELT rubber—untreated or under any of the three treatments. Statistical significance is designated with *, $p < 0.05$.

3.3.4. Compressive Strength

The results of the compressive strength tests for the rubber added samples under three treatments are presented in Figure 19. Furthermore, the figure shows the results with untreated rubber, under the same conditions, i.e., 5% substitution rate, grain size T1 (2.36–4.75 mm), and cement C1.

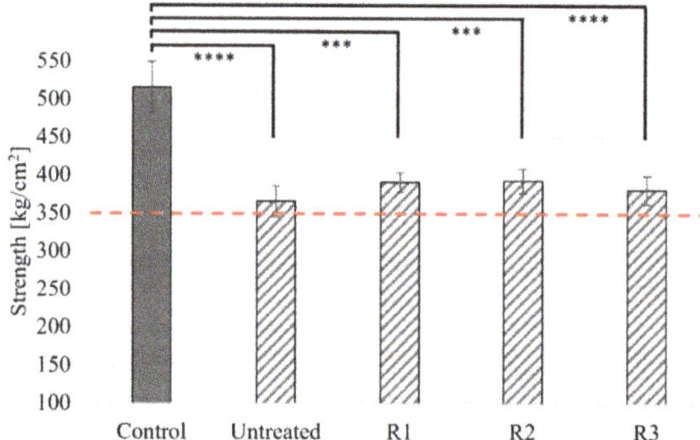

Figure 19. Compressive strength of mortars with 5% replacement rubber addition, grain size T1, using cement C1, after 28 days. *** ($p < 0.001$); **** ($p < 0.0001$).

Compared to the control mortar, in all cases, the compressive strength is lower, with a loss of 29% in the case of mortars with untreated rubber, 24% when the rubber is under treatment R1 (hydration), 24% under treatment R2 (oxidation–sulphonation), and 26% in the case of treatment R3 (contact with hydrogen peroxide). While it is true that the average result of the untreated ELT samples fulfil the design strength, which is not necessarily valid for all samples as it is shown by the standard deviation. On the contrary, when the ELT rubber is treated, the design strength of 350 (kg/cm^2) is always fulfilled, including the standard error. This is valid for the three treatments, but with better results for treatments R1 and R2, similar to the case of flexural strength.

Using one-way analysis of variance (ANOVA) with Dunnett's post hoc, it is possible to visualize that there is a statistically significant difference between the results of the control samples and the ones with ELT rubber, either untreated or with any of the three treatments. However, in terms of statistical significance, those differences are less for treatments R1 and R2 (*** $p < 0.001$) than for treatment R3 and the untreated samples (**** $p < 0.0001$).

In addition to the technical performance related with the compressive strength, it is important to consider, as well, the economical, practical, and environmental aspects of the treatments. In this sense, the R1 treatment clearly has advantages over the R2 one, as the ELT rubber hydration is a very economical and practical procedure, which uses only water, i.e., an eco-friendly alternative.

3.3.5. Contact Angle

- Treatment R1

The contact angle tests were performed in order to observe the variation of the angle formed by the water on the modified and original rubber particle; thus, evaluating the hydrophilicity of the rubber surface by means of the angle value. Figure 20 shows that the unmodified rubber and the rubber modified with the first treatment have an angle value of 161.4°, and 131.3°, respectively, demonstrating that the adhesion between the rubber and the water increases slightly.

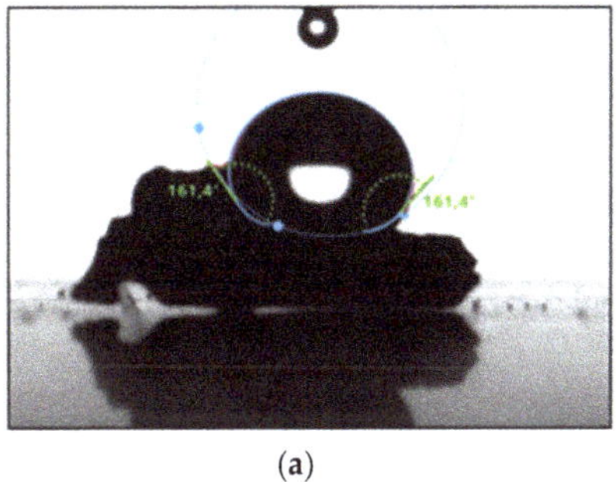

(**a**)

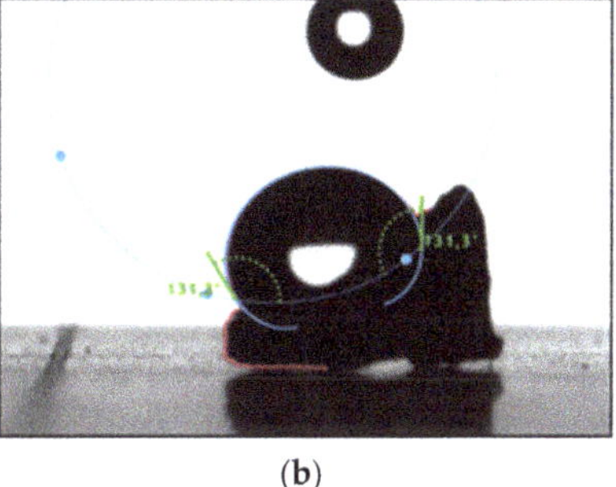

(**b**)

Figure 20. Contact angle: (**a**) unmodified rubber 161.4°, (**b**) rubber under treatment R1, 131.3°.

- Treatment R2

Figure 21 shows the contact angle of the rubber under the second treatment. There is a lower contact angle value compared to the unmodified rubber because the adhesion forces are larger compared to the cohesive forces. This results in the liquid being attracted to the solid and spread out; therefore, it is possible to conclude that this method increases the hydrophilic character of the ELT rubber surface.

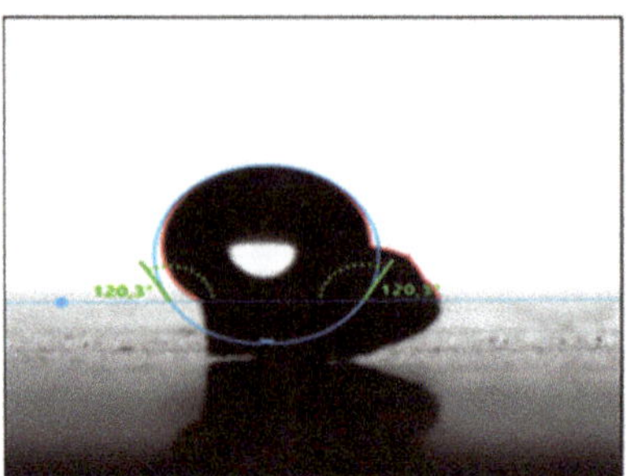

Figure 21. Rubber contact angle under the second treatment (R2), 120.3°.

- Treatment R3

Figure 22 shows that the rubber modified with the third treatment has an angle value of 145.9°. Therefore, it has a low wettability due to the fact that the forces of attraction are lower and the surface tends to repel the liquid. The contact angle value is also affected by the surface roughness, as the rubber particle has small protuberances and the water droplet rests between their peaks, resulting in less contact and, thus, a larger contact angle.

Figure 22. Rubber contact angle under the third treatment (R3), 145.9°.

3.3.6. Scanning Electron Microscope (SEM) Analysis

- Treatment R1

Using the SEM technique, it was possible to obtain images of the surface of the rubber hydrated with water (Figure 23). Differences in the surface can be observed, as the surface of the unmodified rubber is rough and has many pores, while the surface of the modified rubber is smooth and less porous.

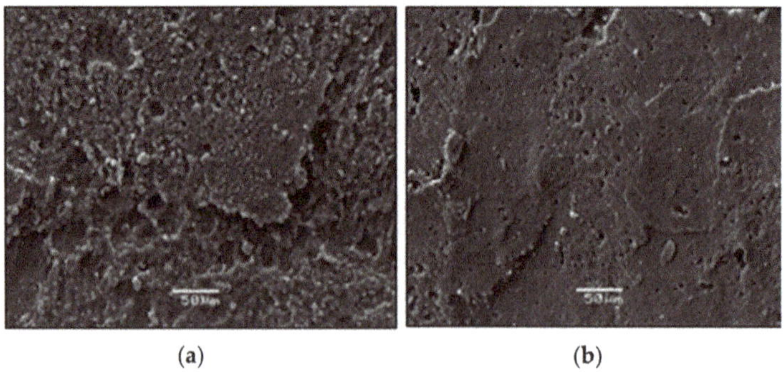

(**a**) (**b**)

Figure 23. Rubber surface: (**a**) unmodified, (**b**) under treatment R1.

- Treatment R2

Figure 24 characterises the morphological aspects of the modified rubber surface under the second treatment, showing a surface with some roughness and pores, but still less than the unmodified rubber.

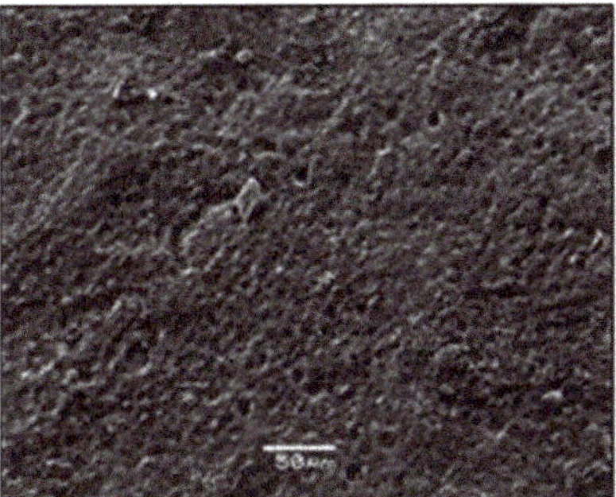

Figure 24. Rubber surface under treatment R2.

- Treatment R3

Figure 25 shows the modified surface for experiment 3, with a rough surface composed of small cavities.

Figure 25. Rubber surface under treatment R3.

When comparing the surfaces of the unmodified rubber with respect to the three treatments, it is possible to observe that these last surfaces are slightly smoother and with fewer pores, especially regarding to the first and second treatment. The wettability of the surfaces not only depends on their materiality, but on their surface at the micro level as well. In this way, the more irregular the surface, the more hydrophobic it is due to the interaction surface-water [43]. Thus, under these treatments, the surface irregularity at the micro level decreases the hydrophobicity of the rubber, allowing a better interaction between the rubber and the cementitious matrix. The contact angle measurements present trends in the same direction. Although the contact angle is not greatly reduced, under the first (R1) and second treatment (R2) it decreases more than R3, which shows a slight improvement in the interaction with water with respect to the untreated rubber. This is more clearly observed with the treatment R2, where the contact angle decreases by more than 30° and is the one with the best result, in terms of mechanical resistance, together with the treatment R1.

Hence, the analysis at the micro level with the contact angle and the SEM images, provide further explanation to the macro performance observed in the cement mortar samples. In this way, the better behaviour of treatments R1 and R2 is confirmed.

4. Discussion

The results indicate that the addition of ELT recycled rubber modifies the behaviour of the cement mortar. The changes include a reduction of compressive and flexural strength as the rubber is added to the mix. Although this effect is undesirable, it is possible to minimise its impact with the application of rubber treatments, such as those shown in this work.

Although the percentage of rubber added in the concrete mix is low, the large quantities of concrete required in the construction industry can lead to a high total volume of ELT rubber used. For that is important to fulfil the strength design, which is the case presented in this work, without adding, as other investigations, a higher amount of cement [44] or additives [45,46] that reduce the loss of strength. Actually, in this paper, only a treatment on the surface of the rubber that is practical, easy to apply, effective, and economically feasible is proposed. In effect, the results of this study show that is possible to use this waste material while maintaining the design strength requirements, which is different from other investigations, where adding ELT rubber decreases the strength under the designed one [44,47]. Considering the importance of the concrete strength, the promissory results presented in this article open the way to numerous real applications in opposition to the limitations found in other investigations [48,49].

Therefore, although two of the treatments may present similar results in terms of mortar strength, when comparing various factors, such as duration, difficulty of application, and costs, the R1 treatment is more favourable. Particularly, this last aspect is fundamental to develop cost-effective alternatives due to the fact that, although adding ELT rubber can improve different concrete properties, there is also a related cost involved.

Although there is a trend in favour of the T1 ELT rubber size, its election is complemented by other aspects, such as the energy and costs required for the grains production, the smaller the size, the longer the crushing time, which implies higher costs and energy. Another important aspect is the ease with which the rubber grain to be treated can be processed. In the laboratory, when the R1 treatment was applied, the smallest rubber grains did not submerge in the water, as was the case with most of the larger grains at the end of the contact time. Hence, the smallest grain sizes have difficulties to receive the hydration treatment.

This stage of the research focusses in cement mortars, which are a particular case of the concrete material. This allows considering a large number of cases, including different relevant variables, as rubber size, percentage of substitution, rubber treatments, and types of cements, and their influences in fundamental properties as the compressive and flexural strength. This is related with an effective and realistic use of the waste end-of-life tyres, which is part of the motivation of this article. Actually, the composite mortar-waste can have excellent behaviour in other properties, but if the composite material strength is significantly reduced, the practical possibilities to be massively used can be very limited. Considering the promissory results obtained in the present phase of the research, it is recommended to continue investigating other composite properties as ductility, long-term durability, and behaviour at elevated temperatures, among others.

Finally, if the final purpose is to effectively use waste, it is important to consider the geo-dependency of the concrete material [29]. This is particularly evident in the case of cement C1, which is abundant in the Chilean market at a similar cost of cement C2. However, cement C1 is capable of delivering mortars that, despite the strength reduction, they can satisfy the design requirements. Furthermore, cement C1 includes fly ash, which reduces the amount of clinker whilst is reusing a local industrial waste. In this sense, the results obtained in this investigation can be useful for other regions as well. For instance, where it is possible to replace part of the cement by fly ash (produced or imported), or other cement substitute, producing similar results. If the substitute is a waste as the fly ash, it will be important not only from a technical perspective, but also as a contribution to a sustainable development.

5. Conclusions

The properties and behaviour of mortars are modified when recycled rubber is used as partial replacement of the aggregate. The tests and measurements carried out show that is possible to add ELT rubber up to 5% with respect to the weight of the aggregate. Of the three treatments analysed, the hydration of the rubber is the best option from a technical, practical, and economic point of view. Regarding to the three rubber sizes evaluated, the T1 (2.36–4.75 mm), which is close to the maximum aggregate size, is recommended. This is due to the favourable results in terms of strength, handling, and less energy involved in crushing the ELT. Additionally, cement C1 offers the best mortar results due to its composition, which includes another waste as the fly ash, one of the main responsible for the strength differences at 28 days. In effect, cement C1 not only gives higher compressive strength than C2, but also allows fulfilling the designed strength with the addition of ELT rubber.

In the fresh state, the workability of the mixture depends on the amount and size of rubber added, as well as the surface treatment applied. In general, with larger grain sizes, the workability tends to remain the same or increase, as the amount of rubber increases. However, it is important to highlight that this workability changes remain in the design range.

As expected, the density of the samples with the addition of ELT recycled rubber decreases as the rubber content increases, due to the lower density of ELT recycled rubber with respect to the aggregate. This behaviour depends as well on the size of the rubber grains, due to the specific surface. Indeed, considering the low compatibility between rubber and water, higher presence of entrapped air in the mix is expected due to a larger and irregular surface area generated when the grain size decreases.

The mechanical strength of the samples studied decreases as the percentage of ELT rubber replacement increases. At early age (7 days), the comparative results between mortars made with both cements and untreated ELT rubber are slightly similar. The results are different at 14 and 28 days, where the behaviour of the mortars with cement C1 are always better. For this reason, the third stage of optimization, only considers the cement C1.

Although the three applied treatments modify the hydrophobic nature of rubber, the best results are obtained with treatment R1 (hydration) and treatment R2 (oxidation–sulphonation). The behaviours at macro level can be explained at the micro level, due to the contact angle measurements and SEM images analysis. These results indicate that the treatments are able to alter the surface roughness and the contact angle between surface and water, improving the interaction between the ELT rubber and the cementitious matrix. This is particularly relevant on treatments R1 and R2, which is coherent with the macro results obtained of the cement mortar samples.

Treatments R1 and R2 provide the best results, being these ones very similar between them. However, treatment R1 is much less difficult to apply in practice, due to the fact that R1 uses elements and substances easily accessible at lower cost. Hence, in addition to the technical performance mainly related with the compressive strength is important to consider, as well, the economical, practical, and environmental aspects of the treatments. In this sense, the R1 treatment clearly has advantages over the R2 one, as the ELT rubber hydration is a very economical and practical procedure, which uses only water, i.e., an eco-friendly alternative.

Finally, when C1 cement in conjunction with the R1 treatment are used, the resulting cement mortars with ELT rubber are capable of exceeding the designed strength without adding, as other investigations, more cement, or using additives to reduce loss of strength. Considering the importance of the concrete strength, the promissory results presented in this article open the way to numerous real applications in opposition to the limitations found in other investigations related with concrete and/or mortar incorporating ELT rubber. Hence, the obtained results open the way to numerous practical applications, and then effective uses of the waste end-of-life tyres.

Author Contributions: Conceptualization, M.P., B.U., V.H.C.-R., C.M., and P.F.; methodology, M.P., B.U., V.H.C.-R., C.M., and P.F.; validation, M.P.; formal analysis, M.P., B.U., V.H.C.-R., and P.F.; investigation, E.G., and B.V.; resources, B.U., C.M., and P.F.; writing—original draft preparation, E.G. and B.V.; writing—review and editing, M.P.; supervision, M.P., B.U., and V.H.C.-R.; project administration, M.P.; funding acquisition, M.P., B.U., C.M., and P.F. All authors have read and agreed to the published version of the manuscript.

Funding: This research was funded by ANID PIA/Apoyo CCTE AFB170007 and VRID Multidisciplinario 219.091.051-M.

Institutional Review Board Statement: Not applicable.

Informed Consent Statement: Not applicable.

Data Availability Statement: Raw data of this paper will be available from corresponding author, M.P., on a reasonable request.

Acknowledgments: The authors acknowledge the contribution of the company "Polambiente" for providing the ELT rubber used in this study.

Conflicts of Interest: The authors declare no conflict of interest.

References

1. Kashani, A.; Ngo, T.D.; Hemachandra, P.; Hajimohammadi, A. Effects of surface treatments of recycled tyre crumb on cement-rubber bonding in concrete composite foam. *Constr. Build. Mater.* **2018**, *171*, 467–473. [CrossRef]
2. Thai, Q.B.; Le, D.K.; Do, N.H.; Le, P.K.; Phan-Thien, N.; Wee, C.Y.; Duong, H.M. Advanced aerogels from waste tire fibers for oil spill-cleaning applications. *J. Environ. Chem. Eng.* **2020**, *8*, 104016. [CrossRef]
3. CINC. Generación y Manejo de Neumáticos Fuera de Uso (NFU). Cámara de la Industria del Neumático de Chile. 2019. Available online: http://cinc.cl/wp-content/uploads/2020/02/200205-Estad%C3%ADsticas-NFU.pdf (accessed on 1 November 2020).
4. Gobierno Regional de Arica y Parinacota. Available online: https://www.goreayp.cl/index.php/noticias/1962-nueva-ley-para-combatir-140-mil-toneladas-de-neumaticos-en-desuso (accessed on 1 December 2020).
5. Ley N°20920. Diario Oficial de la República de Chile, Santiago, Chile, 1 June 2016. Available online: https://mma.gob.cl/wp-content/uploads/2017/03/acuerdo-N-5-ley-20920.pdf (accessed on 1 December 2020).
6. Marinković, S.; Malešev, M.; Ignjatović, I. Life cycle assessment (LCA) of concrete made using recycled concrete or natural aggregates. In *Eco-Efficient Construction and Building Materials*; Woodhead Publishing: Cambridge, UK, 2014; pp. 239–266.
7. Toutanji, H. The use of rubber tire particles in concrete to replace mineral aggregates. *Cem. Concr. Compos.* **1996**, *18*, 135–139. [CrossRef]
8. Fedroff, D.; Ahmad, S.; Savas, B.Z. Mechanical properties of concrete with ground waste tire rubber. *Transp. Res. Rec.* **1996**, *1532*, 66–72. [CrossRef]
9. Shu, X.; Huang, B. Recycling of waste tire rubber in asphalt and portland cement concrete: An overview. *Constr. Build. Mater.* **2014**, *67*, 217–224. [CrossRef]
10. Thomas, B.S.; Gupta, R.C.; Mehra, P.; Kumar, S. Performance of high strength rubberized concrete in aggressive environment. *Constr. Build. Mater.* **2015**, *83*, 320–326. [CrossRef]
11. Karakurt, C. Microstructure properties of waste tire rubber composites: An overview. *J. Mater. Cycles Waste Manag.* **2014**, *17*, 422–433. [CrossRef]
12. Liu, H.; Wang, X.; Jiao, Y.; Sha, T. Experimental Investigation of the Mechanical and Durability Properties of Crumb Rubber Concrete. *Materials* **2016**, *9*, 172. [CrossRef]
13. Callister, W.D. *Introducción a la Ciencia a Ingeniería de los Materiales*; Reverté: Barcelona, España, 1995.
14. Médici, M.; Benegas, O.; Uñac, R.; Vidales, A. The effect of blending granular aggregates of different origin on the strength of concrete. *Phys. A: Stat. Mech. Its Appl.* **2012**, *391*, 1934–1941. [CrossRef]
15. Topçu, I.B.; Bilir, T. Experimental investigation of some fresh and hardened properties of rubberized self-compacting concrete. *Mater. Des.* **2009**, *30*, 3056–3065. [CrossRef]
16. Aiello, M.; Leuzzi, F. Waste tyre rubberized concrete: Properties at fresh and hardened state. *Waste Manag.* **2010**, *30*, 1696–1704. [CrossRef]
17. Freitas, C.; Galvão, J.C.A.; Portella, K.F.; Joukoski, A.; Filho, C.V.G.; Ferreira, E.S. Desempenho físico-químico e mecânico de concreto de cimento Portland com borracha de estireno-butadieno reciclada de pneus. *Química Nova* **2009**, *32*, 913–918. [CrossRef]
18. Albano, C.; Camacho, N.; Reyes, J.; Feliu, J.; Hernández, M. Influence of scrap rubber addition to Portland I concrete composites: Destructive and non-destructive testing. *Compos. Struct.* **2005**, *71*, 439–446. [CrossRef]
19. Batayneh, M.K.; Marie, I.; Asi, I. Promoting the use of crumb rubber concrete in developing countries. *Waste Manag.* **2008**, *28*, 2171–2176. [CrossRef] [PubMed]
20. Bignozzi, M.; Sandrolini, F. Tyre rubber waste recycling in self-compacting concrete. *Cem. Concr. Res.* **2006**, *36*, 735–739. [CrossRef]

21. Skripkiūnas, G.; Grinys, A.; Černius, B. Deformation properties of concrete with rubber waste additives. *Mater. Sci.* **2007**, *13*, 219–223.
22. Pacheco-Torgal, F.; Shasavandi, A.; Jalali, S. Tyre rubber wastes based concrete: A review. In Proceedings of the 1st International Conference WASTES: Solutions, Treatments and Opportunities, Guimarães, Portugal, 12–14 September 2011.
23. Su, H.; Yang, J.; Ling, T.-C.; Ghataora, G.S.; Dirar, S. Properties of concrete prepared with waste tyre rubber particles of uniform and varying sizes. *J. Clean. Prod.* **2015**, *91*, 288–296. [CrossRef]
24. Yu, Y.; Zhu, H. Influence of Rubber Size on Properties of Crumb Rubber Mortars. *Materials* **2016**, *9*, 527. [CrossRef]
25. Shatanawi, K.M.; Biro, S.; Naser, M.; Amirkhanian, S.N. Improving the rheological properties of crumb rubber modified binder using hydrogen peroxide. *Road Mater. Pavement Des.* **2013**, *14*, 723–734. [CrossRef]
26. Mohammadi, I.; Khabbaz, H.; Vessalas, K. In-depth assessment of Crumb Rubber Concrete (CRC) prepared by water-soaking treatment method for rigid pavements. *Constr. Build. Mater.* **2014**, *71*, 456–471. [CrossRef]
27. He, L.; Ma, Y.; Liu, Q.; Mu, Y. Surface modification of crumb rubber and its influence on the mechanical properties of rubber-cement concrete. *Constr. Build. Mater.* **2016**, *120*, 403–407. [CrossRef]
28. Liu, F.; Zheng, W.; Li, L.; Feng, W.; Ning, G. Mechanical and fatigue performance of rubber concrete. *Constr. Build. Mater.* **2013**, *47*, 711–719. [CrossRef]
29. Mansilla, C.; Pradena, M.; Fuentealba, C.; César, A. Evaluation of Mechanical Properties of Concrete Reinforced with *Eucalyptus globulus* Bark Fibres. *Sustainabilty* **2020**, *12*, 10026. [CrossRef]
30. INN. *Cemento—Terminología, Clasificación y Especificaciones Generales. NCh148.1968*; Instituto de Normalización Nacional: Santiago, Chile, 1968.
31. Tapia, M. Caracterización y Comparación Mecánica, Física y Química de Tres Tipos de Cementos Disponibles en el Mercado Regional del Bío Bío. Tesis de Pregrado, Universidad Andrés Bello, Concepción, Chile, 2016.
32. INN. *Áridos para Morteros y Hormigones—Requisitos. NCh163.2013*; Instituto de Normalización Nacional: Santiago, Chile, 2013.
33. INN. *Áridos para Morteros y Hormigones—Determinación de las Densidades real y Neta y de la Absorción de Agua de las Arenas. NCh1239.2009*; Instituto de Normalización Nacional: Santiago, Chile, 2009.
34. INN. *Áridos para Morteros y Hormigones—Tamizado y Determinación de la Granulometría. NCh165.2009*; Instituto de Normalización Nacional: Santiago, Chile, 2009.
35. INN. *Hormigón y Mortero–Agua de Amasado—Clasificación y Requisitos. NCh1498.2012*; Instituto de Normalización Nacional: Santiago, Chile, 2012.
36. Réunion Internationale des Laboratoires d'Essais et de Recherches Sur Les Matériaux et Les Construction (RILEM). *RILEM Technical Recommendations for the Testing and Use of Construction Materials*, 1st. ed.; CRC Press: London, UK, 1994.
37. INN. *Morteros—Determinación de la consistencia—Parte 3: Método del Asentamiento del cono. NCh2257/3.1996*; Instituto de Normalización Nacional: Santiago, Chile, 1996.
38. INN. *Cementos—Ensayo de Flexión y Compresión de Morteros de Cemento. NCh158.1967*; Instituto de Normalización Nacional: Santiago, Chile, 1967.
39. Zabaleta, H.; Egaña, J. *Manual del Mortero*; Instituto Chileno del Cemento y del Hormigón: Santiago, Chile, 1989.
40. Najim, K.B.; Hall, M.R. Crumb rubber aggregate coatings/pre-treatments and their effects on interfacial bonding, air entrapment and fracture toughness in self-compacting rubberised concrete (SCRC). *Mater. Struct.* **2013**, *46*, 2029–2043. [CrossRef]
41. Raffoul, S.; Garcia, R.; Escolano-Margarit, D.; Guadagnini, M.; Hajirasouliha, I.; Pilakoutas, K. Behaviour of unconfined and FRP-confined rubberised concrete in axial compression. *Constr. Build. Mater.* **2017**, *147*, 388–397. [CrossRef]
42. Onuaguluchi, O.; Panesar, D.K. Hardened properties of concrete mixtures containing pre-coated crumb rubber and silica fume. *J. Clean. Prod.* **2014**, *82*, 125–131. [CrossRef]
43. Li, J.; Zhou, Y.; Fan, F.; Du, F.; Yu, H. Controlling surface wettability and adhesive properties by laser marking approach. *Opt. Laser Technol.* **2019**, *115*, 160–165. [CrossRef]
44. Youssf, O.; Mills, J.E.; Benn, T.; Zhuge, Y.; Ma, X.; Roychand, R.; Gravina, R. Development of Crumb Rubber Concrete for Practical Application in the Residential Construction Sector–Design and Processing. *Constr. Build. Mater.* **2020**, *260*, 119813. [CrossRef]
45. Farfán, M.; Leonardo, E. Recycled rubber in the compressive strenght and bending of modified concrete with plasticizing admixtrue. *Rev. Ing. Constr.* **2019**, *33*, 241–250. [CrossRef]
46. Meesit, R.; Kaewunruen, S. Vibration characteristics of micro-engineered crumb rubber concrete for railway sleeper applications. *J. Adv. Concr. Technol.* **2017**, *15*, 55–66. [CrossRef]
47. Rashid, K.; Yazdanbakhsh, A.; Rehman, M.U. Sustainable selection of the concrete incorporating recycled tire aggregate to be used as medium to low strength material. *J. Clean. Prod.* **2019**, *224*, 396–410. [CrossRef]
48. Fraile-Garcia, E.; Ferreiro-Cabello, J.; Mendivil-Giro, M.; Vicente-Navarro, A.S. Thermal behaviour of hollow blocks and bricks made of concrete doped with waste tyre rubber. *Constr. Build. Mater.* **2018**, *176*, 193–200. [CrossRef]
49. Medina, N.F.; Garcia, R.; Hajirasouliha, I.; Pilakoutas, K.; Guadagnini, M.; Raffoul, S. Composites with recycled rubber aggregates: Properties and opportunities in construction. *Constr. Build. Mater.* **2018**, *188*, 884–897. [CrossRef]

MDPI

Article

Mechanical Performance and Microscopic Mechanism of Coastal Cemented Soil Modified by Iron Tailings and Nano Silica

Xinjiang Song [1], Haibo Xu [1], Deqin Zhou [2], Kai Yao [3], Feifei Tao [4], Ping Jiang [2] and Wei Wang [2,*]

[1] Department of Geotechnical Engineering, Anhui and Huaihe River Water Resources Research Institute, Bengbu 233000, China; sxj06@163.com (X.S.); hbxu2006@163.com (H.X.)
[2] School of Civil Engineering, Shaoxing University, Shaoxing 312000, China; zhoudeqinusx@163.com (D.Z.); jiangping@usx.edu.cn (P.J.)
[3] School of Qilu Transportation, Shandong University, Jinan 250002, China; yaokai@sdu.edu.cn
[4] School of Chemistry and Chemical Engineering, Shaoxing University, Shaoxing 312000, China; feifeitao@usx.edu.cn
* Correspondence: wellswang@usx.edu.cn

Abstract: In order to explore the effect of composite materials on the mechanical properties of coastal cement soil, cement soil samples with different iron tailings and nano silica contents were prepared, and unconfined compression and scanning electron microscope tests were carried out. The results show that: (1) The compressive strength of cement soil containing a small amount of iron tailings is improved, and the optimum content of iron tailings is 20%. (2) Nano silica can significantly improve the mechanical properties of iron tailings and cement soil (TCS). When the content of nano silica is 0.5%, 1.5%, and 2.5%, the unconfined compressive strength of nano silica- and iron tailings-modified cement soil (STCS) is 24%, 137%, and 323% higher than TCS, respectively. (3) Nano silica can promote the hydration reaction of cement and promote the cement hydration products to adhere to clay particles to form a relatively stable structure. At the same time, nano silica can fill the pores in TCS and improve the compactness of STCS.

Keywords: coastal cemented soil; nano silica; iron tailings; mechanical properties; microscopic mechanism

Citation: Song, X.; Xu, H.; Zhou, D.; Yao, K.; Tao, F.; Jiang, P.; Wang, W. Mechanical Performance and Microscopic Mechanism of Coastal Cemented Soil Modified by Iron Tailings and Nano Silica. *Crystals* **2021**, *11*, 1331. https://doi.org/10.3390/cryst11111331

Academic Editors: Shujun Zhang, Cesare Signorini, Antonella Sola, Sumit Chakraborty and Valentina Volpini

Received: 6 October 2021
Accepted: 28 October 2021
Published: 31 October 2021

1. Introduction

Cemented soil has the advantages of high compressive strength [1], low permeability, and low price. It has been widely used in projects such as soft soil foundation reinforcement, high-grade highway cushions, and seepage prevention in small reservoirs, among other projects, and has achieved good engineering benefits [2]. However, in practical engineering, it was found that coastal cemented soil (CS) cannot meet the mechanical requirements of some large-scale and special coastal engineering projects. For example, Liu et al. studied the heterogeneity in strength and Young's modulus of cement-admixed clay slabs using random finite-element analyses, considering three sources of variation: namely, a deterministic radial trend in strength and Young's modulus; a stochastic fluctuation component due to non-uniform mixing; and positioning errors arising from off-verticality of the mixing shafts [3]. Li et al. studied the influence of the number of freeze–thaw cycles and curing ages on the mechanical properties of ordinary cemented clay and polypropylene fiber-cemented clay [4]. Liu et al. examined the interaction between the spatial variations in binder concentration and in situ water content, in cement-mixed soil, using field and model data as well as statistical analysis and random field simulation [5]. The results of these studies suggest that it is necessary to add additives to modify the coastal cemented soil. On the other hand, as an industrial by-product, the annual output of iron tailings in China is huge. The accumulation of iron tailings not only takes up farmland and pollutes the environment, but also tends to cause dangerous accidents such as collapses with the increasing of the accumulation height. The use of iron tailings to make composite cemented soil, and at the same time adulterating new admixtures to meet the requirements of engineering mechanics, can not only reduce environmental pollution [6], but also save building materials so as to achieve the dual benefits of economic efficiency and environmental protection, which is a very effective way to achieve sustainable development [7].

Extensive research studies on TCS have been conducted by scholars at home and abroad. Lucas et al. compared the effects of cement, lime, and slag composite iron tailings as roadbed fillers, and found that cement was the most effective stabilizer of iron tailings through unconfined compressive tests [8]. Bastos et al. found through CBR tests that cement composite iron tailings had satisfactory physical and mechanical properties, and thus, were suitable for road bases under most traffic loads with good durability [9]. Through compressive and flexural tests, Hou Rui et al. found through compressive and flexural tests that the compressive and flexural strengths of TCS increased gradually with age, but increased slowly in the later stage. When the iron tailing content was less than 25%, the mechanical properties of TCS were slightly increased compared with that of CS [10].

In recent years, many scholars have tried to develop new engineering materials by compounding various curing agents with iron tailings. Generally, cement can be used to improve the mechanical properties of iron tailings. Tanko et al. studied the performance of cement concrete mixed with fine aggregate of iron tailings [11]. Bao et al. studied the toughening properties of sand cement-based composites of high-performance environmentally friendly tailings [12]. Simonsen et al. studied mine tailings' potential as supplementary cementitious materials based on their chemical, mineralogical and physical characteristics [13]. Long et al. carried out research on the mechanical properties and durability of coal gangue-reinforced cement–soil mixture for foundation treatments [14]. On this basis, lime, slag, fiber, etc. can be added to further improve the mechanical properties of cement-modified iron tailings. Consoli et al. studied fiber-reinforced sand-coal fly ash-lime-NaCl blends under severe environmental conditions [15]. Feng et al. studied lime-and cement-treated sandy lean clay for highway subgrade in China [16]. Tebogo et al. studied mechanical chemically treated and lime-stabilized gold mine tailings using unconfined compressive strength [17]. Kumar et al. studied the use of iron ore tailings, slag sand, ground granular blast furnace slag, and fly ash to produce geopolymer bricks [18]. Falayi et al. conducted a comparative study on the mechanical properties of geological polymers of gold mine tailings modified by fly ash and alkaline oxygen slag [19]. Festugato et al. studied the cyclic shear behavior of fiber-reinforced mine tailings [20]. Chen et al. studied the compressive behavior and microstructural properties of tailings polypropylene fiber-reinforced cemented paste backfill [21]. Cristelo et al. studied the effect of fiber on the cracking behavior of cement-stabilized sandy clay under indirect tensile stress [22]. Lirer et al. studied the strength of fiber-reinforced soils [23]. Among them, cement is the traditional building material with the best modification effect, but it still cannot fully meet the engineering needs. Therefore, it is urgent to find a composite admixture to improve the engineering characteristics of TCS. Nanomaterials have the characteristics of large specific surface area, small particles, and high activity. Therefore, they have been introduced into the engineering exploration of TCS. For example, Ghasabkolaei et al. summarized the geotechnical properties of soil modified by nanomaterials [24]. Wang et al. studied characterization of nano magnesia–cement-reinforced seashore soft soil by direct-shear test [25]. Yao et al. studied effect of nano-MgO on the mechanical performance of cement-stabilized silty clay [26]. It has been found through various studies that nanomaterials can fully exert their own activity in cement-based materials, promote the process of cement hydration reaction, and fully fill the internal pores of cement-based materials at the microscopic scale, thus achieving the purpose of improving the engineering properties of cement-based materials. For example, Zheng et al. studied strength and hydration products of cemented paste backfill from sulfide-rich tailings using reactive MgO-activated slag as a binder [27]. Liu et al. studied the effect of graphite tailings as a substitute for sand on the mechanical properties of concrete [28]. Sarkkinen et al. studied the efficiency of MgO-activated GGBFS and OPC in the stabilization of highly sulfidic mine tailings [29]. Through an indoor unconfined compressive test combined with a microscopic test, Li et al. found that the use of nano clay instead of cement in iron tailings can effectively improve its compressive strength. The addition of nano clay makes the microstructure surface of TCS change from flake to block; the granular structure becomes tightly cemented, and the whole structure tends to be stable [30]. Nano clay can improve the mechanical properties of TCS and can guide the resource application of iron tailings to a certain extent. It is feasible to apply nanomaterials to engineering practice. Due to its unique physical properties, nano silica has also recently been widely used in geotechnical engineering, for example, in modified calcareous sand in the South China Sea [31] and in stabilized coastal silty clay [32].

Therefore, the research on STCS has important theoretical significance and engineering application value, and microstructure has a certain influence on the mechanical properties of materials [33]. In this paper, the optimal iron tailing content was explored through unconfined compressive tests, the modification effect of nano silica on TCS was studied, and the strength growth mechanism of STCS was analyzed in combination with microscopic tests.

2. Test Overview

2.1. Test Materials

The soil used in this experiment was collected from the coastal soft subgrade soil excavated in a certain expressway section in Shaoxing city, Zhejiang. The physical and mechanical indexes are shown in Table 1, and the microscopic characteristics are shown in Figure 1. The main chemical elements of the soil are Si, O, Mg, and Al. The natural state of the soil is a soft plastic state, with high porosity and organic matter content. The iron tailings were taken from Lizhu iron tailings pond in Zhejiang Province. The physical and mechanical indexes are shown in Table 2, and the microscopic characteristics are shown in Figure 2. The main chemical elements of iron tailings are Si, O, Mg, and C, and their particles are small and single. PO 32.5 Conch brand Portland cement, produced in Shangyu, Shaoxing, was used in this test. The physical and mechanical indexes are shown in Table 3, and the microscopic characteristics are shown in Figure 3. The cement's performance meets requirements and it can be used to modify soft soil. Ordinary industrial nano silica was used in this test. The physical and mechanical indexes are shown in Table 4, and the microscopic characteristics are shown in Figure 4. The nano silica used in the test is pure, its microstructure is dense, and it has a large specific surface area.

Table 1. Physical and mechanical indexes of coastal soft soil.

Natural Moisture Content %	Volume Weight g/cm^3	Void Ratio	Saturation Degree %	Plastic Limit ω_P%	Liquid Limit ω_L%	Plasticity Index I_P	Liquidity Index I_l	Organic Content (%)
58	1.63	1.74	98	25	47	22	1.34	6.5

Table 2. Physical and mechanical indexes of iron tailings.

Index	Loss on Ignition (%)	Specific Gravity	Specific Area (m^2/kg)	Liquid Limit (%)	Plastic Limit (%)
Value	7.01	3.06	379	23	17

(**a**) (**b**)

Figure 1. (**a**) SEM, (**b**) EDS. Microscopic characteristics of coastal soft soil.

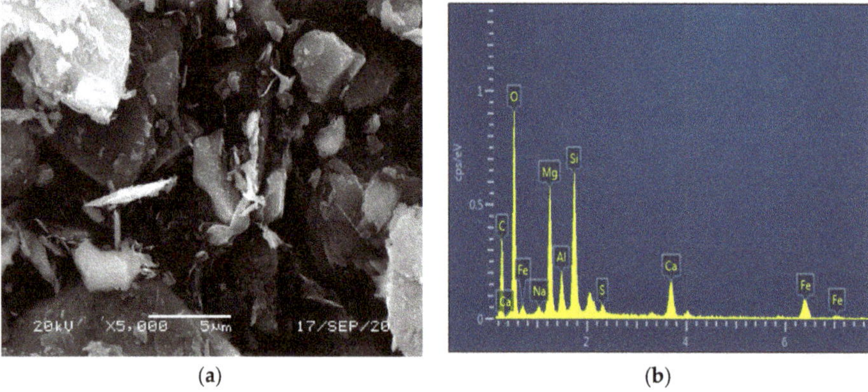

(**a**) (**b**)

Figure 2. (**a**) SEM, (**b**) EDS. Microscopic characteristics of iron tailings.

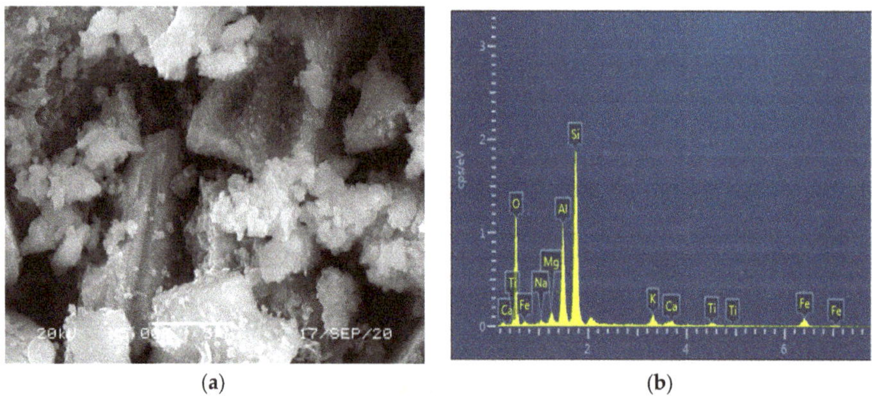

(**a**) (**b**)

Figure 3. (**a**) SEM, (**b**) EDS. Micro characteristics of Portland cement.

Table 3. Physical and mechanical indexes of cement.

Index	Fineness (%)	Initial Setting Time (min)	Final Setting Time (min)	Loss on Ignition (%)	28d Compressive Strength (MPa)	28d Flexural Strength (MPa)
Value	3.5	210	295	1.3	47.0	8.2

Table 4. Physical and mechanical indexes of nano silica.

Index	Loss on Ignition (%)	Bulk Density g/cm^3	Specific Area (m^2/kg)	pH Value
Value	1	3.06	200,000	3.4~4.7

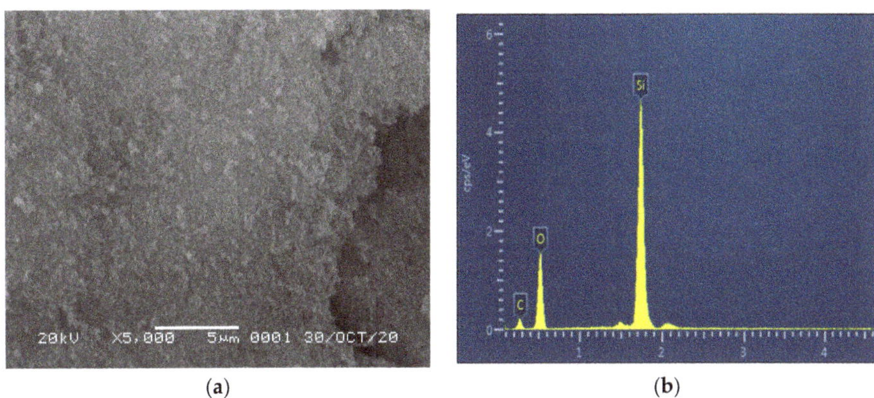

(**a**) (**b**)

Figure 4. (**a**) SEM, (**b**) EDS. Microscopic characteristics of nano silica.

2.2. Test Scheme

According to the application scenarios of cemented soil in actual projects, the moisture content of soil was set at 80%. The test was divided into two parts. First, the mechanical properties of TCS were studied, and the optimal iron tailing content was obtained. The cement content was set at 30% and the iron tailing content was set at 0, 10%, 20%, 30%, and 40%, respectively, to produce the TCS samples. Next, on the basis of the optimal iron tailing content, nano silica at contents of 0.5%, 1.5%, and 2.5% was added continuously to investigate the modification effect of nano silica on TCS. The content of cement, iron tailings and nano silica refers to the mass ratio in relation to dry soil. The test ages of all samples were 7d. The material composition and specific symbols of the samples are shown in Figure 5, and the specific test scheme is shown in Table 5.

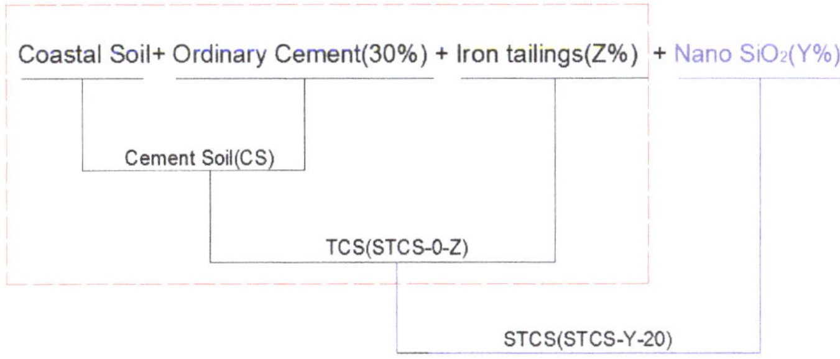

Figure 5. Material composition and specific symbols of the sample.

Table 5. Test scheme for STCS.

Sample Group	Cement Content/%	Iron Tailing Content/%	Nano Silica Content/%	Curing Age/d
STCS-0-0	30	0	0	7
STCS-0-10	30	10	0	7
STCS-0-20	30	20	0	7
STCS-0-30	30	30	0	7
STCS-0-40	30	40	0	7
STCS-0.5-X	30	Optimal content	0.5	7
STCS-1.5-X	30	Optimal content	1.5	7

Note: In STCS-Y-Z, Y represents the content of nano silica, Z represents the content of iron tailings, and the X represents the optimal content.

2.3. Sample Preparation and Maintenance

According to the requirements of the Chinese "GBT 50123-2019" standard, the main process of the sample preparation is shown in Figure 6.

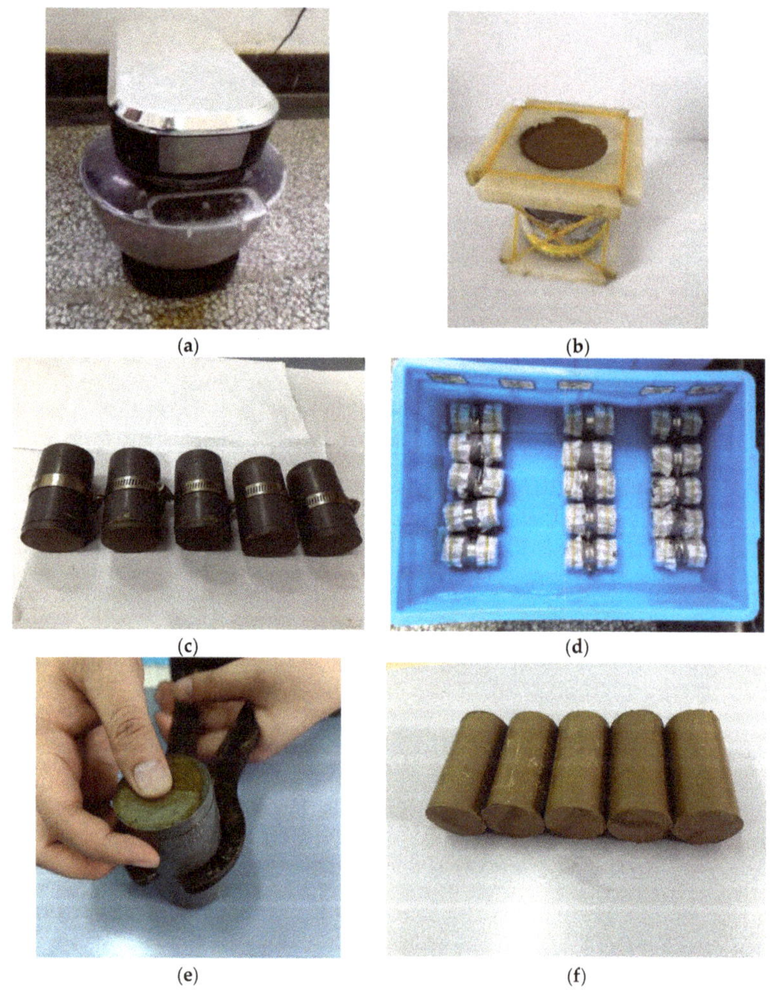

Figure 6. (**a**–**f**) Main processes of sample preparation.

(1) An appropriate amount of wet soil was placed into the mixer, and mixed well; the cement slurry was prepared according to the test mix ratio, then poured into the mixer and mixed evenly, as shown in Figure 6a.

(2) Iron tailings and nano silica were added into the mixer in turn, and were stirred twice, for 4 minutes each time, to ensure that the test materials were fully and evenly mixed. The mixed materials that had been stirred were put into a grouting bag, and were then added to the standard mold three times to ensure the same height each time. The diameter of the mold was 39.1mm, and the height was 80mm. After each feeding, manual vibration was required for about 40 times until it was dense, as shown in Figure 6b.

(3) The sample was left to stand for about 2h. After it was initially set, both ends of the molds were removed and smoothed with a spatula, as shown in Figure 6c. The metal clip outside the test mold was mainly used to apply tightening force to the test mold to ensure the forming of the sample. When demolding the sample, the metal clip was first removed.

(4) Filter paper and rubber bands were used to bind both ends of the sample, and then put in water for curing, as shown in Figure 6d.

(5) After reaching 7d, the sample was removed from the mold using the special demolding tool, as shown in Figure 6e. Then, the sample was placed in a safe place and prepared for subsequent tests, as shown in Figure 6f.

3. Unconfined Compressive Test Analysis of TCS

3.1. Stress–Strain Curve Analysis

Five repeated tests were carried out for each group of samples, and five stress–strain curves were obtained. The stress–strain curves of TCS with different iron tailing contents are summarized in Figure 7, which are all softening curves.

3.2. Curve Normalization

Due to the contingency and error in the unconfined compressive test, based on the research results of Long Hongbo et al. [34], the deviation between the peak points of different stress–strain curves was taken as the research object, and an improved weighted average method was proposed to optimize the curves of five repeated tests by using the weight of each peak point. The detailed calculation steps are as follows:

(1) Determining the standard value. First, the peak point is mean processed to obtain a set of standard values, as shown in Formula (1). N represents the number of tests, in this test, $N = 5$; $i \in [1, N]$ is the peak stress of each curve; q^- is the mean stress.

$$q^- = \frac{1}{N}\sum_{i=1}^{N} q_i \tag{1}$$

(2) Determining the deviation. The peak stress of each peak point is subtracted from the standard stress, and its absolute value is deviation p_i, as shown in Formula (2).

$$p_i = |q_i - q^-| \tag{2}$$

(3) Determining the degree of deviation. The variance M is introduced to describe the degree of deviation between each peak stress and the standard stress, as shown in Formula (3).

$$M = \frac{1}{N}\sum_{i=1}^{N} (q_i - q^-)^2 \tag{3}$$

(4) Selecting the weighted weight. In order to determine the weight of each peak stress, each deviation value is divided by the variance M, respectively, to obtain the weighted weight W of each peak stress, as shown in Formula (4).

$$W = \frac{p_i}{M} \tag{4}$$

(5) Determining the weight function. The weighted weight W of each peak stress and standard stress can be obtained from Formula (4). However, this weight does not conform to the traditional weighting law, which needs to be transformed to a certain extent. Combined with the above analysis, the use of a weight function to transform the initial weight is proposed, as shown in Formula (5).

$$C(x) = \frac{1}{2}\cos^{N}(\pi x + 1) \quad (5)$$

where x is the independent variable and $C(x)$ is the dependent variable; N can be valued according to the actual situation, and its function is to improve the accuracy of the calculation results; in this test, $N = 5$.

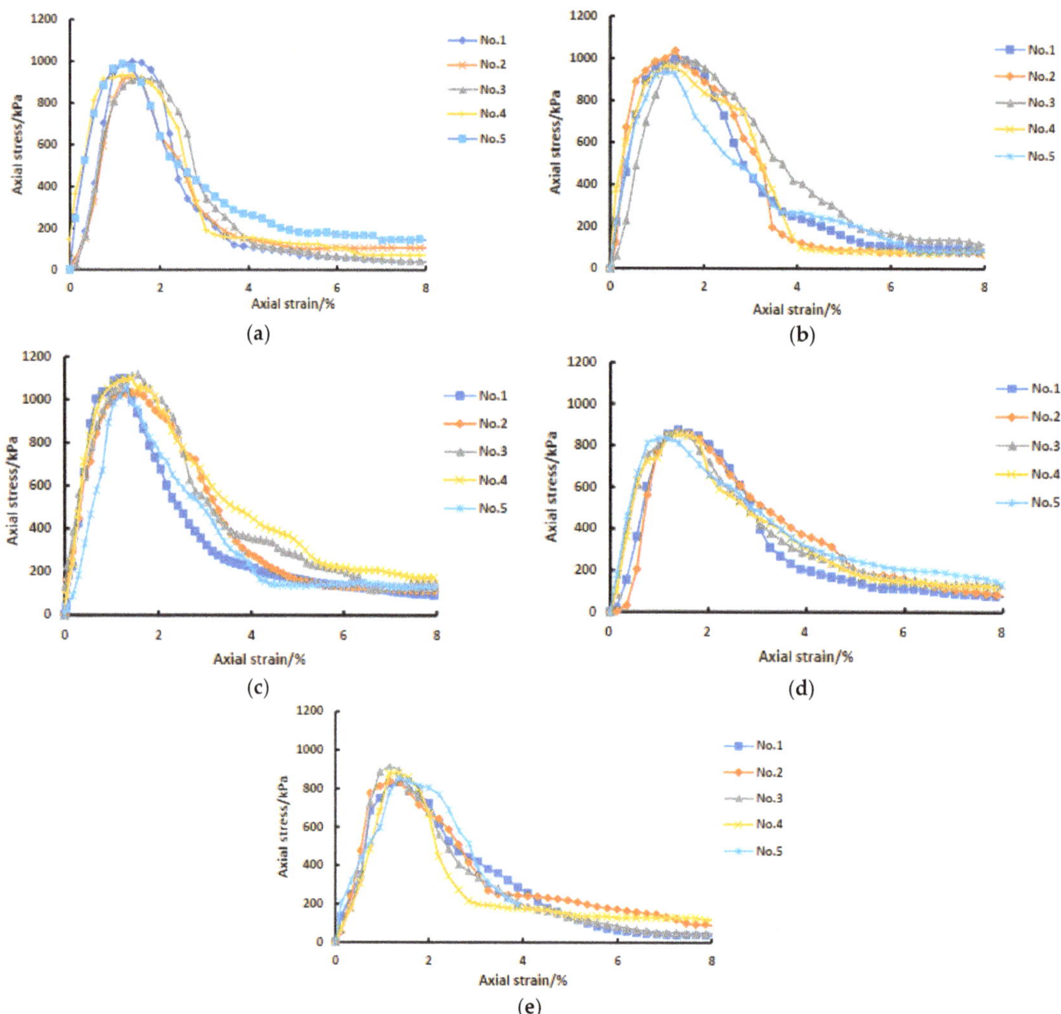

Figure 7. Stress–strain curves of TCS with different iron tailing contents. (**a**) STCS-0-0; (**b**) STCS-0-10; (**c**) STCS-0-20; (**d**) STCS-0-30; (**e**) STCS-0-40.

1. Determining the weight. In order to determine whether Formula (5) is feasible, Formula (4) is substituted into the weight function $G(x)$ to obtain the final weight R, as shown in Formula (6).

$$R = C(W) = \frac{1}{2}\cos^{N}(\pi W + 1) \tag{6}$$

2. Determining the weighting factor. In order to obtain the weighting factor of each peak stress, the weights obtained from Formula (6) are divided by the weight sums, as shown in Formula (7).

$$Y_i = \frac{R}{\sum_{i=1}^{N} R} \tag{7}$$

3. Determining the weighted stress. Each weighting factor obtained in Formula (7) is multiplied by the corresponding peak stress, and the value of each peak stress in the weighted stress can be obtained. The final weighted peak stress q_m can be obtained by adding them together, as shown in Formula (8).

$$q_m = \sum_{i=1}^{N} q_i Y_i \tag{8}$$

It can be seen from above that there were five stress–strain curves for each group of samples, the peak points of the five curves of STCS-0-0 were taken as reference to perform a weighted average, and the obtained specific gravity was multiplied by the standard curve to obtain the q-ε representative curve of the test. In order to verify the applicability of this method, the calculation results were compared with the original curve, and the comparison results are shown in Figure 8. The results show that the representative q-ε curve obtained by the new method has a good correlation with the original q-ε curve.

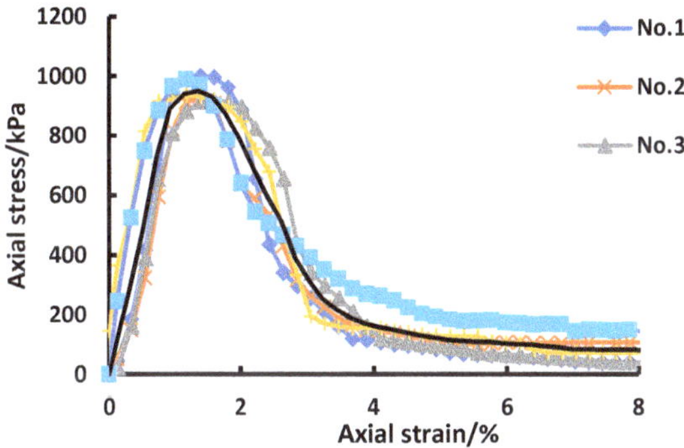

Figure 8. q-ε curve of STCS-0-0.

3.3. Peak Strength and Peak Strain Analysis

The peak strength is the maximum stress value on the stress–strain curve of the unconfined compressive test of the soil. The peak strain is the strain corresponding to the peak strength. According to the method in Section 3.2, the peak strength and peak strain of various types of cemented soil are summarized in Table 6. Table 6 shows that the peak strengths of STCS-0-0, STCS-0-10, and STCS-0-20 were 955kPa, 986kPa, and 1080kPa, respectively. When the iron tailing content was 10% and 20%, the peak strength of TCS was 3% and 13% higher than that of CS, respectively. When the iron tailing content continued to increase, the unconfined compressive strength of TCS began to decrease, and the iron tailings began to show a deterioration effect, which gradually increased with the increase in the iron tailing content. It can be found from Hou Rui's research that when the iron tailing content was less than 25%, iron tailings had a certain improvement effect on the compressive strength of CS, which was consistent with the results in this paper [35].

Table 6. Peak strength and peak strain of each group of TCS.

Sample No.	Peak Strength (kPa)	Average Error of Peak Strength (kPa)	Peak Strain (%)	Average Error of Peak Strain (%)
STCS-0-0	955	35	1.346	0.135
STCS-0-10	986	27	1.348	0.135
STCS-0-20	1080	29	1.379	0.082
STCS-0-30	857	12	1.347	0.132
STCS-0-40	863	26	1.303	0.102

The peak strain of TCS with different iron tailing contents fluctuated between 1.303 and 1.379. When the iron tailing content was 20%, the peak strain increased the most, but only by 2%. The addition of iron tailings had little effect on the peak strain of CS.

Iron tailings can be categorized as a high-silicon type ultra-fine tailing sand. Their particle size is larger than that of clay particles, and thus, they act as a coarse aggregate in soil. The addition of iron tailings in small quantities can improve the gradation of soil particles and reduce the internal pores of the soil. Under the action of cementitious substances formed by cement hydration and consolidation, iron tailings and soil particles clump together to form a whole structure, so as to improve the soil's compressive strength. With the increase in the iron tailing content, the proportion of iron tailings in the composite cemented soil increases, the properties of the composite material begin to approach the sand soil, the internal structure of the soil begins to loosen, the cohesion between particles decreases, and the compressive strength of the soil decreases.

3.4. Residual Strength Analysis

According to "GBT 50123-2019", the sample is defined as damaged when the strain reaches the peak strain in the unconfined compressive stage, and the stress measured at 5% strain, after the selection of the peak strain, is selected as the residual strength. The residual strength of TCS is summarized in Table 7.

Table 7. Residual strength of each group of TCS.

Sample No.	Residual Strength (kPa)	Average Error of Residual Strength (kPa)
STCS-0-0	82	33
STCS-0-10	93	20
STCS-0-20	132	27
STCS-0-30	110	14
STCS-0-40	89	42

It can be seen from Table 7 that iron tailings have a greater impact on the residual strength of TCS. When the iron tailing content is 20%, the residual strength of TCS reaches its peak, which is increased by 60% compared with that of CS. When the iron tailing content continues to increase, the residual strength of TCS begins to decrease, and its mechanism of action is similar to the peak strength.

3.5. Elastic Modulus Analysis

The elastic moduli of TCS are summarized in Figure 9.

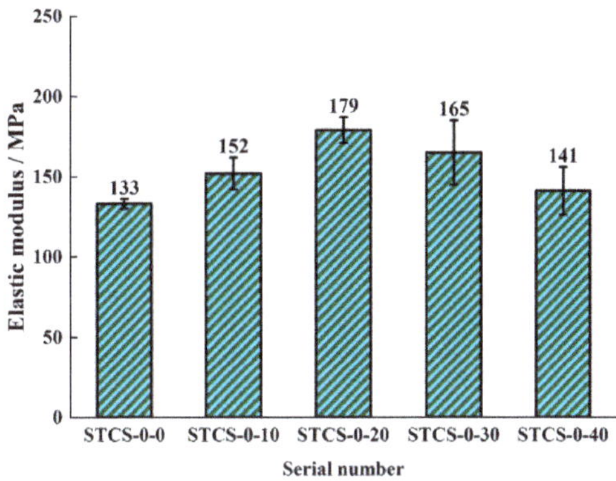

Figure 9. Elastic moduli of TCS.

It can be seen from Figure 9 that after adding 10%, 20%, 30%, and 40% iron tailings, the elastic modulus of TCS increased by 14%, 35%, 24%, and 6%, respectively. When the iron tailing content was 20%, the elastic modulus increased the most.

4. Unconfined Compressive Test Analysis of STCS

4.1. Stress–Strain Curve Analysis

Under the condition that the optimal iron tailing content was 20%, 0.5%, 1.5%, and 2.5% quantities nano silica content were added to modify TCS. The stress–strain curves of STCS with different nano silica contents are summarized in Figure 10, which are all softening curves.

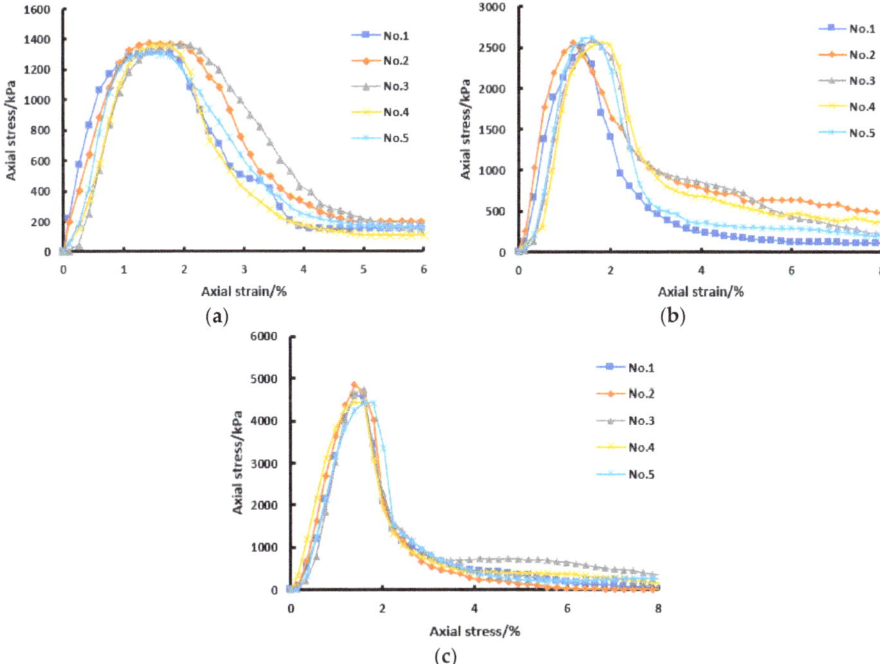

Figure 10. Stress–strain curve of STCS. (**a**) STCS-0.5-20; (**b**) STCS-1.5-20; (**c**) STCS-2.5-20.

4.2. Peak Strength and Peak Strain Analysis

The peak strength and peak strain of STCS are summarized in Figures 11 and 12.

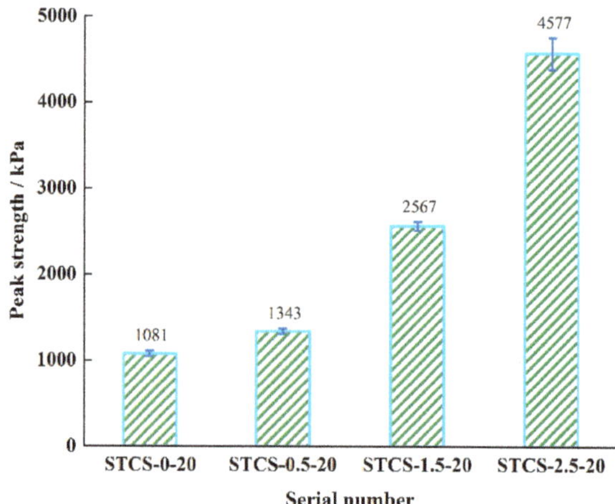

Figure 11. Peak strength of STCS.

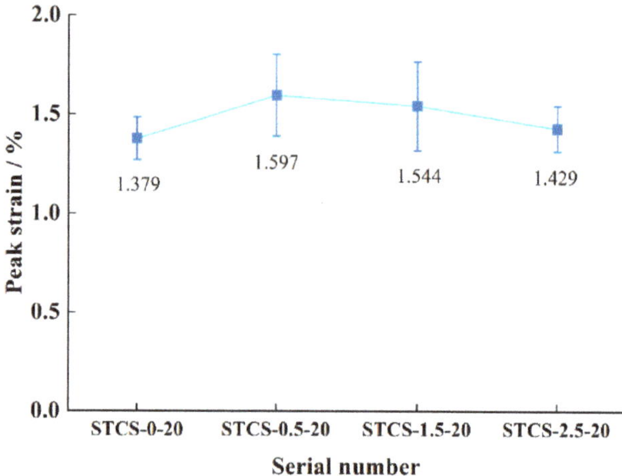

Figure 12. Peak strain of STCS.

It can be seen from Figure 11 that after adding 0.5%, 1.5%, and 2.5% nano silica, the compressive strength of TCS increased by 24%, 137%, and 323%, respectively, which significantly enhanced the compressive strength of TCS.

The peak strain represents the strain variable when the sample reaches the peak stress. The larger the peak strain is, the later the sample reaches failure, which is of great significance in engineering applications. It can be seen from Figure 12 that nano silica increased the peak strain of TCS, but with the increase in the nano silica content, the increment gradually decreased. When the nano silica content was 0.5%, the peak strain of TCS increased by 15%, which indicates that nano silica can help to delay the time of sample failure, but the effect is limited. Because with the increase in nano silica content, the brittleness of the material was also increasing, the increment of peak strain of TCS began to decrease.

4.3. Residual Strength Analysis

The residual strength of STCS is summarized in Table 8.

Table 8. Residual strength of each group of STCS.

Sample No.	Residual Strength (kPa)	Average Error of Residual Strength (kPa)
STCS-0-20	132	27
STCS-0.5-20	152	23
STCS-1.5-20	277	128
STCS-2.5-20	174	144

Table 8 shows that the addition of nano silica effectively improves the residual strength of TCS. When the nano silica content was 0.5%, the residual strength increased by 20kPa; when the nano silica content was 1.5%, the residual strength increased by 142kPa; but when the nano silica content reached 2.5%, the residual strength increased by only 42kPa. The increment of the residual strength of TCS by adding nano silica shows a trend of first increasing and then decreasing, and of reaching the maximum when the nano silica content is 1.5%, which is consistent with the increasing of the peak strength of STCS.

4.4. Elastic Modulus Analysis

The elastic modulus of STCS are summarized in Figure 13.

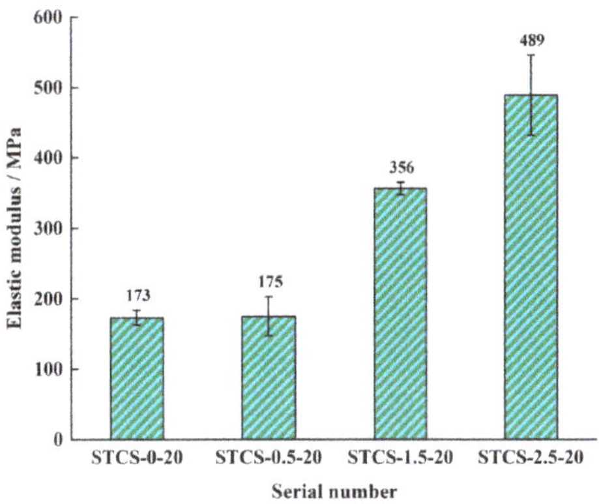

Figure 13. Elastic modulus of STCS.

It can be seen from Figure 13 that after adding 0.5%, 1.5%, and 2.5% nano silica, the elastic modulus of STCS increased by 1%, 106% and 183%, respectively. which significantly enhanced the elastic modulus of TCS.

5. Microscopic Mechanism Analysis

5.1. SEM Test Analysis

In order to better analyze the strength growth mechanism of STCS, the microscopic morphology of various cement soils was observed by SEM. Four groups of unconfined samples of STCS-0-0, STCS-0-20, STCS-0-40, and STCS-2.5-20 were selected for microscopic testing. Via SEM scanning with an electron microscope, four groups of microstructure photos of STCS were obtained at 5000 times magnification, as shown in Figure 14.

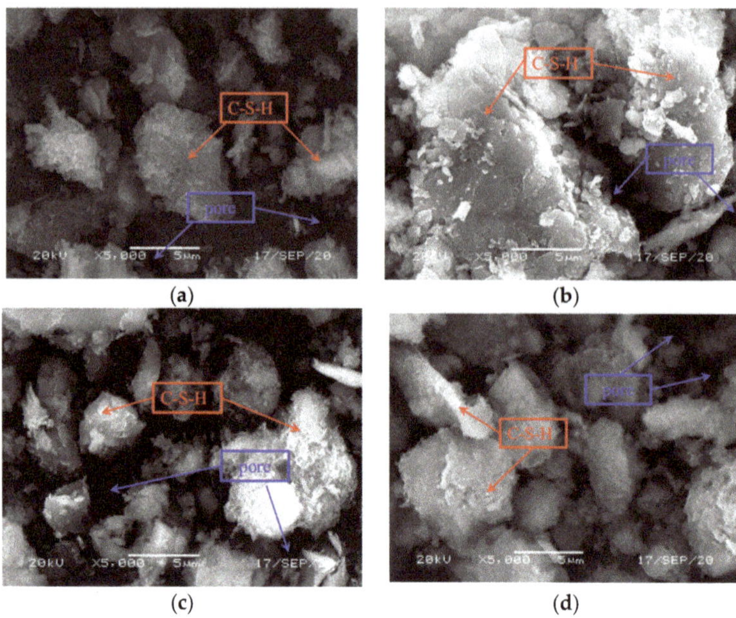

Figure 14. SEM images of different types of STCS. (**a**) STCS-0-0; (**b**) STCS-0-20; (**c**) STCS-0-40; (**d**) STCS-2.5-20.

Figure 14 shows that a large amount of hydration products were produced in each group of samples, including a relatively large amount of flocculated colloid C-S-H, which has strong adsorption capacity. The internal particles of CS are of different sizes and there are many pores. When 20% iron tailings was added, it can be seen from the microscopic image that there were particles of different sizes in the composite cemented soil, forming a good gradation. Some iron tailings particles filled the large particles in the soil, and a large amount of flocculated colloid C-S-H was adsorbed on the soil particles, forming a stable whole structure with iron tailings. The composite cemented soil mixed with 40% iron tailing content was mostly composed of medium and small units, and the gaps between the particles were large and not well filled.

The characteristics of nanomaterials effectively played their role in the improvement of TCS with 2.5% nano silica content. It can be seen from the microscopic pictures of STCS-2.5-20 that the structure was very compact, and the nano silica adequately filled the pores of the composite cemented soil.

5.2. SEM Image Processing

The basic properties of CS depend largely on its particle bonding characteristics at the microscopic level, and the macro-mechanical performance of CS largely depends on the pore size at the microscopic level. In order to quantitatively evaluate the relationship between the compressive strength of CS and the size of microscopic pores, the SEM image obtained at 5000 times magnification was binarized to obtain the quantitative pore size of relevant samples, as shown in Table 9.

Table 9. Porosities of different STCS.

Sample No.	**STCS-0-0**	**STCS-0-20**	**STCS-0-40**	**STCS-2.5-20**
Porosity	0.334	0.225	0.448	0.098

Table 9 shows that the porosity of the four types of STCS samples, from high to low, was STCS-0-40 > STCS-0-0 > STCS-0-20 > STCS-2.5-20. This is consistent with the order of the unconfined compressive strength of each group of STCS samples.

Compared with STCS-0-0, the porosity of STCS-0-20 was reduced by 32%, which shows that the compressive strength of STCS with 20% iron tailing content was improved compared with that of CS. The porosity of STCS-0-40 was higher than that of STCS-0-0, which was because the particle size of iron tailings was larger than that of clay particles; the proportion of fine particles in the mixture decreased due to the increasing of the iron tailing content, which was not enough to fill the pores between the cement and the soil. As a result, the fine aggregate particle gradation became worse after the iron tailing content exceeded 20%. Compared with STCS-0-20, the porosity of STCS-2.5-20 was reduced by 56%. It can be seen that the internal pores of STCS were very small and the structure was compact. Therefore, the unconfined compressive strength of STCS with 2.5% nano silica content reached more than three times that of TCS.

6. Conclusions

The mechanical properties and micro mechanism of TCS and STCs at 7d curing age were studied by unconfined compressive strength test and SEM test. The following conclusions can be drawn.

(1) With the increase in the content of iron tailings, the unconfined compressive strength, peak strain, residual strength, and elastic modulus of TCS first increased and then decreased. The optimal content of iron tailings was 20%.

(2) Nano silica could significantly modify the unconfined compressive strength and elastic modulus of TCS. The unconfined compressive strength of STCS increased with the increasing of the nano silica content. Compared with STCS-0-20, the unconfined compressive strength of STCS-0.5-20, STCS-1.5-20, and STCS-2.5-20 was increased by 24%, 137%, and 323% respectively. At the same time, the addition of nano silica effectively improved the peak strain and residual strength of the TCS. With the increasing of the nano silica content, the peak strain, residual strength and elastic modulus first increased and then decreased.

(3) Analysis of SEM pictures showed that the flocculation colloid C-S-H generated by cement hydration and the porosity of the samples were the main factors affecting the strength of TCS and STCS. An appropriate amount of iron tailings can fill the pores in TCS. With the increase of iron tailings, the porosity of TCS increased, resulting in the decreasing of the strength. In addition, on the one hand, nano silica can promote the hydration reaction of cement and improve the cementation of particles in STCS. On the other hand, nano silica can fill the pores between particles and improve the compactness of STCS.

The above conclusions are based on the test data of 7d curing age. The strength of TCS and STCs will gradually increase with the curing age, and final strength of the composite may be achieved after 2–3 months.

Author Contributions: Conceptualization, X.S. and H.X.; investigation, D.Z. and K.Y.; formal analysis, F.T. and P.J.; writing—review and editing, W.W. All authors have read and agreed to the published version of the manuscript.

Funding: This research was funded by the National Natural Science Foundation of China, grant number 41772311, and the Shandong Provincial Natural Science Foundation, grant number ZR202102240826.

Data Availability Statement: The data presented in this study are available on request from the corresponding author.

Acknowledgments: This authors thank Linxia Wang from the Micro Testing Center of Shaoxing University for her help in the process of micro testing.

Conflicts of Interest: The authors declare no conflict of interest.

References

1. Aline, C.; Gustavo, C.X.; Jonas, A.; Leonardo, G.P.; Afonso, R.G.A.; Carlos, M.F.V.; Sergio, N.M. Environmental durability of soil-cement block incorporated with ornamental stone waste. In *Materials Science Forum*; Trans Tech Publications Ltd.: Kapellweg, Switzerland, 2014; Volume 798, pp. 548–553.
2. Datta, S.G.; Roy, D. Influence of flow velocity and nutrient availability on microbially mediated reduction of erosion suscep-tibility of sand. In *Geotechnics for Sustainable Infrastructure Development*; Springer: Singapore, 2020; pp. 945–950.
3. Liu, Y.; Lee, F.H.; Quek, S.T.; Chen, E.J.; Yi, J.T. Effect of spatial variation of strength and modulus on the lateral compression response of cement-admixed clay slab. *Géotechnique* **2015**, *65*, 851–865. [CrossRef]
4. Li, N.; Zhu, Y.; Zhang, F.; Lim, S.M.; Wu, W.; Wang, W. Unconfined compressive properties of fiber-stabilized coastal cement clay subjected to Freeze–Thaw cycles. *J. Mar. Sci. Eng.* **2021**, *9*, 143. [CrossRef]
5. Liu, Y.; He, L.Q.; Jiang, Y.J.; Sun, M.M.; Chen, E.J.; Lee, F.-H. Effect of in situ water content variation on the spatial variation of strength of deep cement-mixed clay. *Géotechnique* **2019**, *69*, 391–405. [CrossRef]
6. Rybak, J.; Adigamov, A.; Kongar-Syuryun, C.; Khayrutdinov, M.; Tyulyaeva, Y. Renewable-Resource technologies in mining and metallurgical enterprises providing environmental safety. *Minerals* **2021**, *11*, 1145. [CrossRef]
7. Rybak, J.; Gorbatyuk, S.M.; Bujanovna-Syuryun, K.C.; Khairutdinov, A.M.; Tyulyaeva, Y.S.; Makarov, P.S. Utilization of mineral waste: A method for expanding the mineral resource base of a mining and smelting company. *Metallurgist* **2021**, *64*, 851–861. [CrossRef]
8. Muguda, S.; Lucas, G.; Hughes, P.N.; Augarde, C.E.; Perlot, C.; Bruno, A.W.; Gallipoli, D. Durability and hygroscopic behaviour of biopolymer stabilised earthen construction materials. *Constr. Build. Mater.* **2020**, *259*, 119725. [CrossRef]
9. Bastos, L.A.D.C.; Silva, G.C.; Mendes, J.C.; Peixoto, R.A.F. Using iron ore tailings from tailing dams as road material. *J. Mater. Civ. Eng.* **2016**, *28*, 04016102. [CrossRef]
10. Hou, R.; Chen, S.l.; Ma, X.; Hu, Y.T. Study on mechanical properties of iron tailings sand-cement composite soil. *Sino-Foreign Highw.* **2019**, *39*, 206–209.
11. Tang, K.; Mao, X.S.; Xu, W.; Tang, X.L. Performance analysis of cement concrete mixed with fine aggregate of iron tailings. *Ind. Archit.* **2019**, *8*, 153–157.
12. Bao, W.B.; Di, G.H.; Chen, S.l.; Li, W.F. Study on toughening properties of high performance environmental friendly tailings sand cement-based composites. *Funct. Mater.* **2016**, *47*, 7–12. [CrossRef]
13. Simonsen, A.M.T.; Solismaa, S.; Hansen, H.K.; Jensen, P.E. Evaluation of mine tailings' potential as supplementary cementi-tious materials based on chemical, mineralogical and physical characteristics. *Waste Manag.* **2020**, *102*, 710–721. [CrossRef]
14. Long, G.; Li, L.; Li, W.; Ma, K.; Zhou, K.L. Enhanced mechanical properties and durability of coal gangue reinforced cement-soil mixture for foundation treatments. *J. Clean. Prod.* **2019**, *231*, 468–482. [CrossRef]
15. Consoli, N.C.; Godoy, V.B.; Tomasi, L.E.; Paula, T.M.D.; Bortolotto, M.S.; Favretto, E. Fibre-reinforced sand-coal fly ashlime-NaCl blends under severe environmental conditions. *Geosynth. Int.* **2019**, *26*, 525–538. [CrossRef]
16. Feng, R.; Wu, L.; Liu, D.; Wang, Y.; Peng, B. Lime- and Cement-Treated sandy lean clay for highway subgrade in China. *J. Mater. Civ. Eng.* **2020**, *32*, 04019335. [CrossRef]
17. Tebogo, M.; Thandiwe, S. Evaluation of chemically treated and lime stabilized gold mine tailings: Effect on unconfined compressive strength. *Key Eng. Mater.* **2019**, *803*, 366–370. [CrossRef]
18. Kumar, R.; Das, P.; Beulah, M.; Arjun, H.R. Geopolymer Bricks using iron ore tailings, slag sand, ground granular blast furnace slag and fly ash. In *Geopolymers and Other Geosynthetics*; IntechOpen: London, UK, 2020.
19. Falayi, T. A comparison between fly ash and basic oxygen furnace slag-modified gold mine tailings geopolymers. *Int. J. Energy Environ. Eng.* **2019**, *11*, 207–217. [CrossRef]
20. Festugato, L.; Consoli, N.; Fourie, A. Cyclic shear behaviour of fibre-reinforced mine tailings. *Geosynth. Int.* **2015**, *22*, 196–206. [CrossRef]
21. Chen, X.; Shi, X.; Zhou, J.; Chen, Q.; Li, E.; Du, X. Compressive behavior and microstructural properties of tailings polypropylene fibre-reinforced cemented paste backfill. *Constr. Build. Mater.* **2018**, *190*, 211–221. [CrossRef]
22. Cristelo, N.; Cunha, V.M.; Gomes, A.T.; Araújo, N.; Miranda, T.; Lopes, M.D.L. Influence of fibre reinforcement on the post-cracking behaviour of a cement-stabilised sandy-clay subjected to indirect tensile stress. *Constr. Build. Mater.* **2017**, *138*, 163–173. [CrossRef]
23. Lirer, S.; Flora, A.; Consoli, N.C. On the strength of Fibre-Reinforced soils. *Soils Found.* **2011**, *51*, 601–609. [CrossRef]
24. Ghasabkolaei, N.; Janalizadeh, C.A.; Roshan, N.; Ghasemi, S.E. Geotechnical properties of the soils modified with nanomaterials: A comprehensive review. *Arch. Civ. Mech. Eng.* **2017**, *17*, 639–650. [CrossRef]
25. Wang, W.; Zhang, C.; Li, N.; Tao, F.; Yao, K. Characterisation of nano magnesia–cement-reinforced seashore soft soil by direct-shear test. *Mar. Georesour. Geotechnol.* **2019**, *25*, 1–10. [CrossRef]
26. Yao, K.; An, D.L.; Wang, W.; Li, N.; Zhou, A. Effect of nano-MgO on mechanical performance of cement stabilized silty clay. *Mar. Georesour. Geotechnol.* **2020**, *38*, 250–255. [CrossRef]
27. Zheng, J.; Sun, X.; Guo, L.; Zhang, S.; Chen, J. Strength and hydration products of cemented paste backfill from sulphide-rich tailings using reactive MgO-activated slag as a binder. *Constr. Build. Mater.* **2019**, *203*, 111–119. [CrossRef]
28. Liu, H.; Xue, J.; Li, B.; Wang, J.; Lv, X.; Zhang, J. Effect of graphite tailings as substitute sand on mechanical properties of concrete. *Eur. J. Environ. Civ. Eng.* **2020**, *10*, 1–19. [CrossRef]

29. Sarkkinen, M.; Kujala, K.; Gehör, S. Efficiency of MgO activated GGBFS and OPC in the stabilization of highly sulfidic mine tailings. *J. Sustain. Min.* **2019**, *18*, 115–126. [CrossRef]
30. Li, N.; Lv, S.; Wang, W.; Guo, J.; Jiang, P.; Liu, Y. Experimental investigations on the mechanical behavior of iron tailings powder with compound admixture of cement and nano-clay. *Constr. Build. Mater.* **2020**, *254*, 119259. [CrossRef]
31. Wang, W.; Li, J.; Hu, J. Unconfined Mechanical Properties of Nanoclay Cement Compound Modified Calcareous Sand of the South China Sea. *Adv. Civ. Eng.* **2020**, *2020*, 6623710. [CrossRef]
32. Wang, W.; Li, Y.; Yao, K.; Li, N.; Zhou, A.Z.; Zhang, C. Strength properties of nano-MgO and cement stabilized coastal silty clay subjected to sulfuric acid attack. *Mar. Georesour. Geotechnol.* **2020**, *38*, 1177–1186. [CrossRef]
33. Liu, C.; Li, K.; Geng, Y.; Li, Q. Microstructure characteristics of soils with different land use types in the Yellow River Delta. *Trans. Chin. Soc. Agric. Eng.* **2020**, *36*, 81–87, (In Chinese with English abstract).
34. Long, H.; Ye, X.H.; Tan, S.W. Application of normalized weighted average algorithm in measurement. *Electr. Light Control.* **2010**, *17*, 68–70.
35. Chen, S.l.; Hou, R.; Ni, C.L.; Wang, J.X. Study on mechanical properties of cement soil based on triaxial compression test. *Silic. Bull.* **2018**, *37*, 4012–4017.

crystals

MDPI

Article

Residual Repeated Impact Strength of Concrete Exposed to Elevated Temperatures

Raad A. Al-Ameri [1], Sallal R. Abid [2,*], G. Murali [3], Sajjad H. Ali [2] and Mustafa Özakça [1]

[1] Department of Civil Engineering, Gaziantep University, Gaziantep 27310, Turkey; raada.alameri@gmail.com (R.A.A.-A.); ozakca@gantep.edu.tr (M.Ö.)
[2] Department of Civil Engineering, Wasit University, Kut 52003, Iraq; sajhali.wasit@gmail.com
[3] School of Civil Engineering, SASTRA Deemed University, Thanjavur 613401, India; murali@civil.sastra.edu
* Correspondence: sallal@uowasit.edu.iq

Abstract: Portland cement concrete is known to have good fire resistance; however, its strength would be degraded after exposure to the temperatures of fire. Repeated low-velocity impacts are a type of probable accidental load in many types of structures. Although there is a rich body of literature on the residual mechanical properties of concrete after high temperature exposure, the residual repeated impact performance of concrete has still not been well explored. For this purpose, an experimental study was conducted in this work to evaluate the effect of high temperatures on the repeated impact strength of normal strength concrete. Seven identical concrete patches with six disc specimens each were cast and tested using the ACI 544-2R repeated impact setup at ambient temperature and after exposure to 100, 200, 300, 400, 500 and 500 °C. Similarly, six cubes and six prisms from each patch were used to evaluate the residual compressive and flexural strengths at the same conditions. Additionally, the scattering of the impact strength results was examined using three methods of the Weibull distribution, and the results are presented in terms of reliability. The test results show that the cracking and failure impact numbers of specimens heated to 100 °C reduced slightly by only 2.4 and 3.5%, respectively, while heating to higher temperatures deteriorated the impact resistance much faster than the compressive and flexural strengths. The percentage reduction in impact resistance at 600 °C was generally higher than 96%. It was also found that the deduction trend of the impact strength with temperature is more related to that of the flexural strength than the compressive strength. The test results also show that, within the limits of the adopted concrete type and conducted tests, the strength reduction after high temperature exposure is related to the percentage weight loss.

Keywords: repeated impact; ACI 544-2R; high temperatures; fire; residual strength

Citation: Al-Ameri, R.A.; Abid, S.R.; Murali, G.; Ali, S.H.; Özakça, M. Residual Repeated Impact Strength of Concrete Exposed to Elevated Temperatures. *Crystals* **2021**, *11*, 941. https://doi.org/10.3390/cryst11080941

Academic Editors: Cesare Signorini, Antonella Sola, Sumit Chakraborty and Valentina Volpini

Received: 14 July 2021
Accepted: 10 August 2021
Published: 12 August 2021

1. Introduction

Repeated accidental impact is a type of unfavorable load that most building structures are not designed to withstand. Impact loads are a type of short-term dynamic load that exposes the material to unusual and unwanted stresses, which is more effective on brittle materials such as concrete. Repeated impacts occur due to the accidental falling of building materials during the construction period, or other objects during the life span, from higher stories [1–3]. In parking garages, the collision of cars can also be a source of repeated impacts on columns and walls. Other example sources are repeated hits by projectiles or explosive shrapnel in conflict regions [4,5]. The evaluation of the influence of repeated impacts on the material microstructure and the residual performance of the structural members requires an adequate evaluation of the cracking and fracture behavior under such type of loads. Pre-evaluation of material behavior under repeated impacts can be conducted using the ACI 544-2R [6] repeated drop weight impact test, which can be considered as the simplest and cheapest impact test to evaluate the impact performance of concrete [7–9].

This test was used to evaluate the impact performance of several types of concrete mixtures by several recent researchers.

Concrete is a type of cementitious material that is composed mainly of Portland cement and aggregate that are mixed with water to form a moderately low cost construction solution with a sufficient compressive strength [10–13]. Although there have been great advances in the branch of material science and the development of many evolutionary modern types of concrete that possess seriously superior strength and ductility characteristics by incorporating new fiber-reinforced cementitious composites, normal weight, normal strength Portland cement-based concrete is still the most widely used type in reinforced concrete buildings [14,15]. Concrete structures, as with other structure types, are always under the danger of accidental fires that may occur due to electrical issues or inconvenient occupation. With the existence of furniture, fires can reach temperatures as high as 1000 °C in a very short period. Yearly, tens of thousands of fires are reported in countries such as the USA and the UK, 40% of which are classified as structural fires [16]. Although concrete is considered a good fire-resistant material, exposure to high temperatures with a steep temperature increase would seriously deteriorate the concrete structural members [15,17]. Due to the low thermal conductivity of concrete and considering the quick temperature increase, high temperature gradients would form between the exposed surfaces, the cores and the opposite unexposed surfaces [18]. Such gradients would induce internal thermal stresses on the microstructural scale, leading to steep material degradation [19,20]. The non-uniform thermal movements (expansion) of the different material parts owing to the non-uniform thermal gradients, and the chemical decomposition of calcium hydroxide beyond 400 °C [19,21–23] would initiate the effective degradation of the strength until failure.

Three main types of fire test procedures have been used by researchers to simulate the strength reduction in the material under high temperature exposure, two of which simulate the conditions of the concrete strength during fire exposure for compression and flexural members. In the first of these tests, which is the stressed fire test, the setup simulates the performance of compression members such as columns or compression parts of beams where the member is stressed by approximately 20% to 40% of its strength [24–26]. In this test, the specimens are preloaded to a constant stress level and then heated to the desired level of temperature, at which the temperature is kept constant to assure an approximately zero thermal gradient in the material (temperature saturation to the steady state condition). Then, the load is increased to failure. The second test is usually termed as the unstressed condition, which follows the same procedure of the first test setup but without preloading. This test simulates the conditions of beams and slabs where the concrete is usually under low tensile stresses [25]. On the other hand, the residual unstressed test is the third fire test setup. This test setup simulates the post-fire evaluation of the strength, where the specimens are heated unstressed to the required temperature, saturated at this temperature for a constant period and then cooled to ambient temperature. Finally, the cooled specimens are tested at ambient temperature to examine the residual strength of the concrete after the fire is over. The post-fire exposure strength evaluation is essential to decide whether the building is still functional, requires strength rehabilitation due to being unsuitable or must be demolished. Previous studies [25] showed that if the specimens are cooled in water (as in the case of water distinguishing a fire), the strength deteriorates at faster rates than when cooled under air convection. The volume changes resulting from the re-hydration (with cooling water) of dehydrated calcium silicate after exposure to temperatures higher than 400 °C have a destructive influence on the concrete microstructure [26].

Few research works are available in the literature about the evaluation of the concrete impact strength after high temperature exposure, most of which investigate high-strain rate impact tests and blast tests [27–31], while a very limited number of works were found on low-velocity impact tests on reinforced concrete members [18,32]. Most of these works developed numerical analyses to evaluate the dual effect of high temperature and high-strain rate impacts on reinforced concrete and composite structural members [33–35].

Although the combined action of fire exposure and low-velocity repeated impacts is probable in many structures such as parking garages, the authors could not find sufficient literature works that tried to explore the post-fire repeated impact performance of concrete. Mehdipour et al. [18] conducted an experimental study to evaluate the residual mechanical properties of concrete containing recycled rubber as a replacement of coarse aggregate, metakaolin as a partial replacement of Portland cement and steel fibers. Among the investigated mechanical properties, the authors evaluated the residual repeated impact strength using the ACI 544-2R test after exposure to temperatures of 150, 300, 450 and 600 °C. The test results showed that high temperature could significantly affect the residual cracking and failure impact numbers, where only one impact could cause the cracking of specimens heated to 600 °C. They also reported that a multi-surface cracking was observed for heated specimens instead of the central circular fracture of unheated fibrous specimens.

The current research investigated the impact resistance of normal concrete to low-velocity repeated impacts after exposure to high temperatures of accidental fires. As reviewed in the previous sections, such scenario is possible along the span life of concrete structures, while the number of studies on this topic is seriously limited. Aiming to fill the gap in knowledge in this area, a research program was initiated starting with the most usual concrete type in reinforced concrete structures, which is normal weight, normal strength concrete.

2. Experimental Work

As stated previously, the experimental work presented in this study was conducted to be a reference work for future works where the post-fire residual impact performance of fibrous concretes will be assessed. Therefore, a normal plain concrete mixture was adopted in this study to investigate the effect of repeated low-velocity impacts on heated concrete. A cement content of 410 kg/m^3 was used with 215 kg/m^3 of water, while the fine and coarse aggregate contents were 787 and 848 kg/m^3, respectively. Local crushed gravel and river sand from Wasit Province, Iraq, were used as coarse and fine aggregates. The maximum size of the crushed gravel was 10 mm. The chemical composition and physical properties of the used cement are listed in Table 1, while the gradings of both the sand and gravel are shown in Figure 1. All specimens were cured for 28 days in temperature-controlled water tanks.

Table 1. Physical properties of cement.

Oxide (%)	Content
SiO_2	20.08
Fe_2O_3	3.6
Al_2O_3	4.62
CaO	61.61
MgO	2.12
SO_3	2.71
Loss on ignition (%)	1.38
Specific surface (m^2/kg)	368
Specific gravity	3.15
Compressive strength 2 days (MPa)	27.4
Compressive strength 28 days (MPa)	46.8

The impact tests were conducted using the standard ACI 544-2R repeated impact procedure. However, instead of the manual operation of the test, an automatic impact machine was built for this purpose and used in this study to perform the repeated impact tests. The impact machine shown in Figure 2a was built to reduce the efforts and time of the test, where repeated impacts were automatically applied using the standard ACI 544-2R drop weight (4.54 kg) and drop height (457 mm), while the crack initiation and failure were observed using a high-resolution camera, as shown in Figure 2b. The same standard test setup was followed, where a steel ball was placed on the top of the 150 mm-diameter and 64 mm-thick disc specimen, while the drop weight fell directly on the steel ball that transferred it to the center of the specimen's top surface, as shown in Figure 2c. Once the first surface crack appeared, the number of repeated impacts was recorded, which is termed as the cracking number (Ncr). After, the impact was continued until the failure of the specimens, where the failure number (Nf) was recorded.

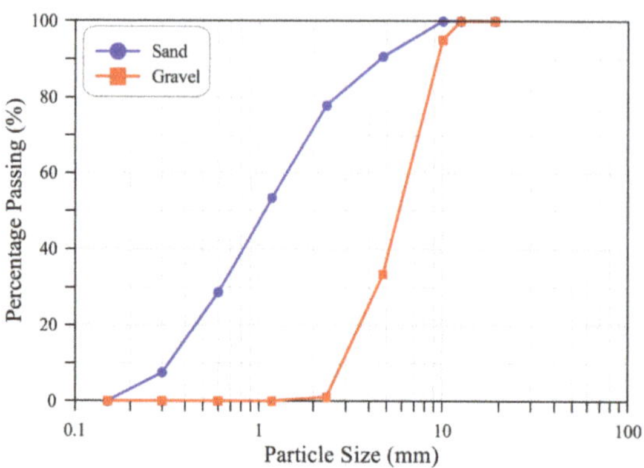

Figure 1. Grading of fine and coarse aggregates.

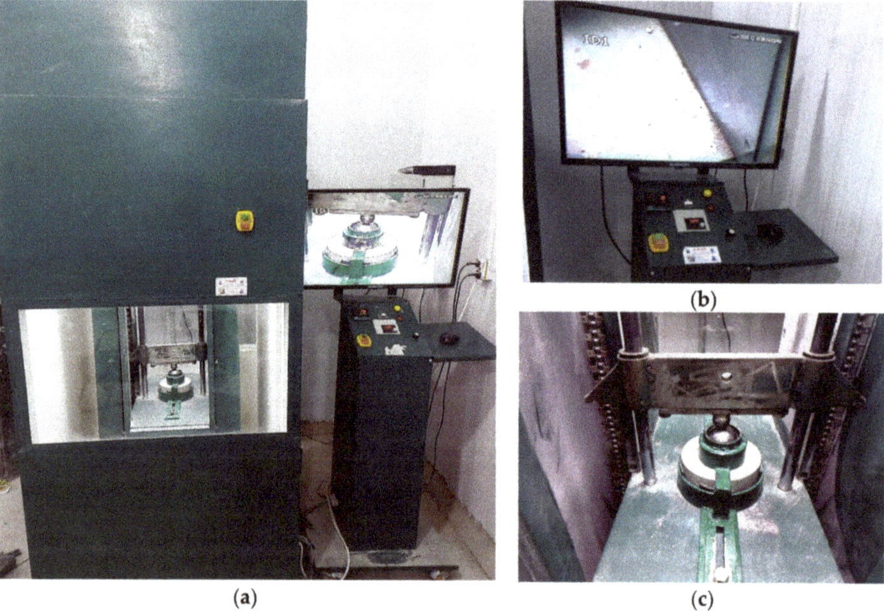

Figure 2. The automatic repeated impact testing machine (**a**) the impact machine; (**b**) crack observation; (**c**) the impact load and holding system.

In addition to the six impact disc specimens, six 100 mm cube specimens and six 100 × 100 × 400 mm prisms were cast from the same concrete patches, as shown in Figure 3. The concrete cubes were used to evaluate the compressive strength, while the prisms were used to evaluate the modulus of rupture. Previous researchers [36] showed that increasing the size of the cube specimen from 100 to 150 mm had no significant effect on the residual compressive strength after exposure to temperatures up to 1200 °C, while others [37] drew a similar conclusion for concrete cylinders having different sizes (diameter × length) of 50 × 100 mm, 100 × 200 mm and 150 × 300 mm. The cylinders were exposed to temperatures up to 800 °C. Several previous studies [38–46] adopted 100 mm cubes to evaluate the residual compressive strength of concrete.

Figure 3. Single group of disc, cube and prism specimens.

Seven concrete patches were cast and cured in water tanks under the same conditions for 28 days, after which the specimens were left to air dry for a few hours and then dried in an electrical oven at 100 °C for 24 hours [15,47–50]. After the specimens were naturally air cooled, six of the seven patches were heated to temperatures of 100, 200, 300, 400, 500 and 600 °C using the electrical furnace shown in Figure 4a, while the seventh patch was left as an unheated reference. As shown in Figure 4b, a steel cage was used to reduce the destructive effects of the concrete explosive failure on the internal walls and heaters of the furnace. The heating process followed the heating and cooling procedure shown in Figure 5, where the specimens where heated steadily at a heating rate of approximately 4 °C/min until the target temperature. After, the specimens were thermally saturated at this temperature for one hour to assure the thermal steady state condition [51,52]. Finally, the furnace door was opened to allow for the natural air cooling of the specimens, which were left in the laboratory environment and tested the next day.

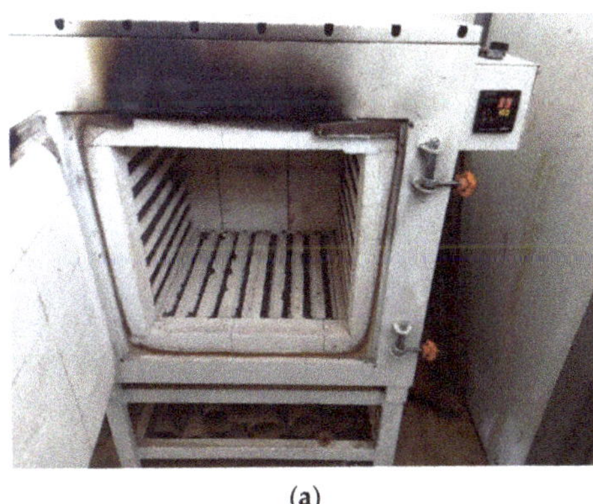

(**a**)

(**b**)

Figure 4. The electrical furnace (**a**) furnace interiors; (**b**) specimens heating.

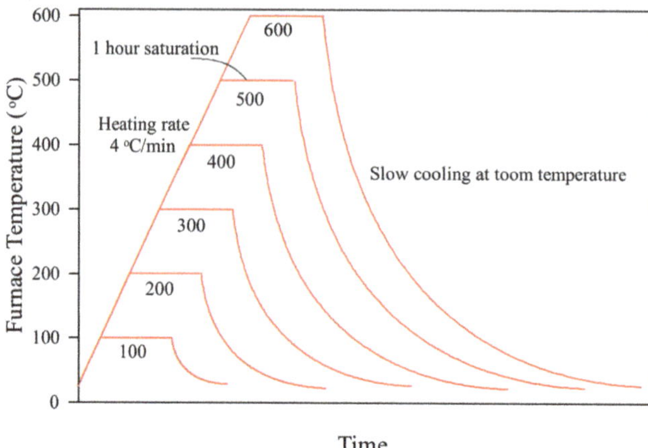

Figure 5. Heating and cooling regime of the electrical furnace.

The heating rate can affect the residual strength of concrete exposed to temperatures exceeding 300 °C [53]. Previous researchers used different heating rates to evaluate the residual properties of concrete, which were as low as 0.5 °C [54] and 1 °C [55] and as high as 30 °C [56]. However, heating rates ranging from 2 to 10 °C were the most common in the literature [57], where a heating rate of 2 °C was adopted by [58,59], while 3 °C was adopted by [60,61], and 4 °C was adopted by [62,63]. On the other hand, 5 °C was used by [64–67], and 6 °C and 7 °C were used by [38,64,68], while other previous researchers [69–71] used a heating rate of 10 °C. In this study, a heating rate within this range, 4 °C, was adopted.

3. Results and Discussion

3.1. Compressive Strength

The residual compressive strength values after exposure to 100, 200, 300, 400, 500 and 600 °C in addition to those at ambient temperature are shown in Figure 6. The figure also shows the percentage reduction in strength due to temperature exposure. It is obvious that a slight strength gain of approximately 1.5% was recorded at 100 °C, where the compressive strength after exposure to this temperature was 43.9 MPa, which is higher than that before heating (43.2 MPa). This initial strength gain was reported by previous researchers [25] and is attributed to the increase in the material density due to the evaporation of free pore water and the increase in hydration products owing to the accelerated pozzolanic reaction [72,73]. The strength gain is represented as a negative percentage reduction in Figure 6. After exposure to 200 °C, the compressive strength recorded 35.1 MPa, with a percentage reduction of approximately 18.8%, while a noticeable percentage strength recovery was recorded at 300 °C. The residual compressive strength at 300 °C was 42.3 MPa, and the percentage reduction was only approximately 2.1%. Hence, more than 16% of strength recovery was gained as the temperature was increased from 200 to 300 °C. As the temperature increased beyond 200 °C, the removal of water from the surfaces of the cement gel particles induced higher attraction surface forces (van der Waals forces), which might increase the ability of the microstructure to absorb higher compression stresses [18,73]. Beyond 300 °C, the compressive strength exhibited a continuous decrease in strength, with an increase in temperature, where the percentage strength reductions were 18.8, 22.4 and 50.0% after exposure to 400, 500 and 600 °C. The significant strength drop after 500 °C is attributed to the volume changes that took place due to the shrinkage of the cement paste and expansion of the aggregate particles, which deteriorate the bond between the two materials [25,74,75]. The dehydration of cement and decomposition of calcium hydroxide also lead to destructive effects [76,77].

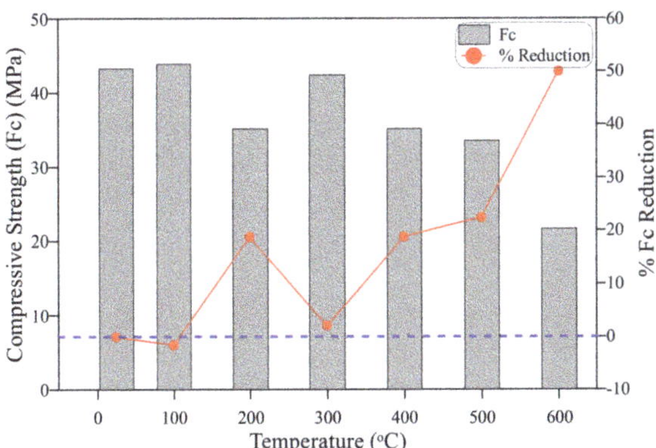

Figure 6. Compressive strength and its percentage reduction at different temperatures.

Figure 7 shows the relationship between the compressive strength and percentage weight loss of the same cube specimens after exposure to high temperatures of 100 to 600 °C. The percentage weight loss was calculated by dividing the weight loss (weight before heating-weight after heating) by the original specimen weight before heating, with the result multiplied by 100. The figure shows that three stages can be recognized in the percentage weight loss relation with temperature. In the first stage, the slope of the percentage weight loss was high after exposure to the sub-high temperature range (100 to 300 °C), which indicates early high weight loss. This weight loss can be attributed to the evaporation of the free pore water before 200 °C and the absorbed water in the cement gel particles. The second stage, which is a semi-stabilization stage with a very small positive slope, is related to the strength recovery and low reduction between 300 and 500 °C, where the microstructure is still not very affected by the chemical and physical changes due to temperature exposure. Finally, the weight loss starts another high-slope reduction region after 500 °C, where the cement matrix was cracked owing to the chemical changes, and the bond between the cement and aggregate was almost lost due to the different thermal movements [78,79]. Excluding the initial strength gain at 100 °C and the strength recovery at 300 °C, it can be said that the strength loss can be related to the loss in weight. The differences between strength loss and weight loss at these regions are attributed to the different behaviors of concrete under the different applied stresses, where it is stated that the residual tensile strength of concrete, for example, has a different behavior with temperature than that of the compressive strength. As weight loss is a stress-free measurement, there would be some expected differences with load tests.

3.2. Flexural Strength

The residual flexural strength records (modulus of rupture) of the tested prisms are visualized in Figure 8, which also visualizes the percentage reduction in flexural strength as a ratio of the unheated strength. The figure explicitly shows that a similar strength gain to that of the compressive strength was recorded for the flexural strength at 100 °C. However, the percentage increase was higher, where the unheated flexural strength was 3.7 MPa, while that after exposure to 100 °C was approximately 4.1 MPa, with a percentage increase of approximately 10.1%, which is depicted in Figure 8 as a negative percentage reduction. The same reason discussed in the previous section for the compressive strength could be the source of this increase in the flexural strength. After, a continuous strength deterioration was recorded as the temperature increased beyond 100 °C, where the residual modulus of rupture values were approximately 2.9, 2.4, 2.2, 1.6 and 0.3 MPa with respective percentage reductions of approximately 22.5, 35.1, 41.6, 57.3 and 91.2 % after exposure to 200, 300, 400, 500 and 600 °C. It can be noticed that the strength recovery recorded for the compressive strength at 300 °C was not recorded for the flexural strength, which reflects the positive effect of physical attraction forces at this temperature to sustain higher compressive stresses, while such effect was insignificant under the flexural tensile stresses. Another result that was noticed is the faster deterioration of the flexural strength compared to the compressive strength, which is directly related to the weak microstructural response of concrete to tensile stresses. A similar flexural strength reduction behavior was reported by many previous works [25,72,80].

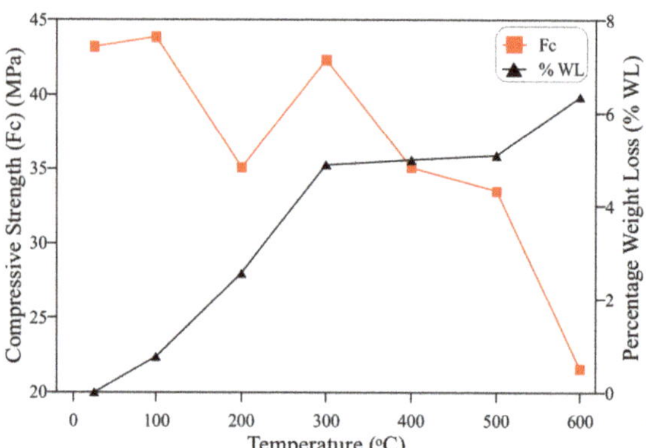

Figure 7. The relation of compressive strength and percentage weight loss of the cube specimens at different temperatures.

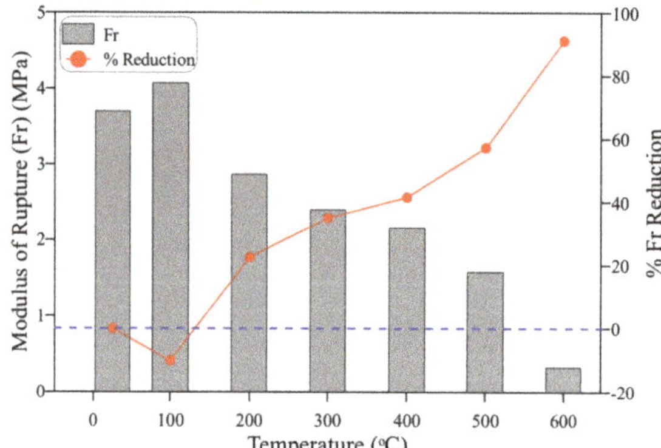

Figure 8. Flexural strength and its percentage reduction at different temperatures.

The weight loss for the same prisms was recorded, from which the percentage weight loss was calculated, which is depicted in Figure 9 against the residual modulus of rupture. The figure shows that a similar three-stage behavior can be recognized for the prism specimens to that recorded for the cube specimens, where the slope of the percentage weight loss was higher before 300 °C and beyond 500 °C, while it was lower between them. Similarly, excluding the high residual flexural strength recorded at 100 °C, the strength reduction curve exhibited a similar three-slope behavior to that of weight loss, where the percentage reduction slope was high from 100 to 300 °C and from 500 to 600 °C, while it was semi-stabilized between 300 and 400 °C for both the flexural strength and weight loss, as shown in Figure 9. This similar trend confirms the strong relation between weight loss and strength reduction after high temperature exposure.

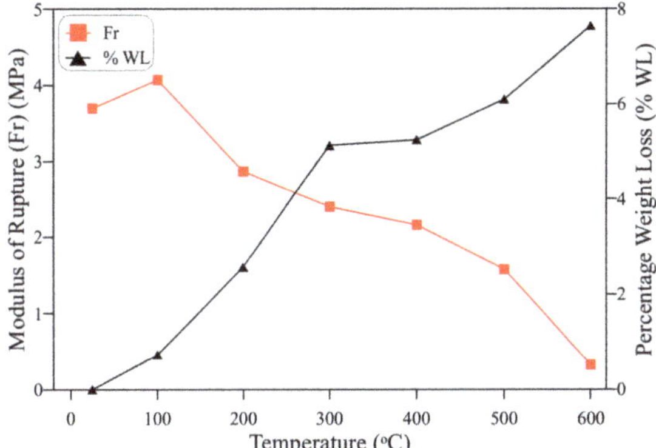

Figure 9. The relation of flexural strength and percentage weight loss of the prism specimens at different temperatures.

3.3. Repeated Impact Strength

This section presents and discusses the obtained results from the conducted drop weight repeated impact tests that were carried out at ambient temperature and after exposure to different levels of high temperatures. The results of all specimens are listed in Table 2 together with the mean, standard deviation (SD) and coefficient of variation (COV) of each of the six disc samples. It is obvious that for all specimens, before and after exposure to high temperature, the recorded failure number was slightly higher than the cracking number, which reveals the brittle nature of normal concrete under impact tests [81–83].

Table 2. Results of repeated impact test.

Temperature	R		100 °C		200 °C		300 °C		400 °C		500 °C		600 °C	
Disc No.	N_{cr}	N_f	N_{cr}	N_f	N_{cr}	N_f	N_{cr}	N_f	N_{cr}	N_f	N_{cr}	N_f	N_{cr}	N_f
1	52	53	28	29	18	19	8	9	4	5	2	3	1	2
2	41	42	76	78	25	26	7	8	2	3	2	3	1	3
3	78	81	41	43	11	12	6	7	4	5	2	3	1	2
4	50	53	90	91	9	10	7	8	3	4	3	4	1	2
5	51	54	39	41	9	10	4	5	2	3	1	2	1	2
6	58	60	48	49	13	14	7	8	3	4	3	3	1	2
Mean	55.0	57.2	53.7	55.2	14.2	15.2	6.5	7.5	3.0	4.0	2.2	3.0	1.0	2.2
SD	12.5	13.0	24.0	24.0	6.3	6.3	1.4	1.4	0.9	0.9	0.8	0.6	0.0	0.4
COV %	22.8	22.8	44.8	43.5	44.3	41.4	21.2	18.4	29.8	22.4	34.7	21.1	0.0	18.8

3.3.1. Cracking and Failure Impact Numbers

The detailed impact numbers are listed in Table 2, while Figures 10 and 11 show the post-high temperature exposure response of the cracking (N_{cr}) and failure (N_f) impact numbers, respectively. Figure 10 shows that after exposure to 100 °C, the cracking number was not noticeably affected, where this number was decreased from 55 to 53.7, with a percentage decrease of only 2.4%. Similarly, N_f was decreased by no more than 3.5%, as illustrated in Figure 11. Considering the known high variation in the ACI 544-2R repeated impact test [6], it can be said that the impact resistance was not affected by heating to 100 °C, which is also confirmed by the comparison of the individual impact numbers in Table 2, where the retained impact numbers for some specimens where higher after exposure to 100 °C than before heating. For instance, a cracking impact number of 90 was recorded for specimens heated to 100 °C, which is higher than all recorded numbers before heating. As discussed in the previous sections, the compressive strength and flexural strength were increased after exposure to a temperature of 100 °C. The strength gain is reported to be due to the shrinkage of the concrete pore holes after the evaporation of the pore water, which resulted in denser media. The slow preheating of specimens in an electrical oven at 100 °C might also help to minimize the reduction in the impact strength at this temperature, where the free pore water was partially evaporated during the initial heating phase, resulting in the stress being relieved during the second (furnace) quick heating phase, which maintained the microstructural deterioration at a minimum. It should be mentioned that the oven's slow preheating is essential to prevent any type of thermal explosive failure during heating, which would be harmful to the furnace and other specimens in the furnace. The preheating process was a typical procedure followed by many experimental studies in the literature, where it was reported that thermal explosive spalling is probable at temperatures in the range of 300 to 650 °C [25].

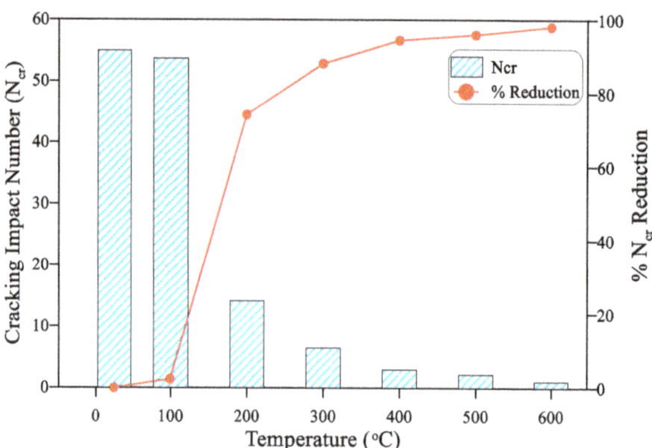

Figure 10. Cracking impact number and its percentage reduction at different temperatures.

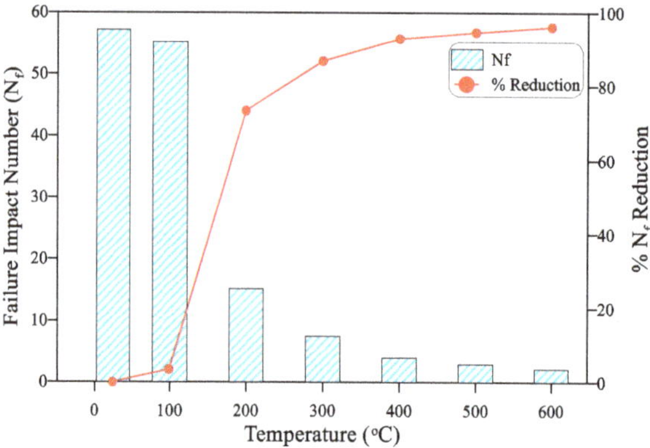

Figure 11. Failure impact number and its percentage reduction at different temperatures.

The strength of the material to resist impact forces dropped sharply after exposure to 200 °C, where the loss in impact resistance in terms of N_{cr} and N_f reached 74.2 and 73.5%, respectively. This reduction is dramatic and is much higher than the losses in the compressive, flexural and tensile strengths according to the literature and the current study results, where previous studies showed that the normal compressive strength range of normal concrete after exposure to 200 °C falls between 70 and 110% of the unheated strength [18,21,72]. Similarly, it is widely addressed in the literature that the modulus of elasticity of normal concrete would be higher than 70% of the unheated values [73–75].

Phan and Carino [25] reported that, for the case of an unstressed residual compressive strength test, which was the heating procedure followed in this research, an initial strength gain or minor loss is the usual trend of normal concrete up to 200 °C. The residual impact strength of the higher temperatures followed the same excessive strength drop, as shown in Figures 10 and 11. The residual cracking impact strength after exposure to 300, 400, 500 and 600 °C was, respectively, 11.8, 5.5, 3.9 and 1.8% of the original unheated strength. Similarly, the residual failure impact strengths were 13.1, 7.0, 5.2 and 3.8% after exposure to 300, 400, 500 and 600 °C, respectively. Two points should be discussed here, namely, the strength reduction behavior with temperature, and the high drop in strength. It is obvious that the impact strength follows a similar trend of reduction with temperature to the flexural strength, where both the flexural and impact strengths showed stable responses at 100 °C, followed by a continuous decrease until the approximate fading of the strength at 600 °C. This might be attributed to the type of stresses caused by the repeated impacts on the concrete, where the received impact forces tend to cause a fracture surface and then transfer this into tensile stresses that try to open the cracks until breakage failure. The higher strength reduction can also be attributed to the nature of the impact loads, where a sudden concentrated loading is induced within a very short time, leading to a higher stress concentration and hence a faster deterioration. As soon as the microstructure is internally fractured by the initiation of microcracks, only a few further concentrated drops are required to induce the surface cracking and failure of the specimen. Since the microstructure of the cement paste, the aggregate particles and the bond between them are negatively affected by the preceding heating phase of the specimens, a quick fracture of these specimens is expected under impact drops.

Figure 12 compares the behaviors of the residual impact strength in terms of the impact numbers of the disc specimens and the percentage weight loss of the same specimens. It can be said that the trend of the impact strength reduction is related to that of the percentage weight loss, where the same minor reduction is clear in the figure after exposure to 100 °C for both the strength and the weight loss, followed by a steep drop from 100 to 200 °C. After, a continuous decrease in the strength and an increase in weight loss are obvious from 300 to 600 °C. Indeed, the percentage strength is much higher than the percentage weight loss at each temperature; however, the behavior of the reduction with the temperature increase is quite similar.

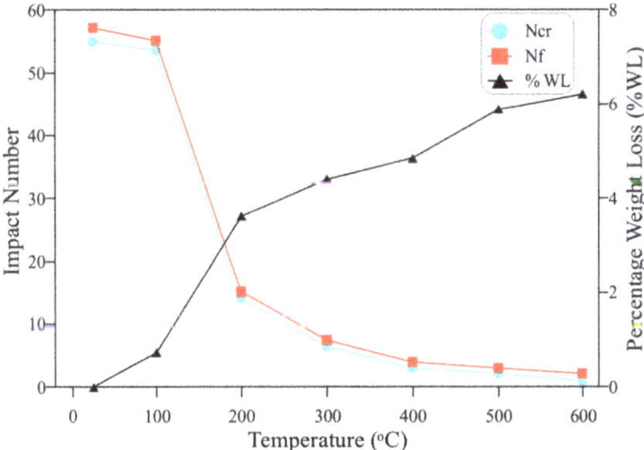

Figure 12. The relation of the impact number and percentage weight loss of the prism specimens at different temperatures.

As disclosed in the previous sections, the ACI 544-2R repeated impact test is known for its high variation in the obtained impact results. Figure 13 shows the effect of high temperature exposure on the variation in the cracking and failure impact numbers in terms of the coefficient of variation (COV). It is clear in the figure that the COV was not so high

(22.8) at ambient temperature, which is a good indication of the better control on the loading parameters using the automatic machine compared to the standard manual apparatus. The COV increased at 100 and 200 °C, recording values in the range of 41.4 and 44.8% for both the cracking and failure numbers. This increase in the result variation can be attributed to the dramatic behaviors at these temperatures, where at 100 °C, the specimens exhibited an increase and a decrease in strength compared to those tested at ambient temperature, as listed in Table 2. Similarly, the high drop in the impact number records at 200 °C resulted in high percentage variations, although there were limited numeral variations. The extremely low records of the impact numbers at the higher temperatures reduced the differences between the tested specimens, which, in turn, reduced the COV, as shown in Figure 13.

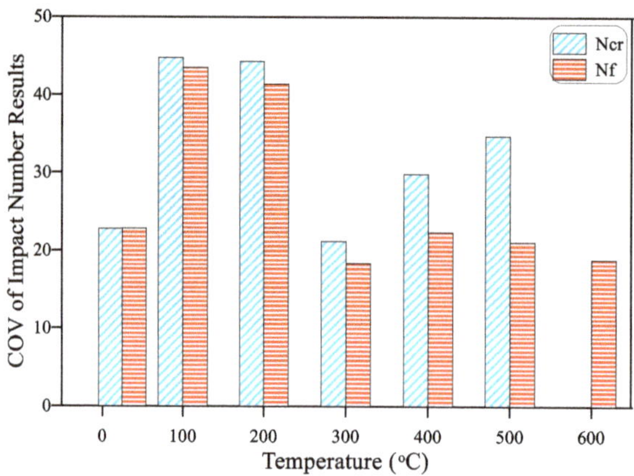

Figure 13. Coefficient of variation of the cracking and failure impact numbers at different temperatures.

3.3.2. Failure Patterns

Pictures for the disc specimens at room ambient temperature (R) and after exposure to 100, 200, 300, 400, 500 and 600 °C are shown in Figure 14a–g, both before and after impact testing. The fracture and failure of the reference unheated specimens align with what has been reported in previous studies [84–88] for plain concrete, where after a number of repeated blows, a small-diameter central fracture zone was created under the concentrated compression impacts via the top surface's steel ball. After a few more blows, the internal cracks propagated to the surface, forming a surface cracking of two or three radial cracks from the central fracture zone, which formed the failure shape that occurred after a few additional blows, as shown in Figure 14a.

Figure 14b–d show that the specimens heated up to 300 °C exhibited a similar cracking and fracture behavior to that of the unheated specimens. However, at 200 and 300 °C, the cracking and failure tended to be softer, where the fracture started at much lower impact numbers, and the central fracture zone was smaller. The fracture of the specimens heated to temperatures above 400 °C is different. The material became so weak and absorbed the applied impact energy after being heated, where the cement paste became softer, the crushing strength of the aggregate particle reduced and the bond between them was almost lost. Due to these effects, internal thermal cracks were formed in the whole volume of the specimens. Therefore, for the specimens exposed to 400, 500 and 600 °C, only three impacts, two impacts and one impact were enough to cause the already existing cracks to appear at the surface. As a result, the central fracture zone was barely formed, and a higher number of major cracks (four or five) were formed accompanied by more hair cracks, as shown in Figure 14e–g.

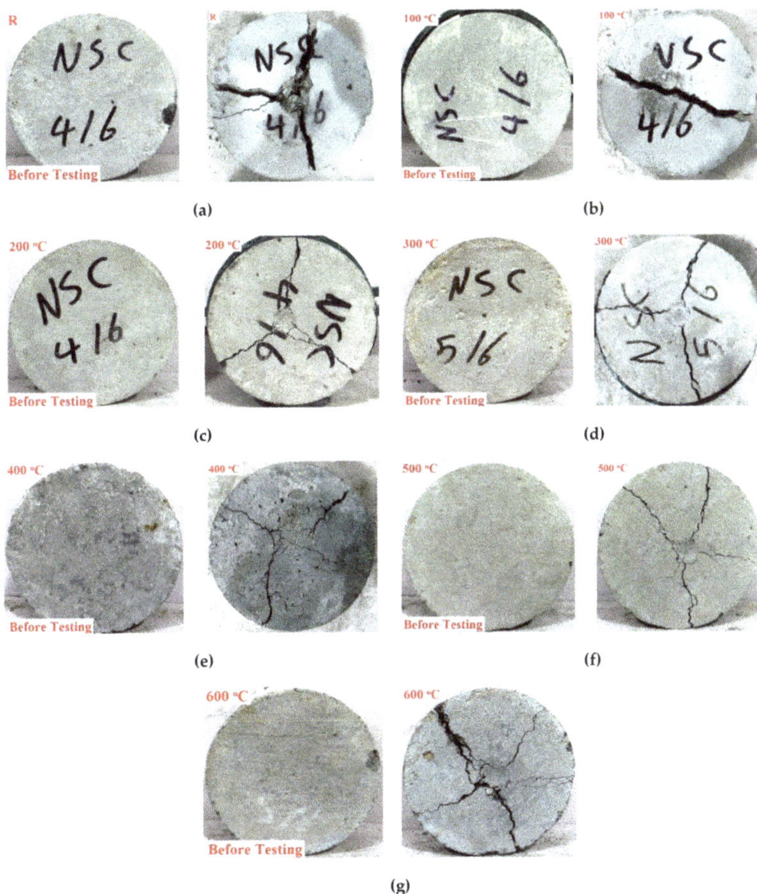

Figure 14. Failure patterns of the impact disc specimens at different temperatures (**a**) ambient temperature; (**b**): 100 °C; (**c**): 200 °C; (**d**) 300 °C; (**e**) 400 °C; (**f**) 500 °C; (**g**): 600 °C.

4. Weibull Distribution

At first, probabilistic methods were used to provide a rationale for the scattering of fracture strength results with a brittle nature. This statistical method has most widely been used to assess the statistical variability of impact test results in recent times [7,9,89]. Barbero et al. [90] investigated the mechanical properties of composite materials using the Weibull distribution. The authors recommended that the Weibull distribution is a pragmatic approach for determining 90% and 95% reliability values. The Weibull distribution is accentuated by two parameters, namely, shape and scale, and these parameters can be evaluated by several methods [91]. The scattering of the failure impact number of concretes was modeled using a two-parameter Weibull distribution. Lastly, the reliability of the concrete in terms of the failure impact number was presented in graphical form. The scattering of the cracking impact numbers was minor and hence was not modeled using the Weibull distribution.

4.1. Mean Standard Deviation Method (MSDM)

This method is more useful when the means and standard deviations are known; if this occurs, the shape parameter (α) and scale parameter (β) are determined using Equations (1) and (2) as follows [92].

$$\alpha = \left(\frac{\sigma}{\overline{N_f}}\right)^{-1.086} \tag{1}$$

$$\beta = \frac{\overline{R}\,\alpha^{2.6674}}{0.184 + 0.816\,\alpha^{2.73855}} \tag{2}$$

where $\overline{N_f}$ is the mean of the failure impact number, and σ is the standard deviation.

4.2. Energy Pattern Factor Method (EPFM)

The EPF is defined by the ratio of the summation of cubes of individual failure impact numbers to the cube of the mean failure impact number. The scale and shape parameters are calculated using Equations (3) and (4) once the *EPF* value is known [93].

$$EPF = \frac{\overline{N_f{}^3}}{\overline{N_f}^3} \tag{3}$$

$$\alpha = 1 + \frac{3.69}{(Epf)^2} \tag{4}$$

The gamma function is defined in Equation (5), expressed as follows.

$$\Gamma(x) = \int_0^\infty t^{x-1}\exp(-t)dt \tag{5}$$

4.3. Method of Moments (MOM)

Numerical iteration is involved in this method, and the mean failure impact number and corresponding standard deviation (σ) are used to find the shape and scale parameters [51].

$$\alpha = \left(\frac{0.9874}{\frac{\sigma}{\overline{R}}}\right)^{-1.086} \tag{6}$$

$$\overline{N_f} = \beta\Gamma\,(1 + 1/\alpha) \tag{7}$$

Table 3 demonstrates the results of the Weibull parameters obtained from three methods of distribution. It is clear from the table that the MSDM and MOM methods showed approximately the same parameter values. However, EPFM showed a lower value compared to MSDM and MOM. To perform the reliability analysis, the mean value of the three methods was used. The reliability of concrete exposed to various temperatures in terms of the failure impact number can be calculated using Equation (8) [94–98].

$$N_f = \beta(-\ln(R_x))^{(1/\alpha)} \tag{8}$$

where R_x is the reliability level, and R is the failure impact number.

Table 3. Results of Weibull parameters (scale and shape parameters).

Temperature	MSDM		EPFM		MOM		Mean	
	α	β	α	β	α	β	α	β
R	5.00	62.34	3.96	63.15	5.02	62.29	4.66	62.59
100 °C	2.47	62.25	2.80	61.99	2.46	62.24	2.58	62.16
200 °C	2.60	17.12	2.88	17.05	2.59	17.12	2.69	17.10
300 °C	6.19	8.06	4.23	8.25	6.23	8.07	5.55	8.13
400 °C	5.05	4.36	4.01	4.41	5.08	4.35	4.71	4.37
500 °C	5.74	3.24	4.08	3.31	5.78	3.24	5.20	3.26
600 °C	6.37	2.36	4.40	2.41	6.41	2.36	5.73	2.38

Using the Weibull parameters (mean values from Table 3), the reliability analysis was performed to estimate the failure impact number. Figure 15 illustrates the failure impact number in terms of the reliability or survival probability. By examining the 0.99 reliability (1% probability of failure), the failure impact numbers for the 100, 200, 300, 400, 500 and 600 °C specimens were 23, 10, 3, 4, 2, 1 and 1, respectively. By examining another probability level of 0.9 (10% probability of failure), the failure impact numbers were 39, 26, 7, 5, 3, 2 and 2, corresponding to the 100, 200, 300, 400, 500 and 600 °C specimens, respectively. Likewise, the failure impact number for the concrete exposed to different temperatures can be obtained from Figure 15. Using the reliability curves, the design engineer has the option to choose the required failure impact number at the desired reliability level (0.5 to 0.99). These values can be used effectively in the design calculations, and the Weibull distribution can be considered as a powerful tool to examine the scattering of the impact strength results. This statistical method and the outcomes are in good agreement with earlier studies [99–103].

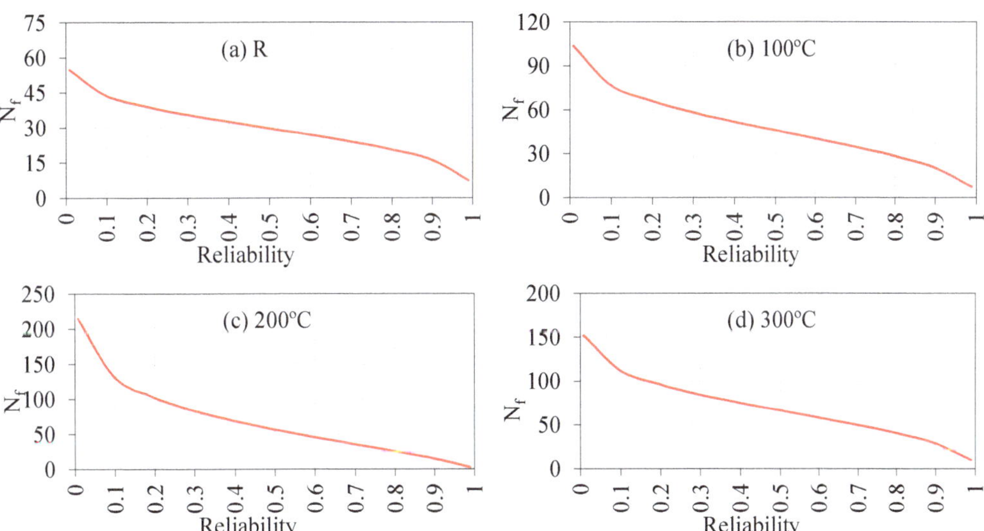

Figure 15. *Cont.*

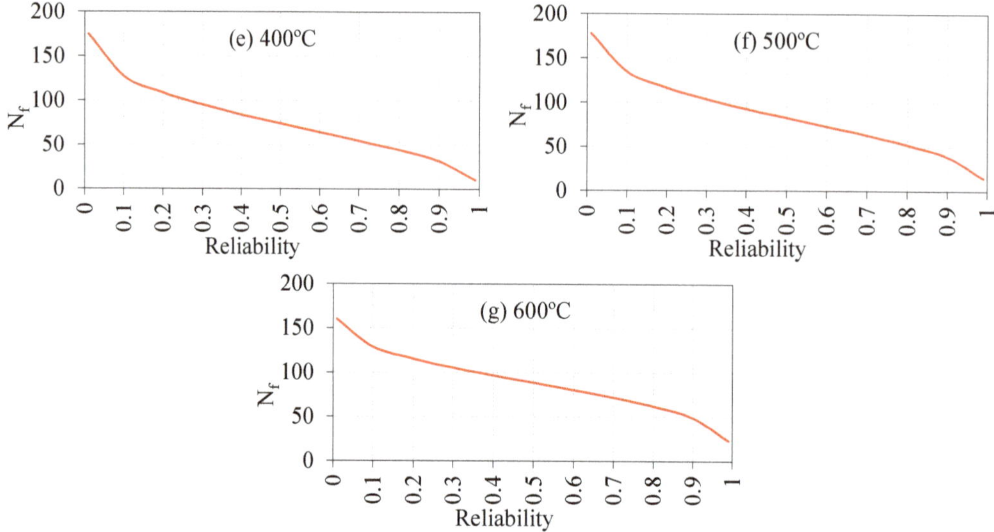

Figure 15. Failure number in terms of reliability (**a**) ambient temperature; (**b**): 100 °C; (**c**): 200 °C; (**d**) 300 °C; (**e**) 400 °C; (**f**) 500 °C; (**g**): 600 °C.

5. Conclusions

Based on the obtained experimental results from the study presented in this work, the following points are the most important conclusions.

1. The compressive strength was increased by less than 2% after exposure to 100 °C, while it decreased by more than 18% after exposure to 200 °C, followed by a partially recovered strength recorded at 300 °C. Finally, the compressive strength exhibited a continuous decrease after 400 °C, reaching a residual strength of approximately 50% of the original strength at 600 °C. This multi-phase behavior is attributed to the early loss of free water, physical and chemical changes in the cement paste after and the different thermal movements of the cement paste and aggregate.
2. Owing to the weak microstructural response of concrete under tensile stresses, the deterioration of the flexural strength after 200 °C was significantly higher than compressive strength, where the residual tensile strength after exposure to 400, 500 and 600 °C was approximately 58, 43 and 9%, respectively.
3. The impact strength, in terms of the cracking and failure impact numbers, was almost unaffected after exposure to 100 °C, where the reduction was only 2.4 and 3.5% in the cracking and failure numbers. However, a sharp drop in the impact strength of 74.2% was recorded after exposure to 200 °C, which was followed be a continuous decrease after exposure to the higher temperatures. The impact strength almost vanished at 600 °C, where the percentage reductions in the cracking and failure numbers were 98.2 and 96.2%, respectively. The higher strength drop under impact loads, compared to the compressive and tensile strengths, is attributed to the high concentrated tensile stresses induced in the material within a short time under the repeated impacts.
4. Comparing the residual strength with the percentage weight loss of the cube, prism and disc specimens, it was found that, by excluding the full or partial recovery regions, the strength follows an approximately similar behavior to that of weight loss after high temperature exposure. However, the percentage reduction in strength was much higher than the percentage increase in weight loss of the same specimens.
5. The failure of the specimens heated to temperatures up to 300 °C was similar to that of the unheated specimens, where a central fracture zone formed due to the concentrated compression impacts beneath the steel ball, followed by cracking and failure along

two or three radial cracks. On the other hand, the deteriorated microstructure of the specimens heated to temperatures of 400 to 600 °C imposed a different fracture behavior, where the specimens cracked quickly and softly along four or five paths accompanied by additional surface hair cracks, which reflects the weak strength of the material and the existence of internal thermal cracks prior to testing.

6. A rational distribution is desirable from a statistical perspective, in line with the relevant impact strength and, most significantly, with the safety of the design calculation. The Weibull distribution was found to be an efficient tool to examine the scattered test results and present the impact strength at the desired levels of reliability.

Author Contributions: Conceptualization, S.R.A.; methodology, S.H.A. and S.R.A.; software, G.M.; validation, S.R.A. and M.Ö.; formal analysis, G.M.; investigation, R.A.A.-A.; resources, R.A.A.-A.; data curation, S.R.A. and R.A.A.-A.; writing—original draft preparation, S.R.A., G.M. and R.A.A.-A.; writing—review and editing, M.Ö.; visualization, S.R.A. and S.H.A.; supervision, S.R.A. and M.Ö.; project administration, S.R.A. and M.Ö.; funding acquisition, S.R.A. and R.A.A.-A. All authors have read and agreed to the published version of the manuscript.

Funding: This research received no external funding.

Data Availability Statement: Not applicable.

Acknowledgments: The authors acknowledge the support from Al-Sharq Lab., Kut, Wasit, Iraq, and Ahmad A. Abbas.

Conflicts of Interest: The authors declare no conflict of interest.

References

1. Nili, M.; Afroughsabet, V. Combined effect of silica fume and steel fibers on the impact resistance and mechanical properties of concrete. *Int. J. Impact Eng.* **2010**, *37*, 879–886. [CrossRef]
2. Wang, W.; Chouw, N. The behavior of coconut fibre reinforced concrete (CFRC) under impact loading. Constr. *Build. Mater.* **2017**, *134*, 452–461. [CrossRef]
3. Abid, S.R.; Gunasekaran, M.; Ali, S.H.; Kadhum, A.L.; Al-Gasham, T.S.; Fediuk, R.; Vatin, N.; Karelina, M. Impact Performance of Steel Fiber-Reinforced Self-Compacting Concrete against Repeated Drop Weight Impact. *Crystals* **2021**, *11*, 91. [CrossRef]
4. Jabir, H.A.; Abid, S.R.; Murali, G.; Ali, S.H.; Klyuev, S.; Fediuk, R.; Vatin, N.; Promakhov, V.; Vasilev, Y. Experimental Tests and Reliability Analysis of the Cracking Impact Resistance of UHPFRC. *Fibers* **2020**, *8*, 74. [CrossRef]
5. Salaimanimagudam, M.P.; Suribabu, C.R.; Murali, G.; Abid, S.R. Impact Response of Hammerhead Pier Fibrous Concrete Beams Designed with Topology Optimization. *Period. Polytech. Civ. Eng.* **2020**, *64*, 1244–1258. [CrossRef]
6. ACI 544-2R. *Measurement of Properties of Fiber Reinforced Concrete*; American Concrete Institute: Indianapolis, IN, USA, 1999.
7. Haridharan, M.; Matheswaran, S.; Murali, G.; Abid, S.R.; Fediuk, R.; Amran, Y.M.; Abdelgader, H.S. Impact response of two-layered grouted aggregate fibrous concrete composite under falling mass impact. *Constr. Build. Mater.* **2020**, *263*, 120628. [CrossRef]
8. Murali, G.; Abid, S.R.; Amran, Y.M.; Abdelgader, H.S.; Fediuk, R.; Susrutha, A.; Poonguzhali, K. Impact performance of novel multi-layered prepacked aggregate fibrous composites under compression and bending. *Structures* **2020**, *28*, 1502–1515. [CrossRef]
9. Murali, G.; Abid, S.R.; Karthikeyan, K.; Haridharan, M.; Amran, M.; Siva, A. Low-velocity impact response of novel prepacked expanded clay aggregate fibrous concrete produced with carbon nano tube, glass fiber mesh and steel fiber. *Constr. Build. Mater.* **2021**, *284*, 122749. [CrossRef]
10. Imbabi, M.S.; Carrigan, C.; McKenna, S. Trends and developments in green cement and concrete technology. *Int. J. Sustain. Built Environ.* **2012**, *1*, 194–216. [CrossRef]
11. Sharma, N.K.; Kumar, P.; Kumar, S.; Thomas, B.S.; Gupta, R.C. Properties of concrete containing polished granite waste as partial substitution of coarse aggregate. *Constr. Build. Mater.* **2017**, *151*, 158–163. [CrossRef]
12. Babalola, O.; Awoyera, P.O.; Le, D.-H.; Romero, L.B. A review of residual strength properties of normal and high strength concrete exposed to elevated temperatures: Impact of materials modification on behaviour of concrete composite. *Constr. Build. Mater.* **2021**, *296*, 123448. [CrossRef]
13. Gagg, C.R. Cement and concrete as an engineering material: An historic appraisal and case study analysis. *Eng. Fail. Anal.* **2014**, *40*, 114–140. [CrossRef]
14. Behera, M.; Bhattacharyya, S.K.; Minocha, A.K.; Deoliya, R.; Maiti, S. Recycled aggregate from C&D waste & its use in concrete–A break through towards sustainability in construction sector: A review. *Constr. Build. Mater.* **2014**, *68*, 501–516.
15. Tufail, M.; Shahzada, K.; Gencturk, B.; Wei, J. Effect of Elevated Temperature on Mechanical Properties of Limestone, Quartzite and Granite Concrete. *Int. J. Concr. Struct. Mater.* **2016**, *11*, 17–28. [CrossRef]

16. Arna'Ot, F.H.; Abid, S.R.; Özakça, M.; Tayşi, N. Review of concrete flat plate-column assemblies under fire conditions. *Fire Saf. J.* **2017**, *93*, 39–52. [CrossRef]
17. Guo, Y.-C.; Zhang, J.-H.; Chen, G.-M.; Xie, Z.-H. Compressive behaviour of concrete structures incorporating recycled concrete aggregates, rubber crumb and reinforced with steel fibre, subjected to elevated temperatures. *J. Clean. Prod.* **2014**, *72*, 193–203. [CrossRef]
18. Mehdipour, S.; Nikbin, I.; Dezhampanah, S.; Mohebbi, R.; Moghadam, H.; Charkhtab, S.; Moradi, A. Mechanical properties, durability and environmental evaluation of rubberized concrete incorporating steel fiber and metakaolin at elevated temperatures. *J. Clean. Prod.* **2020**, *254*, 120126. [CrossRef]
19. Alimrani, N.; Balazs, G.L. Investigations of direct shear of one-year old SFRC after exposed to elevated temperatures. *Constr. Build. Mater.* **2020**, *254*, 119308. [CrossRef]
20. Drzymała, T.; Jackiewicz-Rek, W.; Tomaszewski, M.; Kuś, A.; Gałaj, J.; Šukys, R. Effects of high temperatures on the properties of high performance concrete (HPC). *Procedia Eng.* **2017**, *172*, 256–263. [CrossRef]
21. Chu, H.-Y.; Jiang, J.-Y.; Sun, W.; Zhang, M. Mechanical and physicochemical properties of ferro-siliceous concrete subjected to elevated temperatures. *Constr. Build. Mater.* **2016**, *122*, 743–752. [CrossRef]
22. Roufael, G.; Beaucour, A.-L.; Eslami, J.; Hoxha, D.; Noumowé, A. Influence of lightweight aggregates on the physical and mechanical residual properties of concrete subjected to high temperatures. *Constr. Build. Mater.* **2020**, *268*, 121221. [CrossRef]
23. Deng, Z.; Huang, H.; Ye, B.; Wang, H.; Xiang, P. Investigation on recycled aggregate concretes exposed to high temperature by biaxial compressive tests. *Constr. Build. Mater.* **2020**, *244*, 118048. [CrossRef]
24. Phan, L.T.; Carino, N.J. Code provisions for high strength concrete strength-temperature relationship at elevated temperatures. *Mater. Struct.* **2003**, *36*, 91–98. [CrossRef]
25. Phan, L.T.; Carino, N.J. Review of Mechanical Properties of HSC at Elevated Temperature. *J. Mater. Civ. Eng.* **1998**, *10*, 58–65. [CrossRef]
26. Husem, M. The effects of high temperature on compressive and flexural strengths of ordinary and high-performance concrete. *Fire Saf. J.* **2006**, *41*, 155–163. [CrossRef]
27. Chen, L.; Fang, Q.; Jiang, X.; Ruan, Z.; Hong, J. Combined effects of high temperature and high strain rate on normal weight concrete. *Int. J. Impact Eng.* **2015**, *86*, 40–56. [CrossRef]
28. Zhai, C.; Chen, L.; Xiang, H.; Fang, Q. Experimental and numerical investigation into RC beams subjected to blast after exposure to fire. *Int. J. Impact Eng.* **2016**, *97*, 29–45. [CrossRef]
29. Zhai, C.; Chen, L.; Fang, Q.; Chen, W.; Jiang, X. Experimental study of strain rate effects on normal weight concrete after exposure to elevated temperature. *Mater. Struct.* **2016**, *50*, 40. [CrossRef]
30. Mirmomeni, M.; Heidarpour, A.; Zhao, X.-L.; Packer, J.A. Effect of elevated temperature on the mechanical properties of high-strain-rate-induced partially damaged concrete and CFSTs. *Int. J. Impact Eng.* **2017**, *110*, 346–358. [CrossRef]
31. Kakogiannis, D.; Pascualena, F.; Reymen, B.; Pyl, L.; Ndambi, J.M.; Segers, E.; Lecompte, D.; Vantomme, J.; Krauthammer, T. Blast performance of reinforced concrete hollow core slabs in combination with fire: Numerical and experimental assessment. *Fire Saf. J.* **2013**, *57*, 69–82. [CrossRef]
32. Jin, L.; Bai, J.; Zhang, R.; Li, L.; Du, X. Effect of elevated temperature on the low-velocity impact performances of reinforced concrete slabs. *Int. J. Impact Eng.* **2021**, *149*, 103797. [CrossRef]
33. Pan, L.; Chen, L.; Fang, Q.; Zhai, C. A modified layered-section method for responses of fire-damaged reinforced concrete beams under static and blast loads. *Int. J. Prot. Struct.* **2016**, *7*, 495–517. [CrossRef]
34. Yu, X.; Chen, L.; Fang, Q.; Ruan, Z.; Hong, J.; Xiang, H. A concrete constitutive model considering coupled effects of high temperature and high strain rate. *Int. J. Impact Eng.* **2017**, *101*, 66–77. [CrossRef]
35. Jin, L.; Lan, Y.; Zhang, Z.; Du, X. Impact performances of RC beams at/after elevated temperature: Ameso-scale study. *Eng. Fail. Anal.* **2019**, *105*, 196–214. [CrossRef]
36. Arioz, O. Retained properties of concrete exposed to high temperatures: Size effect. *Fire Mater.* **2009**, *33*, 211–222. [CrossRef]
37. Erdem, T.K. Specimen size effect on the residual properties of engineered cementitious composites subjected to high temperatures. *Cem. Concr. Compos.* **2014**, *45*, 1–8. [CrossRef]
38. Chan, W.; Luo, Y.N.; Sun, X. Compressive strength and pore structure of high performance concrete after exposure to high temperature up to 800 °C. *Cem. Concr. Res.* **2000**, *30*, 247–251. [CrossRef]
39. Xu, Y.; Wong, Y.; Poon, C.S.; Anson, M. Impact of high temperature on PFA concrete. *Cem. Concr. Res.* **2001**, *31*, 1065–1073. [CrossRef]
40. Poon, C.S.; Azhar, S.; Anson, M.; Wong, Y.-L. Comparison of the strength and durability performance of normal- and high-strength pozzolanic concretes at elevated temperatures. *Cem. Concr. Res.* **2001**, *31*, 1291–1300. [CrossRef]
41. Netinger, I.; Kesegic, I.; Guljas, I. The effect of high temperatures on the mechanical properties of concrete made with different types of aggregates. *Fire Saf. J.* **2011**, *46*, 425–430. [CrossRef]
42. Rashad, A.M. Potential Use of Silica Fume Coupled with Slag in HVFA Concrete Exposed to Elevated Temperatures. *J. Mater. Civ. Eng.* **2015**, *27*, 4015019. [CrossRef]
43. Malik, M.; Bhattacharyya, S.K.; Barai, S.V. Microstructural Changes in Concrete: Postfire Scenario. *J. Mater. Civ. Eng.* **2021**, *33*, 4020462. [CrossRef]

44. Hager, I.; Tracz, T.; Choin´ska, M.; Mróz, K. Effect of cement type on the mechanical behavior and permeability of concrete subjected to high temperatures. *Materials* **2019**, *12*, 3021. [CrossRef] [PubMed]
45. Li, Y.; Tan, K.H.; Yang, E.-H. Synergistic effects of hybrid polypropylene and steel fibers on explosive spalling prevention of ultra-high performance concrete at elevated temperature. *Cem. Concr. Compos.* **2018**, *96*, 174–181. [CrossRef]
46. Yonggui, W.; Shuaipeng, L.; Hughes, P.; Yuhui, F. Mechanical properties and microstructure of basalt fibre and nano-silica reinforced recycled concrete after exposure to elevated temperatures. *Constr. Build. Mater.* **2020**, *247*, 118561. [CrossRef]
47. Lau, A.; Anson, M. Effect of high temperatures on high performance steel fibre reinforced concrete. *Cem. Concr. Res.* **2006**, *36*, 1698–1707. [CrossRef]
48. Düğenci, O.; Haktanir, T.; Altun, F. Experimental research for the effect of high temperature on the mechanical properties of steel fiber-reinforced concrete. *Constr. Build. Mater.* **2015**, *75*, 82–88. [CrossRef]
49. Da Silva, J.B.; Pepe, M.; Filho, R.D.T. High temperatures effect on mechanical and physical performance of normal and high strength recycled aggregate concrete. *Fire Saf. J.* **2020**, *117*, 103222. [CrossRef]
50. Mathews, M.E.; Kiran, T.; Naidu, V.C.H.; Jeyakumar, G.; Anand, N. Effect of high-temperature on the mechanical and durability behaviour of concrete. *Mater. Today: Proc.* **2020**, *42*, 718–725. [CrossRef]
51. Al-Owaisy, S.R. Effect of high temperatures on shear transfer strength of concrete. *J. Eng. Sustain. Dev.* **2007**, *11*, 92–103.
52. Tang, C.-W. Residual Mechanical Properties of Fiber-Reinforced Lightweight Aggregate Concrete after Exposure to Elevated Temperatures. *Appl. Sci.* **2020**, *10*, 3519. [CrossRef]
53. Zhang, B.; Bicanic, N.; Pearce, C.J.; Balabanic, G.; Purkiss, J.A. Discussion: Residual fracture properties of normal- and high-strength concrete subject to elevated temperatures. *Mag. Concr. Res.* **2000**, *52*, 123–136. [CrossRef]
54. Yermak, N.; Pliya, P.; Beaucour, A.-L.; Simon, A.; Noumowé, A. Influence of steel and/or polypropylene fibres on the behaviour of concrete at high temperature: Spalling, transfer and mechanical properties. *Constr. Build. Mater.* **2017**, *132*, 240–250. [CrossRef]
55. Pliya, P.; Beaucour, A.-L.; Noumowé, A. Contribution of cocktail of polypropylene and steel fibres in improving the behaviour of high strength concrete subjected to high temperature. *Constr. Build. Mater.* **2010**, *25*, 1926–1934. [CrossRef]
56. Ahmad, S.; Rasul, M.; Adekunle, S.K.; Al-Dulaijan, S.U.; Maslehuddin, M.; Ali, S.I. Mechanical properties of steel fiber-reinforced UHPC mixtures exposed to elevated temperature: Effects of exposure duration and fiber content. *Compos. Part B Eng.* **2018**, *168*, 291–301. [CrossRef]
57. Wu, H.; Lin, X.; Zhou, A. A review of mechanical properties of fibre reinforced concrete at elevated temperatures. *Cem. Concr. Res.* **2020**, *135*, 106117. [CrossRef]
58. Khaliq, W.; Kodur, V. Thermal and mechanical properties of fiber reinforced high performance self-consolidating concrete at elevated temperatures. *Cem. Concr. Res.* **2011**, *41*, 1112–1122. [CrossRef]
59. Khaliq, W.; Kodur, V. High temperature mechanical properties of high-strength fly ash concrete with and without fibers. *ACI Mater. J.* **2012**, *109*, 665–674.
60. Ju, Y.; Wang, L.; Liu, H.; Tian, K. An experimental investigation of the thermal spalling of polypropylene-fibered reactive powder concrete exposed to elevated temperatures. *Sci. Bull.* **2015**, *60*, 2022–2040. [CrossRef]
61. Caetano, H.; Ferreira, G.; Rodrigues, J.P.C.; Pimienta, P. Effect of the high temperatures on the microstructure and compressive strength of high strength fibre concretes. *Constr. Build. Mater.* **2018**, *199*, 717–736. [CrossRef]
62. Zheng, W.; Li, H.; Wang, Y. Compressive behaviour of hybrid fiber-reinforced reactive powder concrete after high temperature. *Mater. Des.* **2012**, *41*, 403–409. [CrossRef]
63. Raza, S.S.; Qureshi, L.A. Effect of carbon fiber on mechanical properties of reactive powder concrete exposed to elevated temperatures. *J. Build. Eng.* **2021**, *42*, 102503. [CrossRef]
64. Luo, X.; Sun, W.; Nin Chan, S.Y. Effect of heating and cooling regimes on residual strength and microstructure of normal strength and high-performance concrete. *Cem. Concr. Res.* **2000**, *30*, 379–383. [CrossRef]
65. Sideris, K.K.; Manita, P. Residual mechanical characteristics and spalling resistance of fiber reinforced self-compacting concretes exposed to elevated temperatures. *Constr. Build. Mater.* **2013**, *41*, 296–302. [CrossRef]
66. Zhang, D.; Dasari, A.; Tan, K.H. On the mechanism of prevention of explosive spalling in ultra-high performance concrete with polymer fibers. *Cem. Concr. Res.* **2018**, *113*, 169–177. [CrossRef]
67. Aslani, F.; Kelin, J. Assessment and development of high-performance fibre-reinforced lightweight self-compacting concrete including recycled crumb rubber aggregates exposed to elevated temperatures. *J. Clean. Prod.* **2018**, *200*, 1009–1025. [CrossRef]
68. Eidan, J.; Rasoolan, I.; Rezaeian, A.; Poorveis, D. Residual mechanical properties of polypropylene fiber-reinforced concrete after heating. *Constr. Build. Mater.* **2018**, *198*, 195–206. [CrossRef]
69. Peng, G.-F.; Yang, W.-W.; Zhao, J.; Liu, Y.-F.; Bian, S.-H.; Zhao, L.-H. Explosive spalling and residual mechanical properties of fiber-toughened high-performance concrete subjected to high temperatures. *Cem. Concr. Res.* **2006**, *36*, 723–727. [CrossRef]
70. Gao, D.; Yan, D.; Li, X. Splitting strength of GGBFS concrete incorporating with steel fiber and polypropylene fiber after exposure to elevated temperatures. *Fire Saf. J.* **2012**, *54*, 67–73. [CrossRef]
71. Canbaz, M. The effect of high temperature on reactive powder concrete. *Constr. Build. Mater.* **2014**, *70*, 508–513. [CrossRef]
72. Sultan, H.K.; Alyaseri, I. Effects of elevated temperatures on mechanical properties of reactive powder concrete elements. *Constr. Build. Mater.* **2020**, *261*, 120555. [CrossRef]
73. Arna'Ot, F.H.; Abbass, A.A.; Abualtemen, A.A.; Abid, S.R.; Özakça, M. Residual strength of high strength concentric column-SFRC flat plate exposed to high temperatures. *Constr. Build. Mater.* **2017**, *154*, 204–218. [CrossRef]

74. Al-Owaisy, S.R. Strength and elasticity of steel fiber reinforced concrete at high temperatures. *J. Eng. Sustain. Dev.* **2007**, *11*, 125–133.
75. Torić, N.; Boko, I.; Peros, B. Reduction of Postfire Properties of High-Strength Concrete. *Adv. Mater. Sci. Eng.* **2013**, *2013*, 712953. [CrossRef]
76. Cheng, F.-P.; Kodur, V.K.R.; Wang, T.-C. Stress-Strain Curves for High Strength Concrete at Elevated Temperatures. *J. Mater. Civ. Eng.* **2004**, *16*, 84–90. [CrossRef]
77. Al-Owaisy, S.R. Post heat exposure properties of steel fiber reinforced concrete. *J. Eng. Sustain. Dev.* **2006**, *10*, 194–207.
78. Varona, F.B.; Baeza-Brotons, F.; Tenza-Abril, A.J.; Baeza, F.J.; Bañón, L. Residual Compressive Strength of Recycled Aggregate Concretes after High Temperature Exposure. *Materials* **2020**, *13*, 1981. [CrossRef]
79. Amin, M.N.; Khan, K. Mechanical Performance of High-Strength Sustainable Concrete under Fire Incorporating Locally Available Volcanic Ash in Central Harrat Rahat, Saudi Arabia. *Materials* **2020**, *14*, 21. [CrossRef]
80. Zhang, P.; Kang, L.; Wang, J.; Guo, J.; Hu, S.; Ling, Y. Mechanical Properties and Explosive Spalling Behavior of Steel-Fiber-Reinforced Concrete Exposed to High Temperature—A Review. *Appl. Sci.* **2020**, *10*, 2324. [CrossRef]
81. Ding, Y.; Li, D.; Zhang, Y.; Azevedo, C. Experimental investigation on the composite effect of steel rebars and macro fibers on the impact behavior of high performance self-compacting concrete. *Constr. Build. Mater.* **2017**, *136*, 495–505. [CrossRef]
82. Abid, S.R.; Abdul-Hussein, M.L.; Ayoob, N.S.; Ali, S.H.; Kadhum, A.L. Repeated drop-weight impact tests on self-compacting concrete reinforced with micro-steel fiber. *Heliyon* **2020**, *6*, e03198. [CrossRef]
83. Mastali, M.; Dalvand, A. The impact resistance and mechanical properties of self-compacting concrete reinforced with recycled CFRP pieces. *Compos. Part B Eng.* **2016**, *92*, 360–376. [CrossRef]
84. Nili, M.; Afroughsabet, V. The effects of silica fume and polypropylene fibers on the impact resistance and mechanical properties of concrete. *Constr. Build. Mater.* **2010**, *24*, 927–933. [CrossRef]
85. Abid, S.R.; Hussein, M.L.A.; Ali, S.H.; Kazem, A.F. Suggested modified testing techniques to the ACI 544-R repeated drop-weight impact test. *Constr. Build. Mater.* **2020**, *244*, 118321. [CrossRef]
86. Murali, G.; Abid, S.R.; Abdelgader, H.S.; Amran, Y.H.M.; Shekarchi, M.; Wilde, K. Repeated Projectile Impact Tests on Multi-Layered Fibrous Cementitious Composites. *Int. J. Civ. Eng.* **2021**, *19*, 635–651. [CrossRef]
87. Abid, S.R.; Murali, G.; Amran, M.; Vatin, N.; Fediuk, R.; Karelina, M. Evaluation of Mode II Fracture Toughness of Hybrid Fibrous Geopolymer Composites. *Materials* **2021**, *14*, 349. [CrossRef]
88. Murali, G.; Abid, S.; Amran, M.; Fediuk, R.; Vatin, N.; Karelina, M. Combined Effect of Multi-Walled Carbon Nanotubes, Steel Fibre and Glass Fibre Mesh on Novel Two-Stage Expanded Clay Aggregate Concrete against Impact Loading. *Crystals* **2021**, *11*, 720. [CrossRef]
89. Ramakrishnan, K.; Depak, S.; Hariharan, K.; Abid, S.R.; Murali, G.; Cecchin, D.; Fediuk, R.; Amran, Y.M.; Abdelgader, H.S.; Khatib, J.M. Standard and modified falling mass impact tests on preplaced aggregate fibrous concrete and slurry infiltrated fibrous concrete. *Constr. Build. Mater.* **2021**, *298*, 123857. [CrossRef]
90. Barbero, E.; Fernandez-Saez, J.; Navarro, C. Statistical analysis of the mechanical properties of composite materials. *Compos. Part B Eng.* **2000**, *31*, 375–381. [CrossRef]
91. Murali, G.; Indhumathi, T.; Karthikeyan, K.; Ramkumar, V.R. Analysis of flexural fatigue failure of concrete made with 100% coarse recycled and natural aggregates. *Comput. Concr.* **2018**, *21*, 291–298.
92. Mohammadi, K.; Alavi, O.; Mostafaeipour, A.; Goudarzi, N.; Jalilvand, M. Assessing different parameters estimation methods of Weibull distribution to compute wind power density. *Energy Convers. Manag.* **2016**, *108*, 322–335. [CrossRef]
93. Akdağ, S.A.; Dinler, A. A new method to estimate Weibull parameters for wind energy applications. *Energy Convers. Manag.* **2009**, *50*, 1761–1766. [CrossRef]
94. Murali, G.; Muthulakshmi, T.; Nycilin Karunya, N.; Iswarya, R.; Hannah Jennifer, G.; Karthikeyan, K. Impact Response and Strength Reliability of Green High-Performance Fibre Reinforced Concrete Subjected to Freeze-Thaw Cycles in NaCl Solution. *Mater. Sci. Medzg.* **2017**, *23*, 384–388.
95. Murali, G.; Santhi, A.S.; Ganesh, G.M. Impact Resistance and Strength Reliability of Fiber-reinforced Concrete in Bending under Drop Weight Impact Load. *Int. J. Technol.* **2014**, *5*, 111. [CrossRef]
96. Murali, G.; Santhi, A.S.; Mohan Ganesh, G. Impact resistance and strength reliability of fiber reinforced concrete using two parameter Weibull distribution. ARPN. *J. Eng. Appl. Sci.* **2014**, *9*, 554–559.
97. Murali, G.; Gayathri, R.; Ramkumar, V.R.; Karthikeyan, K. Two statistical scrutinize of impact strength and strength reliability of steel Fibre-Reinforced Concrete. *KSCE J. Civ. Eng.* **2017**, *22*, 257–269. [CrossRef]
98. Murali, G.; Asrani, N.P.; Ramkumar, V.R.; Siva, A.; Haridharan, M.K. Impact Resistance and Strength Reliability of Novel Two-Stage Fibre-Reinforced Concrete. *Arab. J. Sci. Eng.* **2018**, *44*, 4477–4490. [CrossRef]
99. Asrani, N.P.; Murali, G.; Parthiban, K.; Surya, K.; Prakash, A.; Rathika, K.; Chandru, U. A feasibility of enhancing the impact resistance of hybrid fibrous geopolymer composites: Experiments and modelling. *Constr. Build. Mater.* **2019**, *203*, 56–68. [CrossRef]
100. Abirami, T.; Loganaganandan, M.; Murali, G.; Fediuk, R.; Sreekrishna, R.V.; Vignesh, T.; Januppriya, G.; Karthikeyan, K. Experimental research on impact response of novel steel fibrous concretes under falling mass impact. *Constr. Build. Mater.* **2019**, *222*, 447–457. [CrossRef]

101. Prasad, N.; Murali, G. Exploring the impact performance of functionally-graded preplaced aggregate concrete incorporating steel and polypropylene fibres. *J. Build. Eng.* **2021**, *35*, 102077. [CrossRef]
102. Murali, G.; Karthikeyan, K.; Ramkumar, V.R. Reliability Analysis of Impact Failure Energy of Fibre Reinforced Concrete Using Weibull Distribution. *J. Appl. Sci. Eng.* **2018**, *21*, 163–170.
103. Gayathri, R.; Murali, G.; Parthiban, K.; Haridharan, M.K.; Karthikeyan, K. A Four Novel Energy Pattern Factor Method for Computation of Weibull Parameter in Impact Strength Reliability of Fibre Reinforced Concrete. *Inter. J. Eng. Tech.* **2018**, *7*, 272–280.

 MDPI

Article

Statistical Damage Constitutive Model for High-Strength Concrete Based on Dissipation Energy Density

Liangliang Zhang [1], Hua Cheng [1,2,*], Xiaojian Wang [1], Jimin Liu [1] and Longhui Guo [1]

[1] School of Civil Engineering and Architecture, Anhui University of Science and Technology, Huainan 232001, China; zllaust@163.com (L.Z.); xjwang@aust.edu.cn (X.W.); jimliu@aust.edu.cn (J.L.); guolonghui7864@163.com (L.G.)

[2] School of Resources and Environmental Engineering, Anhui University, Hefei 230022, China

* Correspondence: hcheng@aust.edu.cn

Abstract: To study the energy evolution law and damage constitutive behavior of high-strength concrete based on the conventional triaxial compression tests of C60 and C70 high-strength concrete subjected to five different confining pressures, the failure characteristics of high-strength concrete are analyzed at different confining pressures, and the evolution of the input energy density, elastic strain energy density, and dissipation energy density with axial strain and confining pressure are quantified. Combined with a continuous damage theory and non-equilibrium statistical method, the ratio of dissipation energy density of concrete to dissipation energy density corresponding to peak stress is used as the mechanical parameter. Assuming that the mechanical parameter obeys the Weibull distribution laws, the statistical damage variable describing the damage characteristics of concrete were derived. According to the Lemaitre strain equivalent principle, the damage variable is introduced to the generalized Hooke law to establish the statistical damage constitutive model for high-strength concrete. The results show that: (1) the input energy density and dissipation energy density increases with the increase of axial strain, while the elastic strain energy density increases first and then decreases as a function of the axial strain and reaches the maximum value at the peak stress; (2) the input, elastic strain, and dissipated energy densities corresponding to the peak stress of the two high-strength concretes all increase as a function of confining pressure, and the elastic strain energy density corresponding to the peak stress increases linearly as a function of the confining pressure; (3) the statistical damage constitutive model results of C60 and C70 high-strength concrete are in good agreement with the test results, and the average relative standard deviations are only 3.64% and 3.99%. These outcomes verify the rationality and accuracy of the model.

Keywords: dissipation energy density; high-strength concrete; Weibull distribution; damage mechanics; constitutive model

Citation: Zhang, L.; Cheng, H.; Wang, X.; Liu, J.; Guo, L. Statistical Damage Constitutive Model for High-Strength Concrete Based on Dissipation Energy Density. *Crystals* **2021**, *11*, 800. https://doi.org/10.3390/cryst11070800

Academic Editors: Cesare Signorini, Antonella Sola, Sumit Chakraborty, Valentina Volpini and Ing. José L. García

Received: 24 May 2021
Accepted: 28 June 2021
Published: 8 July 2021

1. Introduction

With the application of high-strength concrete in civil engineering, transportation, water conservancy, municipal engineering, and other engineering fields, it is of great significance to study the deformation law and failure characteristics of high-strength concrete subject to complex stress states aiming to improved scientific designs of concrete buildings (structures) and to the guarantee of their safety [1,2]. High-strength concrete is a type of composite material composed of sand, stone, cementing material, and water, mixed based on specific analogies. Owing to the incompleteness of vibration, incomplete hydration reaction, and temperature effects, there are a large number of discontinuity and irregular shape cracks and joints in the interior of high-strength concrete. The deformation and mechanical characteristics of high-strength concrete are obviously nonlinear and discontinuous. This leads to the complexity of micro-structure and macro-strength evolution of high-strength concrete at different ages and complex stress conditions. It is difficult to effectively judge the strength change and failure behavior of concrete materials by using classical elastic–plastic theory. Numerous research studies have shown that the failure of materials is a state instability phenomenon driven by energy, and the transmission and transformation of energy are the fundamental reasons for the deformation and failure of materials [3,4]. Therefore, the energy method based on thermodynamic theory is an effective way to study the constitutive relationships and failure behaviors of concrete materials.

The constitutive relationship constitutes the basis for the study of the relationship between material deformation and load. Ever since the proposition of the parallel bar system (PBS) model for the concrete uniaxial tensile process in 1982 by Krajcinovic [5], statistical damage theory has gradually become a new research hotspot of concrete damage mechanics. Yang et al. [6] proposed a statistical damage constitutive model of multi-size polypropylene fiber concrete under impact load and obtained the statistical parameters based on the particle swarm optimization algorithm. Bai et al. [7] assumed that there are two damage mechanisms of fracture and yield in the meso-structures of quasi-brittle materials, and established a triaxial orthotropic statistical damage model for concrete that can predict its constitutive behavior at complex loading environments. Based on Weibull and lognormal statistical distribution theory and Lemaitre's strain equivalent principle, Liang Hui et al. [8] established a sectional uniaxial compression damage constitutive model of concrete materials by introducing the strain-rate factor. Cervera [9] established a rate-independent isotropic damage constitutive model of concrete, and used the model to conduct seismic analysis of concrete dams. Zhou [10] investigated the compression behavior of coral aggregate concrete (CAC) at uniaxial and triaxial loading, and proposed a constitutive model for coral aggregate concrete subjected to uniaxial and triaxial compression, wherein the suggested models correlated well with the test results. Wu [11] established the plastic damage constitutive relation with the internal variables based on the continuum damage mechanics, and proposed an energy release rate-based plastic-damage model for concrete. The aforementioned studies showed that the constitutive behavior of concrete materials can be studied from the perspective of statistics, but at present, most studies use the mechanical or deformation parameters of concrete materials as the basis for establishing statistical damage constitutive equations, and few studies have introduced the energy dissipation density parameter into the constitutive relationship.

To study the energy evolution law and damage constitutive behavior of high-strength concrete subjected to complex stress states, conventional triaxial compression tests at different confining pressures were conducted with ZTCR-2000 rock triaxial testing system. The evolution law of input energy, elastic strain energy and dissipation energy with axial strain and confining pressure were analyzed. Based on the continuum damage theory and non-equilibrium statistical method, a statistical damage constitutive model was established for high-strength concrete based on the use of the ratio of dissipation energy density of concrete to the dissipation energy density that corresponded to peak stress.

2. Conventional Triaxial Compression Tests of C60 and C70 High-Strength Concrete

2.1. Test Materials and Equipment

According to the Specifications for Mix Proportion Design of Ordinary Concrete (JGJ55-2011), high-strength concrete with strength grade of C60 and C70 was prepared. The cement was grade 52.5 ordinary Portland cement, with siliceous river sand with a fineness modulus of 2.73, and crushed basalt with particle sizes in the range of 5–10 mm. The admixture is the NF-F high-efficiency admixture compound, wherein the slag and silicon powder accounted for 73% and 20%, respectively. The mix proportions of high-strength concrete (C60 and C70) are listed in Table 1.

Table 1. Mixture proportions of C60 and C70 high-strength concrete.

Strength Grade	Cement (kg/m^3)	Admixture (kg/m^3)	Cementing Materials (kg/m^3)	Sand (kg/m^3)	Gravel (kg/m^3)	Water (kg/m^3)	Sand Rate	Water Cement Ratio
C60	415	135	550	620	1105	175	0.36	0.42
C70	425	145	570	615	1095	170	0.36	0.4

The concrete mix was poured into 150 × 150 × 150 mm plastic molds. The obtained concrete specimens were demolded after 16 h and immediately transferred into a curing box at 20 ± 2 °C and at a relative humidity of 95% for 28 days. Three specimens were subsequently randomly selected for uniaxial compressive strength testing at 28 days. The remaining concrete specimens were finished by coring, cutting, and grinding according to the Standard for Tests Method of Engineering Rock Masses (GB/T50266-2013). The obtained standard cylindrical specimens measured $\varphi 50 \times 100$ mm. The upper and lower faces of the specimens were ground to be plane and parallel, and ensured a uniform stress distribution in the final cylindrical specimens.

Conventional triaxial compressive tests were conducted with a ZTCR-2000 rock triaxial testing system (Figure 1). The equipment is mainly used for uniaxial, conventional triaxial and creep tests of concrete and rock materials, which is mainly composed of an axial pressure system, confining pressure system, servo oil source, temperature control system, and computer control system. The computer system controls the whole process of the test equipment and operates the test steps, and automatically collects and processes the test data. The axial compression system includes an axial loading frame, three-axis cavity lifting device, controller, sensor, and electro-hydraulic servo valve. The confining pressure system consists of a pressure chamber, pressurization device, pressure transmitter, digital display meter, low temperature liquid filling oil source, air pump, guide rail, controller, and sensor. The computer system can draw the curve of each parameter in real time. The main technical parameters of the test equipment are shown in Table 2.

The circumferential and axial deformation was measured by a linear variable differential transducer (LVDT) attached to a chain wrapped tightly around the sample and an axial LVDT, respectively. Cylindrical samples of C60 and C70 high strength concrete were tested by applying compressive stress σ_1 and the constant confining pressures $\sigma_2 = \sigma_3 = 0, 5, 10, 15, 20$ MPa, respectively. The samples were hydrostatically compressed until the level of desired circumferential pressure was reached. The sample was then vertically compressed at a rate of 0.05 mm/min until failure.

Table 2. Main technical parameters of ZTCR-2000 rock triaxial test system.

Project	Technical Parameters
Maximum axial test force	2000 kN
Test force accuracy	±1%
Test force resolution	1/180,000
Maximum displacement of piston	120 mm
Displacement accuracy	±0.5% FS
Displacement resolution	5 μm
Deformation measurement accuracy	±1%
Deformation resolution	1/180,000
Maximum confining pressure	50 MPa
Confining pressure accuracy	±1%
Confining pressure resolution	1/180,000
Minimum control temperature	−20 °C
Temperature control accuracy	±0.5 °C

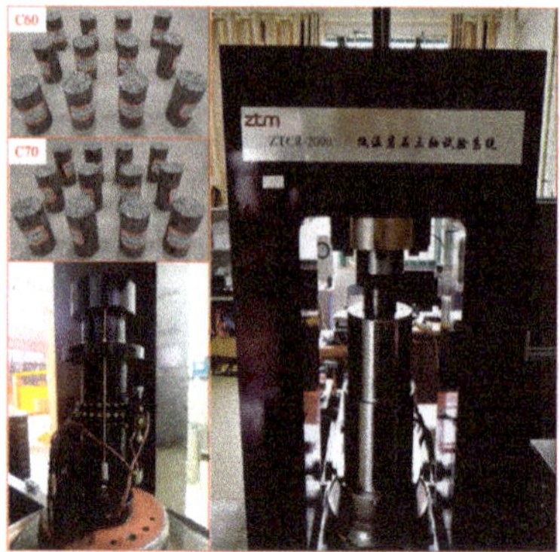

Figure 1. ZTCR-2000 rock triaxial testing system.

2.2. Analysis of Test Results

The stress–strain curves of high-strength concrete specimens (C60 and C70) at the confining pressures of 0, 5, 10, 15, and 20 MPa are shown in Figure 2.

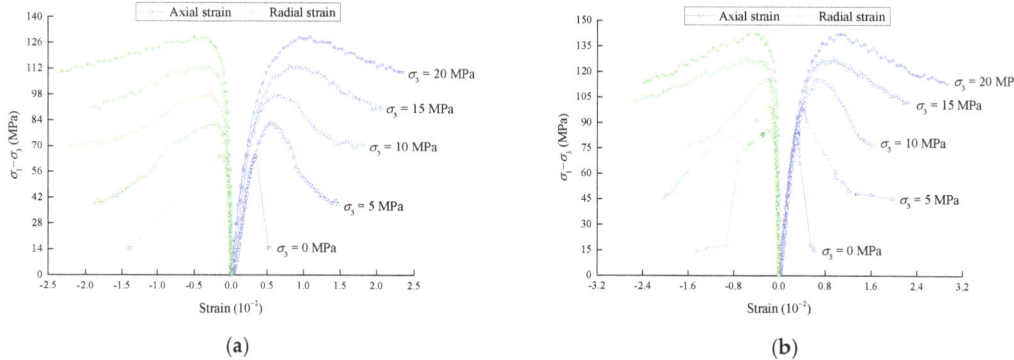

Figure 2. Conventional triaxial stress–strain curves of high-strength concrete. (**a**) C60 and (**b**) C70.

According to the test results, the basic mechanical parameters of C60 and C70 high-strength concrete at different confining pressures are shown in Table 3.

Table 3. Mechanical parameters of high-strength concrete (C60 and C70) at different confining pressures.

σ_3 (MPa)	$\sigma_1-\sigma_3$ (MPa)		Elastic Modulus (GPa)		Poisson's Ratio	
	C60	C70	C60	C70	C60	C70
0	65.38	84.41	28	31	0.31	0.29
5	81.96	100.22	29	32	0.28	0.26
10	98.26	115.98	31	33	0.29	0.26
15	114.01	128.22	33	35	0.28	0.28
20	129.12	141.67	34	36	0.26	0.26

During the setting process of concrete, many microcracks and holes will be formed owing to the drying shrinkage and water evaporation of cement slurry as well as the excessive interfacial microdefects (a) between the cement paste and aggregate, (b) among various phases of cement paste, and (c) among hydration products and un-hydrated cement particles. When subjected to uniaxial loading, the original cracks propagate along the interface and their direction is basically the same as the loading direction. The upper and lower parts of the cracks are compressive stress concentration areas, and the side is a tensile stress area. Because the tensile strength is far lower than the compressive strength, the microcracks first produce tensile failure. At this time, owing to the lack of lateral restraints, the rapid development of microcracks eventually leads to the loss of concrete strength [12]. When the confining pressure is 5, 10, 15, and 20 MPa, the high-confining pressure counteracts the tensile stress on the side of the crack, and increases the compressive stress required for the fracture to connect with each other. The macroscopic performance is characterized by the fact that the peak stresses of C60 and C70 high-strength concrete increase as a function of the confining pressure. Therefore, the confining pressure can effectively limit the propagation speed of micro-cracks in concrete samples, slows down the damage degree of samples, and improves the bearing capacity and deformation capacity of concrete samples. Because the conventional triaxial test is carried out in a closed cavity, it is impossible to directly observe the generation and development process of cracks, but it can be studied by means of SEM technology [13].

The failure modes of C60 and C70 high-strength concrete specimens at different confining pressures are shown in Figure 3. When the confining pressure is 0 MPa, the failure mode of the two high-strength concrete specimens is tensile failure. This is because the transverse tensile stress generated by the Poisson effect at axial loading is greater than the tensile strength of concrete, and results in cracks parallel to the maximum principal stress direction of concrete samples. When the confining pressures are 5, 10, 15, and 20 MPa, the failure mode is oblique shear failure. This is attributed to the fact that the larger confining pressure inhibits the expansion of vertical cracks. The shear stress on the inclined section of concrete subject to high-triaxial stress is greater than its shear strength, and the inclined shear failure occurs along a weak plane.

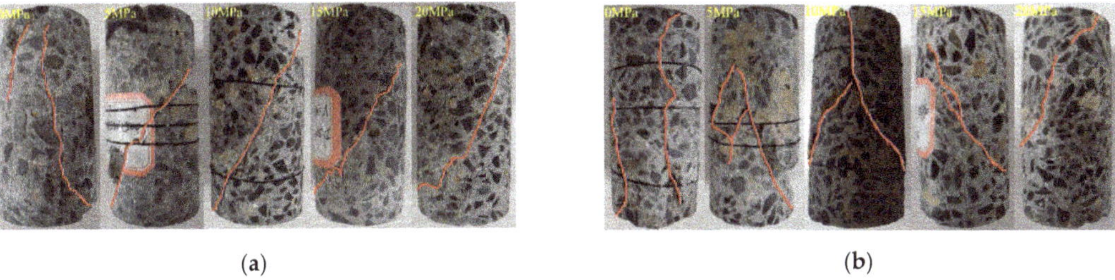

(a) (b)

Figure 3. Failure modes of C60 and C70 high-strength concrete at different confining pressures. (**a**) C60 and (**b**) C70.

3. Energy Analysis of High-Strength Concrete during Compression

3.1. Energy Analysis

This section may be divided by subheadings. It should provide a concise and precise description of the experimental results, their interpretation, as well as the experimental conclusions that can be drawn.

According to the first law of thermodynamics, regardless of the influence of external temperature change and material exchange on the test system, the input energy W_F is equal to the sum of the elastic strain energy W_E and dissipation energy W_D in the test process [14,15].

$$W_F = W_E + W_D \tag{1}$$

The input energy mainly includes the work done by the axial force and confining pressure when the concrete sample deforms, and the elastic strain energy is the energy accumulated in the interior of concrete sample when the elastic deformation occurs because (a) the elastic deformation is reversible, and (b) because the elastic strain energy is also reversible. The dissipation energy mainly includes the (a) surface energy consumed during the initiation, development and penetration of cracks, (b) plastic strain energy for irreversible plastic deformation of concrete samples, (c) heat energy generated by friction and slip between cracks, and various radiation energy [16,17].

The energy input to the concrete specimen by the external force during the test can be expressed as [18],

$$W_F = \frac{\pi}{4}D^2H\left(\int_0^{\varepsilon_1}\sigma_1 d\varepsilon_1 + 2\int_0^{\varepsilon_3}\sigma_3 d\varepsilon_3\right) = VU_F \tag{2}$$

where σ_1, σ_3 are the maximum and minimum principal stresses, respectively, ε_1 and ε_3 are axial and lateral strains, respectively, U_F is the input energy density, V is the volume of the concrete sample, and D and H are the diameter and height of the concrete sample, respectively.

Similarly, the elastic strain energy and dissipation energy are obtained as follows,

$$\begin{cases} W_E = \frac{\pi}{4}D^2HU_E = VU_E \\ W_D = \frac{\pi}{4}D^2HU_D = VU_D \end{cases} \tag{3}$$

where U_E and U_D are the elastic strain and dissipation energy densities, respectively.

Substituting Equations (2) and (3) into Equation (1), we obtain,

$$U_F = U_E + U_D \tag{4}$$

According to the elastic theory (Gong et al., 2018), the elastic-strain energy density is obtained as follows,

$$U_E = \frac{1}{2}(\sigma_1\varepsilon_1^e + 2\sigma_3\varepsilon_3^e) = \frac{1}{2E}\left[\sigma_1^2 + 2(1-\mu)\sigma_3^2 - 4\mu\sigma_1\sigma_3\right] \tag{5}$$

By substituting Equations (2) and (5) into Equation (4), the dissipation energy density is obtained as follows,

$$U_D = \int_0^{\varepsilon_1}\sigma_1 d\varepsilon_1 + 2\int_0^{\varepsilon_3}\sigma_3 d\varepsilon_3 - \frac{1}{2E}\left[\sigma_1^2 + 2(1-\mu)\sigma_3^2 - 4\mu\sigma_1\sigma_3\right] \tag{6}$$

where E is the elastic modulus, and μ is the Poisson's ratio.

3.2. Relationship between Energy Density and Axial Strain

During the loading process of high-strength concrete samples, the changes of input energy, elastic-strain energy, and dissipated energy are always accompanied by the changes of initial hole compaction, elastic deformation, new crack propagation, and penetration. The energy coexisting in the specimen is not isolated from each other, but transformed with the concrete deformation process, and finally changed from the equilibrium state before the test to the new equilibrium state after the test. Based on the conventional triaxial compression test results of high-strength concrete (C60 and C70) at different confining pressures, the curves of input energy density, elastic-strain energy density and dissipation energy density with axial strain at different confining pressures are obtained according to Equations (2), (5), and (6). The results are shown in Figure 4.

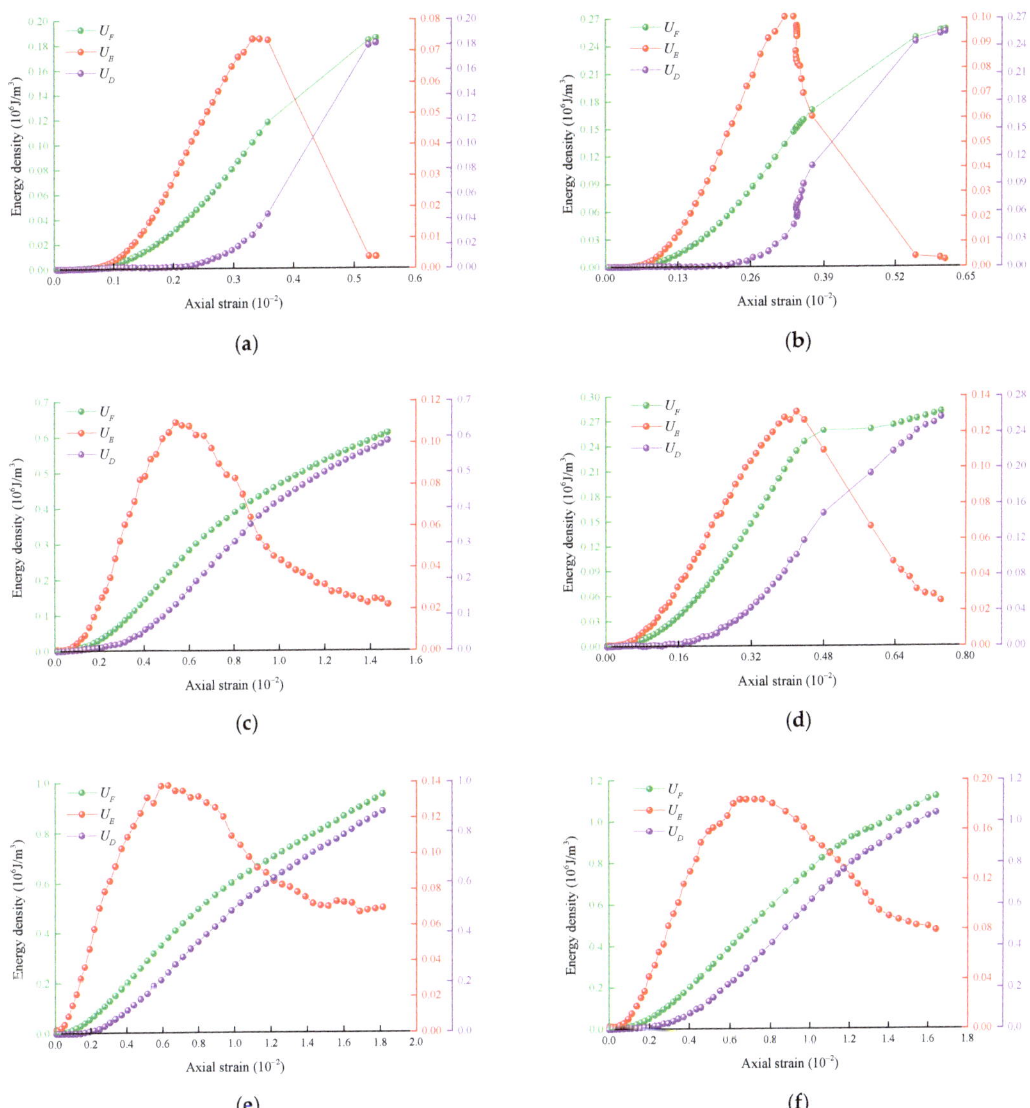

Figure 4. *Cont.*

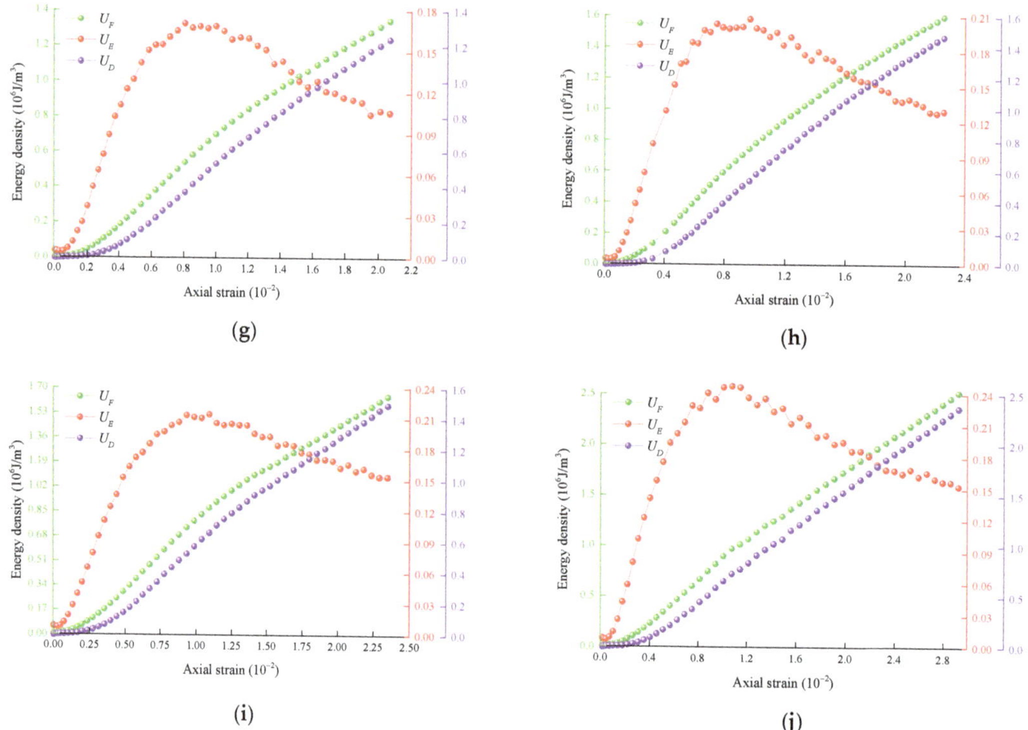

Figure 4. Relationships between energy density and axial strain of high-strength concrete (C60 and C70) at different confining pressures. (**a**) C60, $\sigma_3 = 0$ MPa; (**b**) C70, $\sigma_3 = 0$ MPa; (**c**) C60, $\sigma_3 = 5$ MPa; (**d**) C70, $\sigma_3 = 5$MPa; (**e**) C60, $\sigma_3 = 10$ MPa; (**f**) C70, $\sigma_3 = 10$ MPa; (**g**) C60, $\sigma_3 = 15$ MPa; (**h**) C70, $\sigma_3 = 15$ MPa; (**i**) C60, $\sigma_3 = 20$ MPa; (**j**) C70, $\sigma_3 = 20$ MPa.

It can be observed from Figure 4 that the input energy density and dissipation energy density of the two strength concrete samples at different confining pressures increase as a function of the axial strain. This indicates that the external force always inputs energy to the concrete samples during the entire test process, and at the same time, the energy is gradually dissipated. At the end of the test, the increasing trend of the input energy density slows down owing to the expansion of the sample and the negative work done by confining pressure offset part of the positive work done by axial stress. The elastic-strain energy density increases first and then decreases, as a function of the axial strain, and reaches the maximum value at the peak stress. This indicates that the pre-peak stage is mainly associated with the storage process of elastic-strain energy, and the post-peak stage is mainly associated with the release process of the elastic-strain energy.

Comparing the elastic-strain energy density curves of the two types of high-strength concrete and the stress–strain curves at the same confining pressure, it is found that the change trend of elastic-strain energy density with axial strain is similar to those of concrete samples. At the beginning of the test, the micro-holes and cracks in the concrete gradually close subject to the action of load, the stiffness of the concrete increases, and the curve becomes concave. At this time, most of the work done by the external force is converted into elastic-strain energy, it is stored in the sample, and the dissipation energy is almost zero. The elastic-strain energy density curve basically coincides with the input energy density curve. When the load exceeds the elastic limit of the concrete sample, new cracks will appear in the concrete. The generation and diffusion of the new cracks need to dissipate part of the surface energy, and the crack tip produces acoustic emission energy owing to the stress concentration effect accompanied by irrecoverable plastic strain energy and various radiation energy sources. As a result, the slope of the elastic-strain energy density curve slows down, and the dissipation energy density increases as a function of the axial strain. When the load exceeds the compressive strength of concrete, the concrete will be destroyed. At this time, the elastic-strain energy stored in pre-peak is released rapidly, and most of the input energy of external force is dissipated rapidly by the action of crack initiation, propagation and penetration, as well as the friction of fracture surface, thus resulting in an abrupt increase of dissipation energy density after the peak as a function of strain.

The results show that the storage of elastic-strain energy before the peak of concrete specimen is mainly elastic-strain energy, and constitutes the primary source of concrete failure. At the same time, with the dissipation of energy, the dissipation energy will gradually reduce the bearing capacity of concrete samples. After the peak, the elastic-strain energy is mainly released, and the released elastic-strain energy is transformed into various forms of energy dissipation, so the post-peak dissipation energy accounts for a large proportion of the input energy.

3.3. Relationship between Energy Density Corresponding to Peak Stress and Confining Pressure

Figure 5 shows the relationship between the confining pressure and input energy density U_{FP} and dissipated energy density U_{DP} corresponding to the peak stress of the two types of high-strength concrete samples. It can be observed from Figure 5a,b that the input energy density U_{FP} and the dissipation energy density U_{DP} corresponding to the peak stress of the two types of high-strength concrete samples increase as a function of the confining pressures, and the U_{FP} and U_{DP} of the C70 high-strength concrete samples subjected to the same confining pressure are higher than those of the C60 high-strength concrete. This is attributed to the fact that as the strength of the concrete increases, the system needs to input more energy to cause its failure. At the same time, the compressive capacity of high-strength concrete gradually decreases, and thus needs to dissipate more energy.

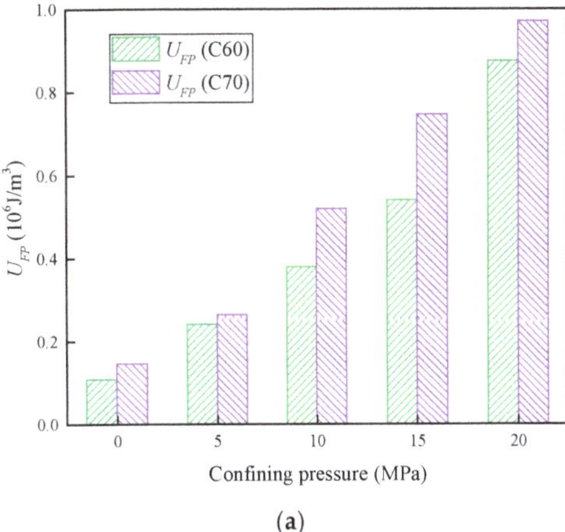

(a)

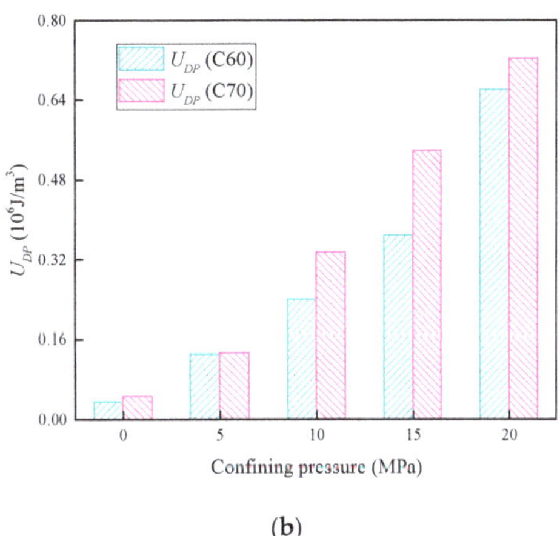

(b)

Figure 5. Relationships between confining pressure and U_{FP} and U_{DP}. (**a**) Relationship between U_{FP} and σ_3; (**b**) Relationship between U_{DP} and σ_3.

Figure 6 shows the relationship between the confining pressure and elastic-strain energy density U_{EP} corresponding to peak stress for two types of high-strength concrete samples. As it can be observed from the figure, U_{EP} and the confining pressure are linearly related, and the correlation coefficients are all above 0.99. It can be observed from Figure 6 that as U_{EP} increases, the energy release at high-confining pressures is more rapid and abrupt compared with low-confining pressures.

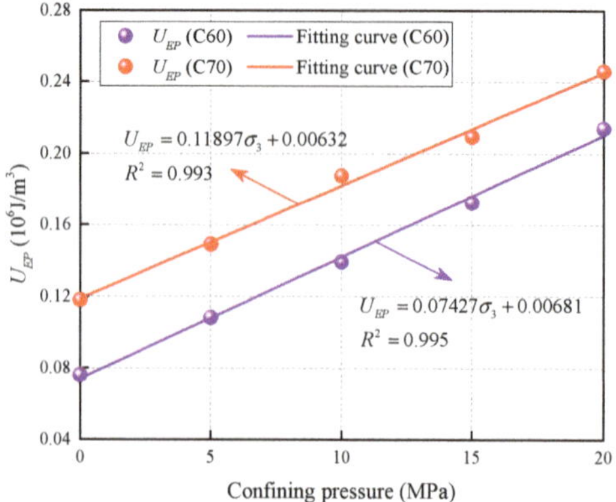

Figure 6. Relationship between confining pressure and U_{EP}.

3.4. Relationship between Final Energy Density and Confining Pressure

Figure 7 shows the relationship between the confining pressure and final elastic-strain energy density U_{ER} of two types of high-strength concrete specimens after complete failure. It can be observed from Figure 7 that the U_{ER} values of the two types of high-strength concrete samples increase as a function of the confining pressure. Considering the C60 high-strength concrete sample as an example, when the confining pressure is 0, 5, 10, 15, and 20 MPa, the U_{ER} values are 0.0038×10^6 J/m^3, 0.0244×10^6 J/m^3, 0.0665×10^6 J/m^3, 0.10150665×10^6 J/m^3, and 0.15330665×10^6 J/m^3, respectively. This shows that in the post-peak failure stage, the elastic-strain energy stored before the peak is not completely released, and a small part of the elastic strain energy is still stored in the sample. Accordingly, the larger the confining pressure is, the greater the residual elastic-strain energy is. This indicates that the confining pressure limits the release of the elastic-strain energy to a certain extent in the post-peak stage. Macroscopically, the concrete sample still has a finite bearing capacity and residual strength after failure. In addition, the larger the elastic strain energy remaining in the specimen is, the greater the residual strength of the concrete sample.

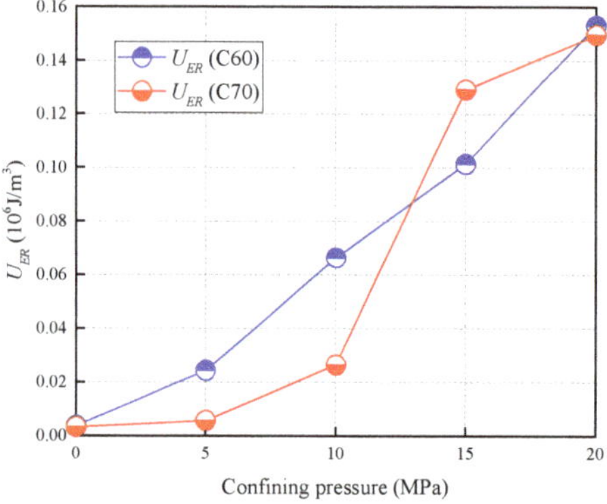

Figure 7. Relationship between confining pressure and final elastic-strain energy density.

4. Statistical Damage Constitutive Model of High-Strength Concrete

4.1. Establishment of Constitutive Model

According to Lemaitre's strain equivalent principle [19], the strain response produced by the nominal stress acting on the damaged material is equivalent to the strain response produced by the effective stress that acts on the undamaged material. Therefore, the constitutive relationship of the damaged material can be obtained by replacing the nominal stress with the effective stress.

$$\sigma = \sigma^*(1 - D) = E\varepsilon(1 - D) \tag{7}$$

where σ and σ^* are nominal and effective stresses, respectively, and D is the damage variable.

In the conventional triaxial compression test of concrete, σ_1, σ_3, and ε_1, can be measured, and the corresponding effective stresses are σ_1^* and σ_3^*. According to Equation (7) and the generalized Hooke's law [20,21], we can obtain,

$$\begin{cases} \varepsilon_1 = \frac{1}{E}(\sigma_1^* - 2\mu\sigma_3^*) \\ \sigma_1^* = \frac{\sigma_1}{1-D} \\ \sigma_3^* = \frac{\sigma_3}{1-D} \end{cases} \tag{8}$$

The high-strength concrete sample is equivalent to the macrostructure composed of innumerable tiny microelements, wherein the microelements are small enough at the macroscale, and the microelements are large enough at the microscale. The change of mechanical properties of each microelement is equivalent to the change of the mechanical properties of the high-strength concrete sample [22]. The microelement is a linear elastic body before failure, and the stress–strain relationship satisfies Hooke's law. Because the concrete is a mixture of various mineral materials and cementitious materials, it has an obvious heterogeneity, and its mechanical properties show the characteristics of random distribution. The resulting damage is also randomly distributed in concrete materials. Therefore, the mechanical properties of concrete microelements can be described mathematically by a statistical method, based on the assumption that the ratio of dissipation energy density U_D to dissipated energy density U_{DP} at peak stress obeys the Weibull statistical law, and the probability density function can be expressed as follows,

$$\begin{cases} \chi = \frac{U_D}{U_{DP}} \\ P[\chi] = \frac{k}{m}\left(\frac{\chi}{m}\right)^{k-1} \exp\left[-\left(\frac{\chi}{m}\right)^k\right] \end{cases} \tag{9}$$

where $P[\chi]$ is the function of probability density, m is the scale parameter, and k is the shape parameter of the distribution.

The dissipation energy increases as a function of the axial strain. When the dissipation energy density reaches a certain level, the number of damaged microelements in concrete samples can be expressed as,

$$N_D = N\int P[\chi]\mathrm{d}(\chi) = N\int \frac{k}{m}\left(\frac{\chi}{m}\right)^{k-1} \exp\left[-\left(\frac{\chi}{m}\right)^k\right]\mathrm{d}(\chi) = N\left\{1 - \exp\left[-\left(\frac{\chi}{m}\right)^k\right]\right\} \tag{10}$$

where N_D is the number of damaged microelements, and N is the number of total microelements.

The damage variable D is the ratio of damaged microelements to the total number of microelements [23], and is expressed as follows,

$$D = \frac{N_D}{N} \tag{11}$$

By substituting Equation (11) into Equation (10), the statistical damage variables describing the damage characteristics of concrete are obtained as follows,

$$D = 1 - \exp\left[-\left(\frac{\chi}{m}\right)^k\right] \tag{12}$$

The statistical damage constitutive relation of high-strength concrete based on Weibull distribution can be obtained by substituting Equation (12) into Equation (8),

$$\sigma_1 - 2\mu\sigma_3 = E\varepsilon_1 \exp\left[-\left(\frac{\chi}{m}\right)^k\right] \tag{13}$$

4.2. Verification of Constitutive Model

The key to establish the statistical damage constitutive relation of high-strength concrete is to determine the Weibull distribution parameters m and k. Applying the logarithms twice on Equation (13), the following equation is obtained:

$$\ln\left[-\ln\left(\frac{\sigma_1 - 2\mu\sigma_3}{E\varepsilon_1}\right)\right] = k\ln(\chi) - k\ln m \tag{14}$$

This is a linear equation with a slope coefficient equal to k and an intercept equal to $-k\ln m$. Therefore, parameters m and k can be easily determined based on a linear regression analysis on a set of triaxial test data of concrete samples, as shown in Table 4. The statistical damage constitutive relation curves of high-strength concrete are obtained by substituting m and k into Equation (13). The results are shown in Figure 8.

Table 4. Weibull distribution parameters m and k at different confining pressures.

Confining Pressure (MPa)	**Weibull Distribution Parameters**				**R^2**	
	C60		**C70**		**C60**	**C70**
	m	k	m	k		
0	3.576	1.538	3.311	2.747	0.962	0.978
5	2.493	1.225	1.973	2.146	0.984	0.988
10	1.778	0.997	1.766	1.056	0.991	0.982
15	1.718	0.947	1.453	0.916	0.994	0.992
20	1.146	0.919	1.233	0.762	0.989	0.985

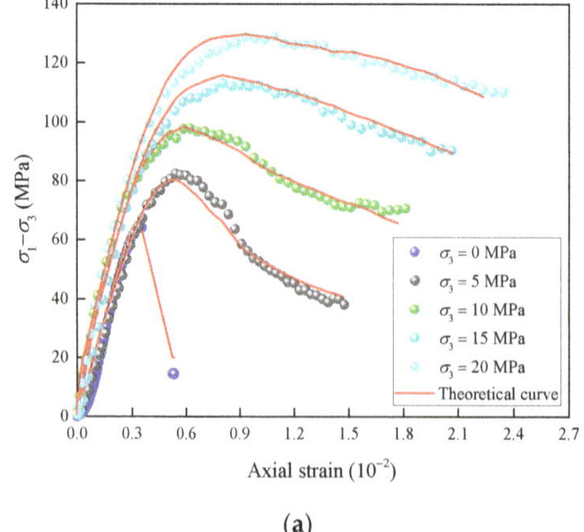

(**a**)

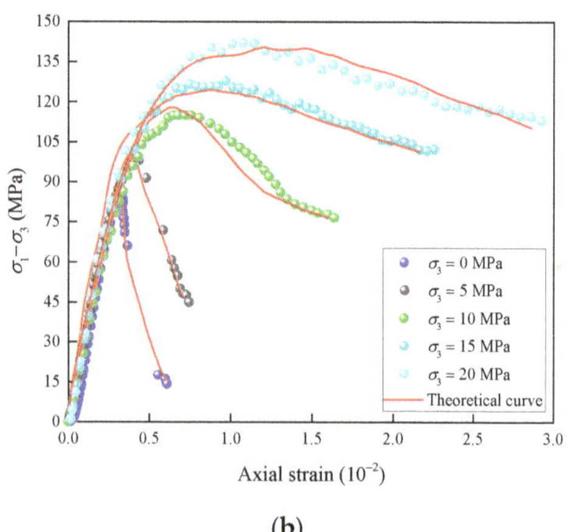

(**b**)

Figure 8. Comparison of constitutive model and test results. (**a**) C60 and (**b**) C70.

According to the comparison results of the statistical damage constitutive model curves and test curves of high-strength concrete (C60 and C70) at different confining pressures in Figure 8, it can be observed that the theoretical curves of the statistical damage constitutive model of high-strength concrete are in good agreement with the experimental curves in both the pre-peak elastic strain energy storage stage and the post-peak elastic strain energy release stage, and the correlation coefficient is above 0.96. Conversely, this constitutive model overcomes the defect of low correlation between concrete constitutive model and test results in the post-peak stage, and improves the accuracy of the model. Conversely, it is verified that the statistical damage constitutive model established from the energy theory and statistical damage theory is suitable to describe the constitutive behavior of high-strength concrete. Although the theoretical curve of the constitutive model established in this paper is in good agreement with the experimental curve, because the model parameters are obtained by fitting the experimental curve, whether the model can predict the total stress–strain curve of high-strength concrete needs further study.

Standard deviation and relative standard deviation can measure the deviation between the experimental and theoretical model results [24]. To verify the accuracy of the statistical damage constitutive model of high-strength concrete, the standard deviation and relative standard deviation of triaxial stress–strain test curve and the theoretical curve of two types of high-strength concrete under five confining pressures are calculated based on Equation (15). The results show that the relative standard deviations between the statistical damage constitutive model results and the test results of C60 high strength concrete are 7.56%, 3.54%, 2.45%, 2.40%, and 2.26%, respectively, at the confining pressures of 0, 5, 10, 15, and 20 MPa, and the average relative standard deviation is only 3.64%. Additionally, the relative standard deviation between the statistical damage constitutive model results and the test results of C70 high-strength concrete are 6.49%, 2.27%, 3.84%, 4.30%, and 3.03%, respectively, and the average relative standard deviation is only 3.99%. The error analysis further shows that the statistical damage constitutive model of high-strength concrete is reasonable and feasible.

$$\begin{cases} \eta = \sqrt{\frac{\sum_{i=1}^{n} (\sigma_s - \sigma_l)^2}{n}} \\ f = \frac{\eta}{\sigma_c} \end{cases} \tag{15}$$

where η is the standard deviation, f is the relative standard deviation, σ_s, σ_l are the test values and theoretical values, respectively, σ_c is the compressive strength, and n is the data volume.

4.3. Damage Analysis of High-Strength Concrete

The Weibull distribution parameters m and k in Table 3 are substituted into Equation (12) to obtain the relationship between the damage variable and axial strain, as shown in Figure 9.

It can be observed from Figure 9 that the damage evolution curves of C60 and C70 high-strength concrete at different confining pressures are similar to the "S" curves. At the initial stage of loading, the damage weakening is not obvious. Owing to the continuous closure of micro-pores and micro-cracks in concrete samples subjected to pressure, the concrete gradually transforms from discontinuous medium to a quasi-continuous medium. With the increase of axial strain, the concrete microdefects are further compacted and closed, and the concrete enters the elastic deformation stage. Because the stress level at this stage is not enough to cause the crack to begin to expand, the microdefects will not decrease after their closure, that is, the damage cannot occur in the real linear elastic stage. However, in the elastic deformation stage of concrete, the damage variable is still increasing slowly. This indicates that the concrete at this time not only includes the elastic deformation, but also the mutual sliding of closed cracks that shows the existence of nonlinearity at low-stress levels. When the stress of concrete exceeds a certain level or its deformation reaches a certain value, new micro-cracks begin to sprout and expand slowly between the relatively weak particle boundaries, the concrete yields and produces plastic deformation, and the damage of concrete begins to evolve and expands steadily. With the increase of the stress level, the micro-cracks in the concrete are concentrated, expanded and penetrated locally, thus forming macro-cracks, and the concrete damage develops rapidly. The main fracture surface is formed by the ladder connection of the macro-cracks that leads to the sudden release of stress and the rapid decrease of concrete strength, thus resulting in damage. However, owing to the incomplete release of elastic strain energy, the damage variable is slightly less than one. With the increase of confining pressure, the rate of change of damage with strain decreases. This indicates that the increase of confining pressure can effectively inhibit the release of elastic-strain energy and the development of damage and will improve the stress state of concrete.

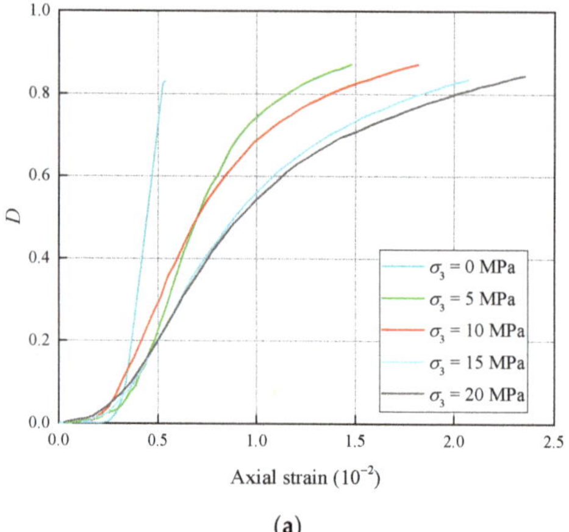

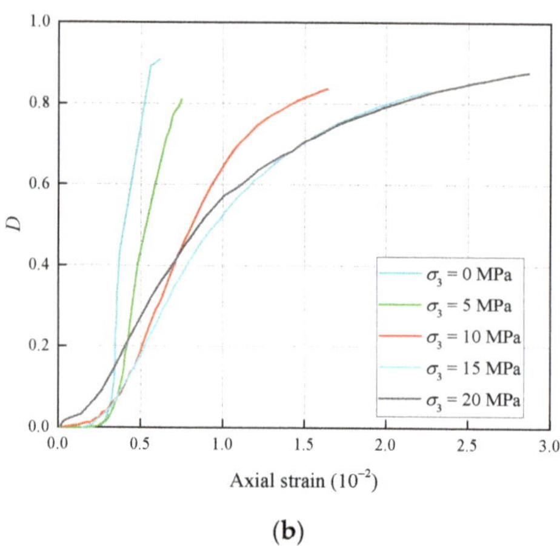

Figure 9. Relationship between damage variable D and axial strain. (**a**) C60 and (**b**) C70.

5. Conclusions

(1) At different confining pressures, the input energy density and dissipation energy density of the C60 and C70 high-strength concrete samples increased as a function of the axial strain, while the elastic strain energy density increased first and then decreased as a function of the axial strain, and reached the maximum value at the peak stress. Before the concrete sample entered the plastic stage, the dissipated energy density increased slowly as a function of the axial strain, and the input energy was basically transformed into elastic-strain energy. After the specimen was destroyed, the elastic-strain energy was released rapidly and the input energy was basically transformed into dissipation energy.

(2) The input, elastic, and the dissipated energy densities corresponding to the peak stresses of the C60 and C70 high-strength concrete increased as a function of the increase of confining pressure, and the elastic-strain energy density corresponding to the peak stress increased linearly as a function of the confining pressure. As the elastic strain energy density of the concrete sample at the moment of failure increased, the energy release at high-confining pressures became more rapid and abrupt compared with those at low-confining pressures.

(3) At both the pre-peak elastic strain energy storage stage and the post-peak elastic strain energy release stage, the statistical damage constitutive model curves of C60 and C70 high-strength concrete at different confining pressures were in good agreement with the experimental curves, and the average relative standard deviations were only 3.64% and 3.99%. Conversely, this constitutive model overcame the defect of low correlation between the concrete constitutive model and the test results in the post-peak stage and

improved the accuracy of the model. It was also verified that the statistical damage constitutive model established from the energy theory and statistical damage theory was suitable in describing the constitutive behavior of high-strength concrete.

Author Contributions: Conceptualization, L.Z. and J.L.; methodology, L.Z.; validation, X.W. and L.G.; writing—original draft preparation, L.Z.; writing—review and editing, H.C.; funding acquisition, H.C. All authors have read and agreed to the published version of the manuscript.

Funding: This research was funded by the National Natural Science Foundation of China (51874005).

Institutional Review Board Statement: Not applicable.

Informed Consent Statement: Not applicable.

Data Availability Statement: The data used to support the findings of this study are available from the corresponding author upon request.

Conflicts of Interest: The authors declare no conflict of interest.

References

1. Li, W.S.; Wu, J.Y. A consistent and efficient localized damage model for concrete. *Int. J. Damage Mech.* **2018**, *27*, 541–567. [CrossRef]
2. Li, R.T.; Li, X.K. A coupled chemo-elastoplastic-damage constitutive model for plain concrete subjected to high temperature. *Int. J. Damage Mech.* **2010**, *19*, 971–1000.
3. Nishiyama, T.; Chen, Y.; Kusuda, H.; Ito, T.; Kita, H. The examination of fracturing process subjected to triaxial compression test in Inada granite. *Eng. Geol.* **2002**, *66*, 257–269. [CrossRef]
4. McSaveney, M.J.; Davies, T.R. Surface energy is not one of the energy losses in rock comminution. *Eng. Geol.* **2009**, *109*, 109–113. [CrossRef]
5. Krajcinovic, D.; Silva, M.A.G. Statistical aspects of the continuous damage theory. *Int. J. Solids Struct.* **1982**, *18*, 551–562. [CrossRef]
6. Yang, X.; Liang, N.H.; Liu, X.R.; Zhong, Z. A study of test and statistical damage constitutive model of multi-size polypropylene fiber concrete under impact load. *Int. J. Damage Mech.* **2019**, *28*, 973–989. [CrossRef]
7. Bai, W.F.; Chen, J.Y.; Fan, S.L.; Lin, G. The statistical damage constitutive model for concrete materials under uniaxial compression. *J. Harbin Inst. Technol. (New Ser.)* **2010**, *17*, 338–344.
8. Liang, H.; Zou, R.H.; Peng, G.; Tian, W.; Chen, X.Q. Rate-dependent constitutive model of concrete based on the statistical theory of Weibull. *J. Yangtze River Sci. Res. Inst.* **2016**, *33*, 111–114.
9. Cervera, M.; Oliver, J.; Manzoli, O. A rate-dependent isotropic damage model for the seismic analysis of concrete dams. *Earthq. Eng. Struct. Dyn.* **1996**, *25*, 987–1010. [CrossRef]
10. Zhou, W.; Feng, P.; Lin, H.W. Constitutive relations of coral aggregate concrete under uniaxial and triaxial compression. *Constr. Build. Mater.* **2020**, *251*, 118957. [CrossRef]
11. Wu, J.Y.; Li, J.; Faria, R. An energy release rate-based plastic-damage model for concrete. *Int. J. Solids Struct.* **2006**, *43*, 583–612. [CrossRef]
12. Zhang, H.M.; Lei, L.N.; Yang, G.S. Characteristic and representative model of rock damage process under constant confining stress. *J. China Univ. Min. Technol.* **2015**, *44*, 59–63. [CrossRef]
13. Golewski, G.L. The beneficial effect of the addition of fly ash on reduction of the size of microcracks in the ITZ of concrete composites under dynamic loading. *Energies* **2021**, *14*, 668. [CrossRef]
14. Chen, W.; Konietzky, H.; Tan, X.; Thomas, F. Pre-failure damage analysis for brittle rocks under triaxial compression. *Comput. Geotech.* **2016**, *74*, 45–55. [CrossRef]
15. You, M.Q.; Hua, A.Z. Energy analysis of failure process of rock specimens. *Chin. J. Rock Mech. Eng.* **2002**, *21*, 778–781.
16. Zhao, Z.H.; Xie, H.P. Energy transfer and energy dissipation in rock deformation and fracture. *J. Sichuan Univ. (Eng. Sci. Ed.)* **2008**, *40*, 26–31.
17. Xie, H.P.; Li, L.Y.; Peng, R.D.; Yang, J. Energy analysis and criteria for structural failure of rocks. *J. Rock Mech. Geotech. Eng.* **2009**, *1*, 11–20. [CrossRef]
18. Zuo, J.P.; Huang, Y.M.; Xiong, G.J.; Liu, J.; Li, M.M. Study of energy-drop coefficient of brittle rock failure. *Rock Soil Mech.* **2014**, *35*, 321–327.
19. Gong, F.Q.; Luo, S.; Li, X.B.; Yan, J. Linear energy storage and dissipation rule of red sandstone materials during the tensile failure process. *Chin. J. Rock Mech. Eng.* **2018**, *37*, 352–363.
20. Lemaitre, J. Evaluation of dissipation and damage in metals submitted to dynamic loading. In Proceedings of the 1st International Conference on Mechanical Behavior of Materials, Kyoto, Japan, 15–20 August 1971; pp. 540–549.
21. Deng, J.; Gu, D.S. 2011 On a statistical damage constitutive model for rock materials. *Comput. Geosci.* **2011**, *37*, 122–128. [CrossRef]
22. Fu, Q.; Xie, Y.J.; Song, H.; Zhou, H. Model for mechanical properties of cement and asphalt mortar. *J. Chin. Ceram. Soc.* **2014**, *42*, 1396–1403.

23. Wang, Z.L.; Li, Y.C.; Wang, J.G. A damage-softening statistical constitutive model considering rock residual strength. *Comput. Geosci.* **2007**, *33*, 1–9. [CrossRef]
24. Guo, J.Q.; Liu, X.R.; Zhao, Q. Theoretical research on rock unloading mechanical characteristics. *Rock Soil Mech.* **2017**, *38*, 123–130.

Article

Crack Resistance of Insulated GRC-PC Integrated Composite Wall Panels under Different Environments: An Experimental Study

Dong Chen [1], Pengkun Li [1], Baoquan Cheng [1,2,*], Huihua Chen [2], Qiong Wang [3] and Baojun Zhao [3]

[1] BIM Engineering Center of Anhui Province, Anhui Jianzhu University, Hefei 230601, China; chendong@ahjzu.edu.cn (D.C.); lpk@ahjzu.edu.cn (P.L.)
[2] School of Civil Engineering, Central South University, Changsha 410083, China; chh@csu.edu.cn
[3] Shenzhen Hailong Construction Technology Company Limited, Shenzhen 518000, China; hlwangqiong@cohl.com (Q.W.); zhaobj@cohl.com (B.Z.)
* Correspondence: curtis_ch@csu.edu.cn

Abstract: GRC-PC wall is a new type of integrated composite exterior wall with decorative and structural functions. It is formed by superimposing GRC surface layer on the outer leaf of prefabricated PC wall. Due to the complexity of indoor and outdoor environment and the difference of shrinkage performance between concrete and GRC materials, GRC surface layer in GRC-PC wall is prone to shrinkage and cracking, among which, the connection modes between GRC layer and PC layer and change of temperature and humidity have the greatest influence. Therefore, GRC material formula was adjusted, and seven experimental panels were produced. In view of the temperature and the humidity changes in different indoor and outdoor environments, the influences of different connection modes between GRC layer and PC layer on the material shrinkage performance were studied, and a one year material shrinkage performance experiment was conducted. The results show that, in indoor environment, the shrinkage of GRC layer and PC layer is relatively gentle due to the small range of temperature and humidity change. Compared with the indoor environment, the changes of outdoor temperature and humidity are more drastic. The shrinkage changes of GRC layer and PC layer show great fluctuations, but the overall strain value is still within a reasonable range, and there is no crack. At the same time, this suggests that smooth interface is more conducive to crack resistance of GRC surface layer compared with different interface types between GRC layer and PC layer. The research provides an experimental basis for the large-scale application of the wall panel, and it has great advantages in improving the efficiency of prefabricated building construction.

Keywords: GRC-PC; integrated wall panels; composite method; shrinkage properties

Citation: Chen, D.; Li, P.; Cheng, B.; Chen, H.; Wang, Q.; Zhao, B. Crack Resistance of Insulated GRC-PC Integrated Composite Wall Panels under Different Environments: An Experimental Study. *Crystals* **2021**, *11*, 775. https://doi.org/10.3390/cryst11070775

Academic Editors: Cesare Signorini, Antonella Sola, Sumit Chakraborty, Valentina Volpini and Tomasz Sadowski

Received: 27 May 2021
Accepted: 29 June 2021
Published: 2 July 2021

1. Introduction

A glass-fiber-reinforced cement (GRC) is a type of composite building material made of cement and glass fiber as the main components while also including white sand, metakaolin, and fly ash [1]. Because of its excellent plasticity and durability, a GRC material can be used as an exterior leaf decorative material on the walls of buildings and is preferred over lacquer materials. The application of GRC not only protects the environment but also plays the role of architectural decoration in accordance with the concept of green buildings [2,3]. A GRC-PC exterior wall is a new type of prefabricated wall formed by a composite GRC surface layer on the outer leaf layer of a precast concrete (PC) wall panel, as shown in Figure 1, which not only ensures the structural bearing capacity of the members but also plays a decorative role. The use of this wall panel significantly reduces pollution, shortens the construction period, and improves the construction efficiency, thereby integrating the building, the structural, and the decorative elements through fabricated systems. It also has great advantages in improving the structural optimization of the conventional PC industry [4,5].

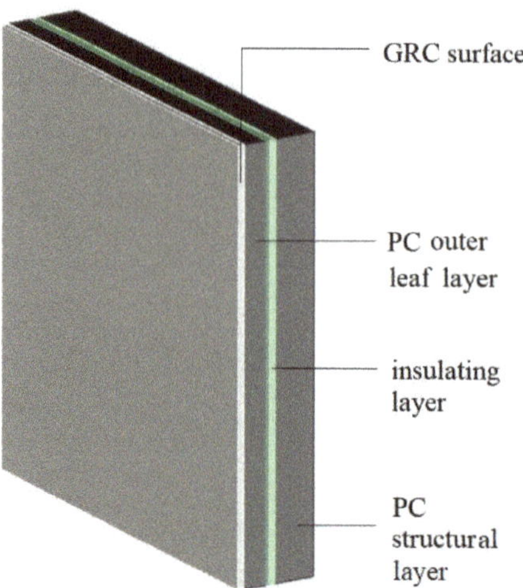

Figure 1. Schematic of a GRC-PC external wall structure.

In practical engineering, the concrete components without glass fiber are prone to cracking, which shortens the service life of components, affects the aesthetic appearance, and reduces the construction quality [6,7]. Previous studies showed that the basic factors causing cracks in concrete cementitious materials are determined by mechanical properties, shrinkage properties, and ductility. Recently, in the development of new concrete materials, materials that incorporate fibers to improve the crack resistance of concrete, such as glass-fiber-cement (GRC) and steel-fiber-cement (SFRC), are gradually being used in engineering applications [8–10].

In a study on the mechanical properties of GRC, Liu and Wu [11] pointed out that adding glass fibers helped reduce the elastic modulus of concrete. Zhao et al. [12] showed that, with the increase in the glass fiber content, compressive strength, splitting tensile strength, and flexural strength of concrete increased first but then decreased, i.e., there is an optimal fiber content. Shen et al. [13] found that an alkali-resistant glass fiber could significantly improve the tension–compression ratio and Poisson's ratio of concrete and enhance its toughness and brittleness. Qian and He [14] studied the influencing factors and the development patterns of glass fibers and fly ash composite cements and concluded that the material strength was most influenced by age and cement content, followed by the glass fiber, and least by the fly ash.

In a study on the shrinkage performance of GRC, Lura et al. [15] found that the shrinkage deformation of materials was the primary factor causing cracks in the process of condensate sclerosis, regardless of whether it was ordinary cement or GRC. The dominant shrinkage mechanism was found to be temperature autogenous shrinkage, which is the shrinkage deformation of a material under the combined action of the hydration heat of cement and the external temperature change [16]. Shrinkage deformation causes shrinkage stress in a material, and cracks are induced when the stress exceeds the maximum tensile stress that the material can withstand [17,18]. In addition, the shrinkage of GRC is affected by curing temperature, humidity, and environment [19,20].

To alleviate the shrinkage deformation of the GRC material and improve its crack resistance, previous studies were mainly carried out from two aspects. The first is the reasonable selection of GRC aggregates. Ye et al. [21] and Nguyen et al. [22] concluded that the incorporation of alkaline aggregates could help increase the amplitude of drying shrinkage of the GRC and the cracking sensitivity of cementation materials by measuring the cracking time and the cracking degree of cement with different alkalinities. Kumarappa et al. [23] reported that the addition of alkaline materials affected the reaction degree and the surface tension of pore solutions based on shrinkage, heat flow, and surface tension of cements blended with Na_2O and SiO_2, thus affecting the shrinkage performance of cement. Wu et al. [24] studied the effect of cementation material composition on the shrinkage properties of GRC materials and concluded that GRC materials prepared with sulphate aluminate cement underwent the least amount of shrinkage, whereas GRC materials prepared with silicate cement shrunk to a greater extent. Additionally, the incorporation of fly ash and silica fume could effectively reduce drying shrinkage and self-shrinkage of GRC materials. Chylík et al. [25] studied the effect of modified gum powder on the shrinkage of GRC materials and concluded that the incorporation of gum powder could help reduce the internal gel pores and the macropores in the material, improve the hydrophilicity of cement, and thus improve flow and toughness of the GRC. Guo et al. [26] studied the effect of swelling agents on the shrinkage of GRC materials and concluded that GRC materials with swelling agents had fewer bonding cracks between the hydration products and the aggregates, better interfacial transition zone of concrete, and fewer cracks due to drying shrinkage.

On the other hand, it is necessary to control the content and the composition of glass fibers in a GRC material. Fiber is added to increase toughness and ductility of the cement base, improve the tensile strength of cement, reduce cracks, and prevent cracks from developing [27]. He et al. [28] studied the influence of fiber geometry on the cracking resistance of GRC and found that, with the increase in the fiber length and the decrease in the fiber diameter, the total plastic shrinkage cracking area of a cement mortar showed a downward trend, and the cracking resistance improved correspondingly. Shen et al. [29] found that the shrinkage strain of fiber-reinforced concrete decreased with the increase in the fiber volume percentage and put forward a prediction model for the early self-shrinkage strain of fiber-reinforced concrete. Kasagani and Rao [30] studied the effect of fiber grading on the crack resistance of GRC, suggesting that short fibers mainly controlled the expansion of microcracks and improved the ultimate strength, while longer fibers inhibited macrocracks and alleviated the deformation of concrete. Consequently, the combination of long and short fibers could help prevent microscopic and macroscopic cracks from developing, thus improving the crack resistance of concrete.

In summary, previous studies on the mechanics and shrinkage performance of GRC mainly focused on pure GRC prefabricated components, and most of the experiments were carried out in a relatively constant temperature and humidity environment. The GRC-PC composite wall panel studied in this paper was to be used both inside and outside the building, with great changes in temperature and humidity. In addition, the shrinkage rate of concrete was less than that of GRC; in this scenario, the concrete layer would hinder the shrinkage of GRC layer when they were combined, which would increase the tensile stress in the GRC layer, causing GRC layer cracking. To improve the crack resistance of the GRC-PC composite wall panel, the research idea of this paper was as follows. First of all, the GRC material formula was adjusted, and the compressive strength and the elastic modulus of the material were measured. Secondly, according to the different interface types of GRC layer and PC layer, seven wall panels of 1 m × 1 m were prepared, and the shrinkage experiment was carried out for 365 days in environments with different temperature and humidity. Among them, the crack resistance of the GRC layer was the core of the test. Finally, according to experimental results, the reasonable interface type between the GRC layer and the PC layer was determined. Findings from this research contribute to application of GRC-PC composite wall panels and promotion of prefabrication.

2. Materials and Methods

2.1. Experimental Raw Materials and Equipment

The raw materials required for the experiment were C30 concrete and GRC mortar. Among them, C30 concrete was produced by a concrete factory, and the GRC material was prepared by mixing in the experimental site; GRC materials are prepared by mixing the raw materials listed in Figure 2 in the experimental site.

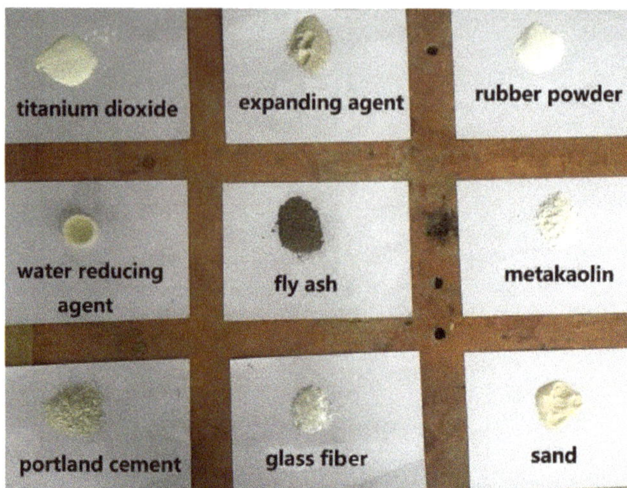

Figure 2. GRC raw materials.

Portland cement, sand, fly ash, water reducing agent, and glass fiber are the traditional GRC formulations. In order to improve the crack resistance of the GRC material, the formula was adjusted, that is, we added rubber powder, expanding agent, metakaolin, and titanium dioxide.

The role of rubber powder was to reduce GRC internal porosity and improve the hydrophilicity of cement so as to increase its mobility and toughness [25]. The function of the expanding agent was to reduce the bond crack between hydration products and aggregate [26]. The role of metakaolin was to improve the pore structure of the cement mortar and improve uniformity and compactness of the mortar structure [31]. Titanium dioxide was used to improve the brightness to achieve the effect of decoration.

Table 1 lists the mix proportion of mortar of GRC, in which the cementing material and the sand were 8:9, and the water–binder ratio was 0.28.

Table 1. Mix proportion of mortar of GRC.

Cement (kg/m^3)	Sand (kg/m^3)	Water (kg/m^3)	Fly Ash (kg/m^3)	Metakaolin (kg/m^3)	Water Reducing Admixture (kg/m^3)	Glass Fiber (kg/m^3)	Rubber Powder (kg/m^3)	Expansion Agent (kg/m^3)
888	1248	322	56	166	31	34	28	64

Figure 3 shows the instruments and the equipment used in the experiment, including a compression testing machine, which was used to measure the compressive strength and the elastic modulus of the raw materials (C30 concrete and GRC materials). An embedded strain sensor DH1204 and a surface strain sensor DH1205 were used to measure the strains of the GRC and the PC layers, respectively. A DH3818Y static strain tester was used for strain data collection and recording.

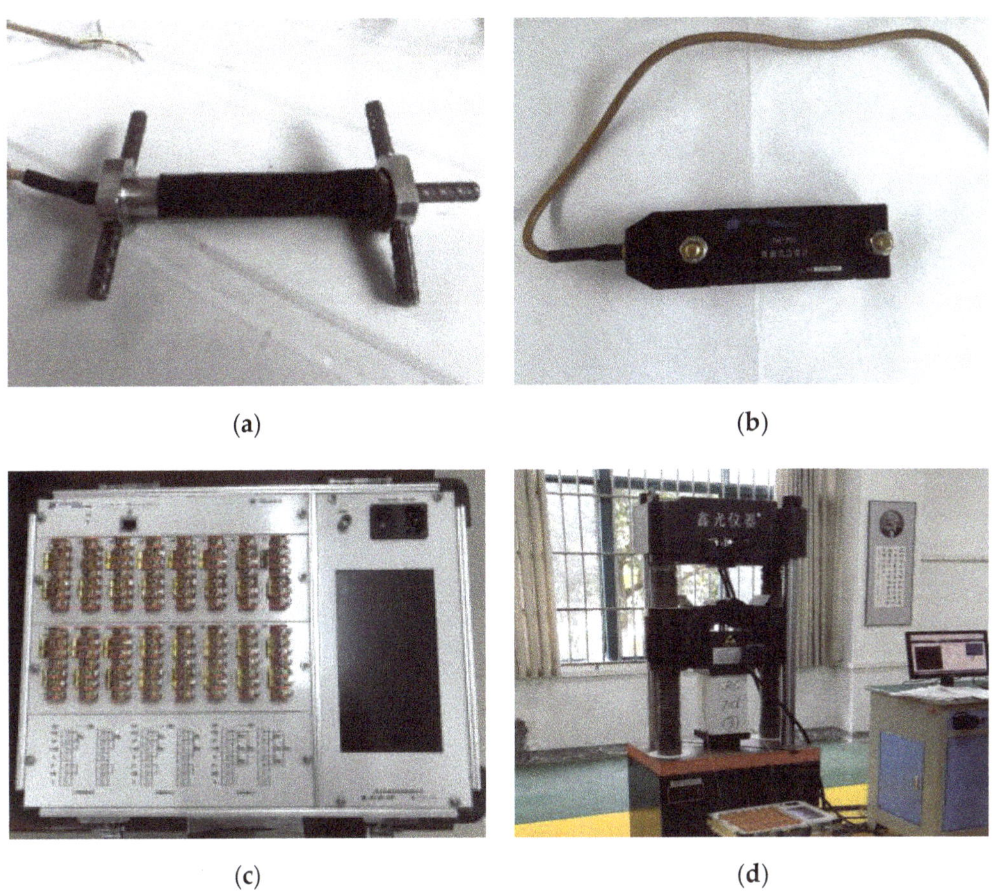

Figure 3. Equipment used in the experiment: (**a**) DH1204 embedded strain sensor; (**b**) DH1205 surface strain sensor; (**c**) DH3818Y static strain tester; (**d**) compression testing machine.

2.2. Experiment on Mechanical Properties of Materials

The compressive strength and the modulus of elasticity are the basic mechanical properties of materials. Tests were carried out on the mechanical properties prior to the fabrication of the test elements to ensure that the material strength met the requirements. The twelve test blocks for the modulus of elasticity and compressive strength tests were divided into four groups of three blocks each based on the curing time listed in Table 2.

Table 2. Parameters of material test block.

Type of Material Property Test	Test Block Size (mm)	Type and Number of Test Blocks	
Compressive strength test	150 × 150 × 150	GRC-12	PC-12
Elastic modulus test	150 × 150 × 150	GRC-12	PC-12

After curing each group of test blocks for the corresponding days, we installed strain sensors on their surface and connected the static strain tester. We then employed the compression testing machine to carry out a pressure test on the test blocks until they were destroyed. Figure 4 shows the experimental process.

Figure 4. Experimental process of mechanical properties of materials: (**a**) Manufacture of concrete test block; (**b**) manufacture of GRC test block; (**c**) grouping and pasting of test blocks; (**d**) loading of test block.

The compressive strength was calculated using the following formula:

$$F_{cu} = \frac{F}{A}, \quad (1)$$

where F_{cu} is the compressive strength (MPa) of C30 concrete and GRC cube specimens; F is the failure load (N) of the specimen; A is the bearing area (mm^2) of the specimen.

The measurement and the calculation formula of the elastic modulus were:

$$E_c = \frac{F_a - F_0}{A} \times \frac{L}{\Delta n}, \quad (2)$$

where E_c is the elastic modulus of the specimen; Fa is the load at which the stress reaches one third of the axial compressive strength value; F_0 is the initial load (N) when the stress is 0.5 MPa; L is the measuring gauge distance (mm); A is the bearing area of the specimen (mm^2); Δn is the average value (mm) of the deformation on both sides of F_a from F_0 loading.

2.3. Shrinkage Experiment of GRC-PC Composite Wall Panels

2.3.1. Specimen Design

A total of seven groups of components (S0 to S6) were designed for the shrinkage experiment. S0 and S1 were panels made of GRC without and with glass fiber, respectively; S2 was a panel made of concrete; S3 to S6 were GRC-PC composite wall panels made according to Figure 1, which omitted the concrete structure layer and the insulation layer, as shown in Figure 5. Table 3 lists the specific parameters. Because of the limitations of the experimental site, the length and the width of the wall panel were designed to be 1000 mm × 1000 mm, and the thicknesses of the GRC and the PC layers were set to 15 mm and 60 mm, respectively. For the interface between GRC and PC, two commonly used concrete surface processes were adopted: smooth surface (surface smoothing) and rough surface (surface grabbing).

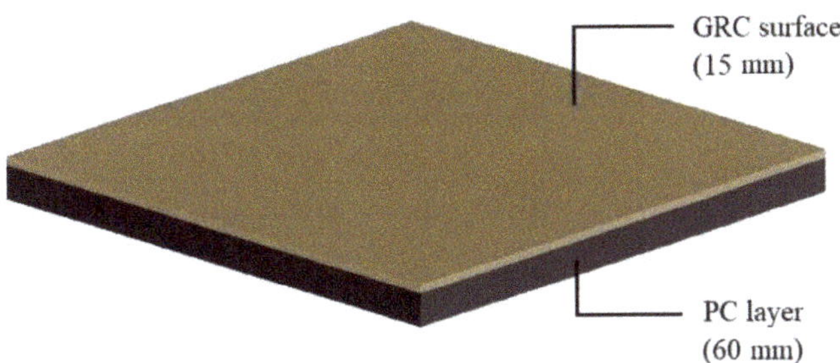

Figure 5. Three-dimensional model of a composite wall panel.

Table 3. Dimensional parameters of GRC-PC composite wall panels.

Specimen Number	Specimen Size (Length × Width × Height) (mm)	GRC Thickness (mm)	Concrete Thickness (mm)	Type of Interface	Environment	Period
S0	1000 × 1000 × 15	15	–	–	indoor	1 June 2019 to 1 September 2019
S1	1000 × 1000 × 15	15	–	–	indoor	1 October 2019 to 1 September 2020
S2	1000 × 1000 × 60	–	60	–	indoor	
S3	1000 × 1000 × (15 + 60)	15	60	smooth	indoor	
S4	1000 × 1000 × (15 + 60)	15	60	rough	indoor	
S5	1000 × 1000 × (15 + 60)	15	60	smooth	outdoor	
S6	1000 × 1000 × (15 + 60)	15	60	rough	outdoor	

The design idea of the components was as follows: S0 panel was a member without glass fiber, which was mainly used to observe the way of crack development so as to determine the location of the strain sensor. Other panels (S1 to S6) took the interface type and the environment as variables to determine the applicable structural type of GRC-PC wall panel.

2.3.2. Process of Experiment

Before collecting data from the shrinkage experiment, we first completed the fabrication of each group of wall panels (Table 3) and the installation of strain sensors, as shown in Figure 6. The placement of the strain sensor, as shown in Figure 7, was determined by the distribution of S0 cracks, which is explained in the next chapter.

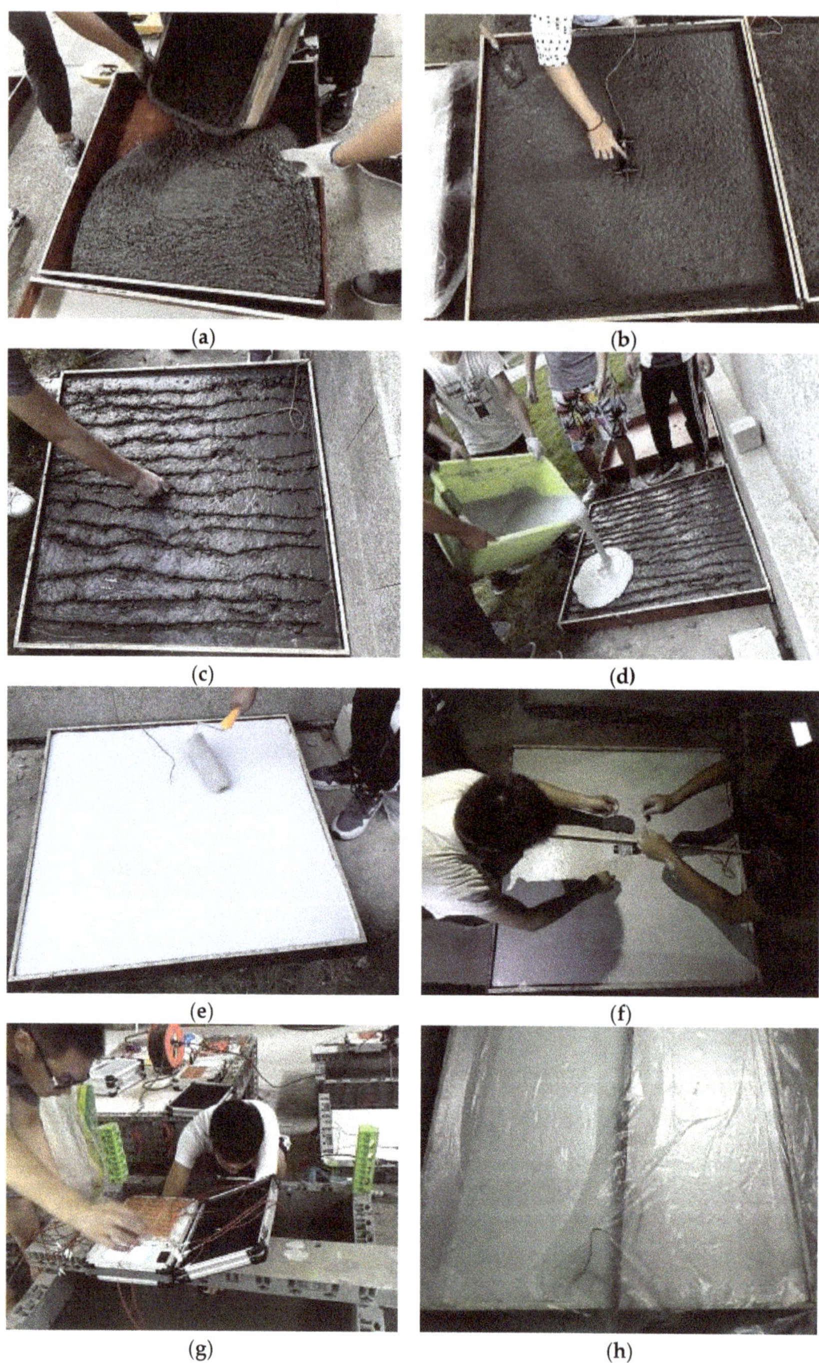

Figure 6. Experimental process: (**a**) pouring concrete; (**b**) placing embedded strain sensor; (**c**) surface processing; (**d**) pouring GRC material; (**e**) leveling the GRC surface; (**f**) placing surface strain sensors; (**g**) connecting the static strain tester; (**h**) covering film maintenance.

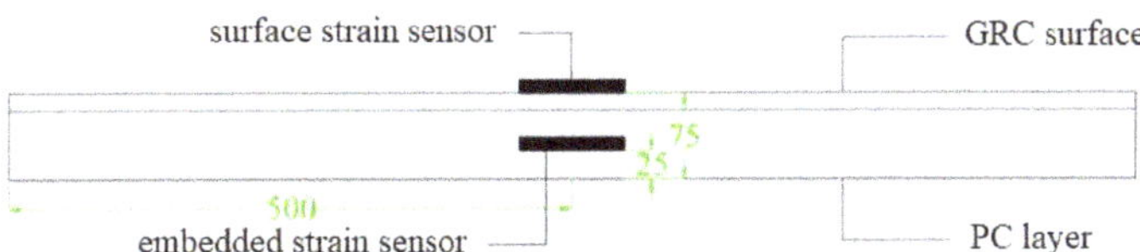

Figure 7. Strain sensor location in the composite wall panel.

The specific process was as follows.

(1) Brush the surface of the template with release oil and then pour the concrete to the height of the specified scale of the template;

(2) Vibrate the concrete and bury the embedded strain sensor at the center of the concrete;

(3) Machine the surfaces of wall panels of different types;

(4) Pour the mixed GRC material into the initial setting concrete and level the surface with a roller;

(5) Install and fix the surface strain sensor at the center of the GRC layer;

(6) Switch on the static strain tester and cover with a film for maintenance.

Since the shrinkage deformation of the wall panel is influenced by the ambient temperature and the humidity, in the experiment, we recorded daily indoor and outdoor temperatures and humidity in the morning, the afternoon, and the evening during the test period while taking the shrinkage strain measurement. The average value of the temperature in the three periods was taken and plotted as the temperature and humidity curve, as shown in Figure 8. As shown, the temperature amplitude was lower in the indoor environment than in the outdoor environment, whereas the air humidity was higher in the indoor environment than in the outdoor environment. Since cement-based cementitious materials are more suitable for maintenance and use in an environment with small temperature difference and high humidity [32], the outdoor environment, compared with the indoor environment, is more severe and places higher requirements on GRC-PC composite wall panels to resist cracking.

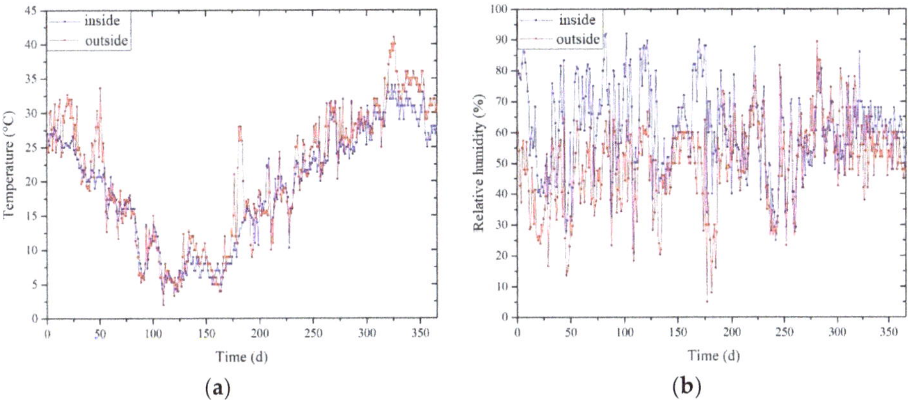

Figure 8. Temperature and humidity curves during the test period: (**a**) Indoor and outdoor temperatures vs. time curves; (**b**) indoor and outdoor humidity vs. time curves.

3. Results and Discussion

3.1. Experimental Results of Mechanical Properties of Materials

Tables 4 and 5 list the measured compressive strength and elastic modulus, respectively. From Table 4, we found that the compressive strengths of three GRC and concrete test blocks reached the standard compressive strength value on the 28th day, and the average value was within the error range.

Table 4. Measurement results of compressive strength experiment.

Group	Age (d)	Materials	Compressive Strength (MPa)			Average Compressive Strength (MPa)
			Specimen 1	Specimen 2	Specimen 3	
I	7	GRC	52.57	60.25	55.78	56.20
		Concrete	24.36	25.42	23.78	24.52
II	14	GRC	59.68	60.52	57.24	59.15
		Concrete	26.59	27.15	28.46	27.40
III	21	GRC	67.50	63.62	60.47	63.86
		Concrete	28.00	28.57	29.61	28.73
IV	28	GRC	67.27	65.86	66.85	66.66
		Concrete	31.02	31.01	31.07	30.68

Table 5. Measurement results of elastic modulus experiment.

Group	Age(d)	Materials	Elastic Modulus Values (GPa)			Average Modulus of Elasticity (GPa)
			Specimen 1	Specimen 2	Specimen 3	
I	7	GRC	27.20	28.5	27.45	27.72
		Concrete	23.40	21.87	22.45	22.57
II	14	GRC	27.60	28.64	29.63	28.62
		Concrete	25.71	23.43	23.87	24.34
III	21	GRC	28.32	28.90	29.80	29.01
		Concrete	27.25	28.21	28.96	28.14
IV	28	GRC	30.63	30.71	33.35	31.56
		Concrete	29.32	31.47	30.63	30.47

3.2. Shrinkage Experiment Results and Discussion

3.2.1. Cracking Analysis of Experimental Panels

The S0 panel was completed on 1 June 2019 and placed in an indoor environment for shrinkage experiments. As a member made of a single material, the S0 was subjected to free shrinkage. Three months later, cracks emerged in the S0 plate. As shown in Figure 9a, the cracks located in the middle of the panel in a cross distribution. These phenomenon indicated that different parts of the panel had different shrinkage and deformation. Thus, the shrinkage stress was generated in the panel. The distribution of cracks indicated that the shrinkage deformation in the middle of the panel was more limited, which led to the shrinkage stress exceeding the tensile limit of the material and finally cracking. Therefore, it was reasonable to place the strain gauge in the center of the panel (Figure 9b). At the same time, S1 to S6 did not crack during the monitoring period, and it can be concluded that the new GRC-PC composite wall panel met the crack resistance requirements.

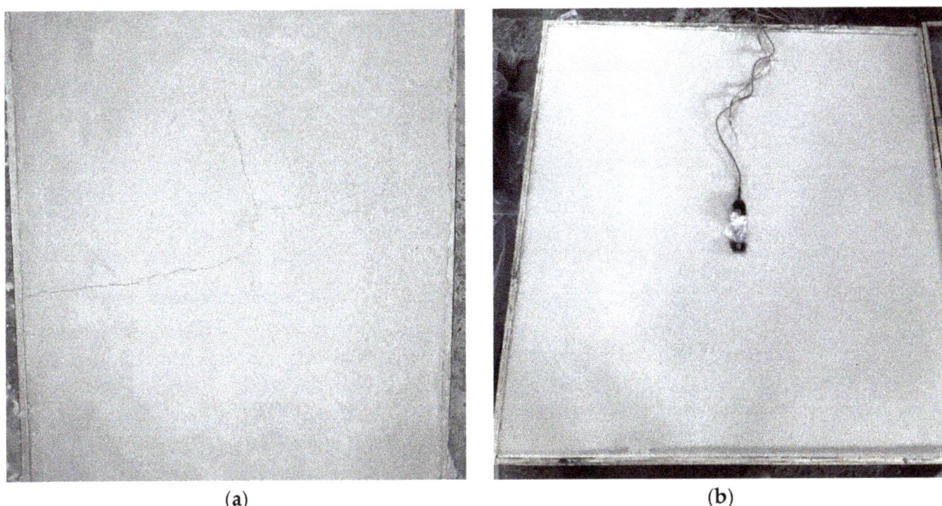

(**a**) (**b**)

Figure 9. Cracking condition of S0 and S1: (**a**) S0; (**b**) S1.

In order to have a visual display of the shrinkage deformation of all specimens, the strain curves of the S1 to S6 wall panels, shown in Figure 10, were plotted based on the data collected by the strain sensor during the experimental period. Figure 10a shows the strain of GRC layer of wall panels collected by surface strain sensor, and Figure 10b shows the strain of PC layer of wall panels collected by embedded strain sensor. The strain curves of S1 and S2, which represent pure GRC and pure PC panels, respectively, were used as the standard free shrinkage curves for the material. The other wall panels were classified in terms of environment and type of interface, and each set of strain curves was compared with the standard free shrinkage curve of the material as the crack resistance curve. The analysis was judged by the degree of adaptability, i.e., the more closely the crack resistance curve fit the standard free shrinkage curve of the material, the closer was the shrinkage of the corresponding composite wall panel to the standard free shrinkage, the lower was the resulting shrinkage stress, and the lower was the likelihood of panel cracking. Table 6 lists the maximum strain values of each group of wall panels.

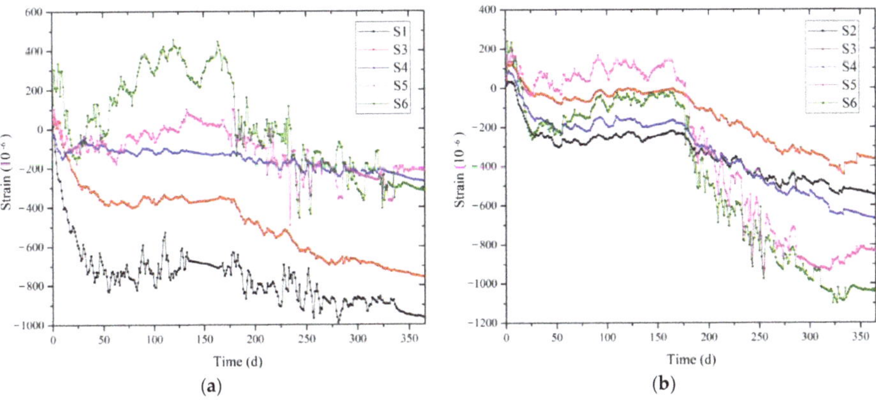

(**a**) (**b**)

Figure 10. Strain curve of wall panels: (**a**) Strain curve of GRC layer for each panel; (**b**) strain curve of PC layer for each panel.

Table 6. Maximum strain values of the GRC and PC layers for each group of wall panels.

Number	Maximum Strain Value of GRC (MPa)	Maximum Strain Value of PC (MPa)
S1	988.934×10^{-6}	-
S2	-	546.987×10^{-6}
S3	759.234×10^{-6}	441.238×10^{-6}
S4	270.548×10^{-6}	668.534×10^{-6}
S5	490.728×10^{-6}	975.871×10^{-6}
S6	454.864×10^{-6}	1100.23×10^{-6}

3.2.2. Shrinkage Analysis of Wall Panels with Different Interface Types

Because of the significant influence of environmental factors on the shrinkage of the composite wall panels, the shrinkage of wall panels with different types of interfaces under two environments, indoor and outdoor, were analyzed separately. Figure 11a,b show the strain curves of panels with different interfaces in an indoor environment. Figure 11c,d show the strain curves of panels with different interfaces in an outdoor environment.

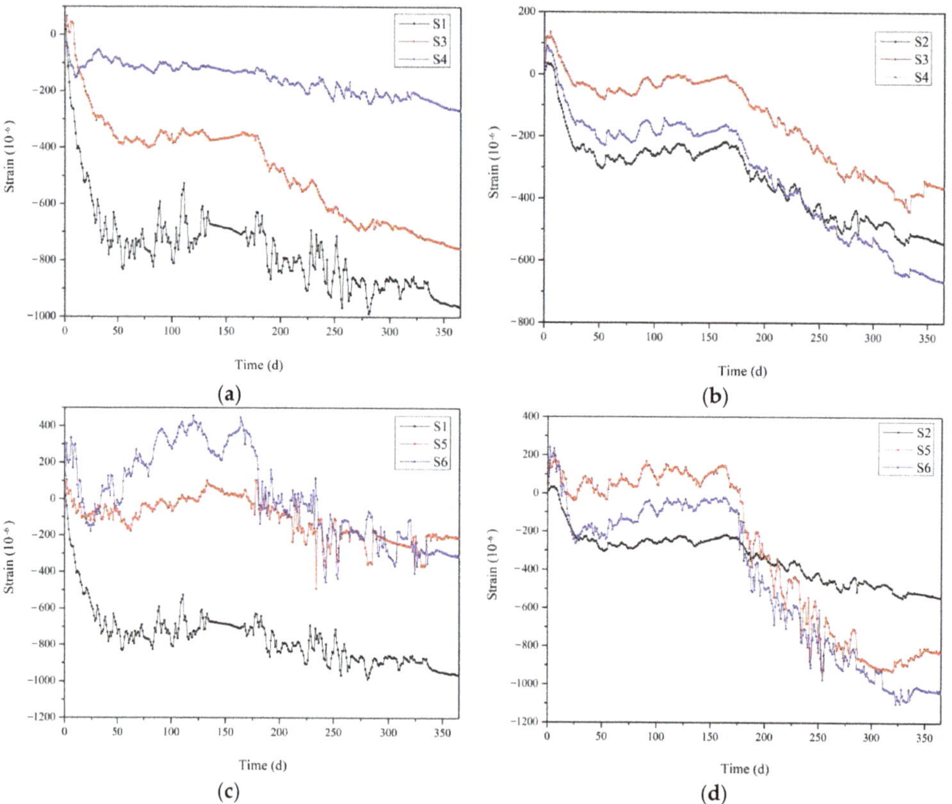

Figure 11. Strain curves of wall panels with different types of interfaces: (**a**) GRC strain curves of S1, S3, and S4; (**b**) PC strain curves of S2, S3, and S4; (**c**) GRC strain curves of S1, S5, and S6; (**d**) PC strain curves of S2, S5, and S6.

As shown in Figure 11a,b, the strain values of the GRC and the PC layers of S3 and S4 followed approximately the same strain curve trend over the monitoring duration. At the beginning of the experiment, the concrete and the GRC materials expanded in volume and were pulled under the effect of hydration heat. With the hydration reaction gradually weakening until disappearing, the GRC material began to shrink, the GRC-PC strain value decreased to a negative value, and the panel began to be under pressure. In the middle and the later stages of the test, the strain showed a small wave change, which indicated that the shrinkage of the GRC materials tended to be stable. By comparing the three curves, we found that the GRC strain of the composite wall panel with a smooth interface in the indoor environment was closer to S1 strain, whereas the PC of the composite wall panel with a rough interface changed to S2 strain, indicating that the interface type of the composite wall panel significantly influenced the shrinkage. From the data listed in Table 6, we found that the shrinkage strains of the GRC material and the concrete with a smooth interface decreased by 23% and 19%, respectively, in the indoor environment, and the shrinkage strain of the GRC material with a rough interface decreased by 72% and that of the concrete increased by 22%. Therefore, the use of a smooth interface is more conducive to improving the shrinkage performance of GRC-PC composite wall panels installed in indoor environments.

Figure 11c,d show a fluctuation in the strain curve of the composite wall panel in the outdoor environment. This was attributed to the significant changes in the temperature and the humidity of the outdoor environment, and the shrinkages of both the PC and the GRC layers were significantly affected. The overall trend in the outdoor strain was similar to that in the indoor strain: both types of layers were in a state of tension in the early stages and began to contract under pressure as the hydration reaction diminished. From the data listed in Table 6, we found that the shrinkage strains of the GRC material and the concrete with a smooth interface decreased by 50% and 75%, respectively, in the outdoor environment, and the shrinkage strain of the GRC material with a rough interface decreased by 54% and that of concrete increased by 50%. Therefore, the use of a smooth interface is more conducive to improving the shrinkage performance of GRC-PC composite wall panels installed in outdoor environments.

In summary, composite wall panels with a smooth interface exhibit better shrinkage performance in both indoor and outdoor environments. It can be concluded that the rough PC surface increases the constraint on the GRC layer, which is not conducive to the free shrinkage of the GRC material, and consequently, the possibility of cracking of the GRC layer increases.

3.2.3. Shrinkage Analysis of Wall Panel under Different Environments

The composite wall panels with the same type of interface were used to compare and analyze their shrinkage patterns in both indoor and outdoor environments.

Figure 12a,b show the strain curves of the composite wall panel with a smooth interface in different environments. Figure 12c,d show the strain curves of the composite wall panel with a rough interface in indoor and outdoor environments. Figure 10a shows a similar overall trend in the GRC strains of S1, S3, and S5, with the GRC shrinkage strain of S3 being significantly lower than that of S5. Figure 12b shows that the strain curves of S2, S3, and S5 followed a similar trend in the early stage, whereas the S5 curve exhibited a downtrend in the later stage. The range of variation in the PC shrinkage strains for S2 and S3 was roughly similar, and the maximum shrinkage strains of PC for S2 and S3 were significantly less than those for S5. From the data listed in Table 6, the shrinkage strain of the GRC was reduced by 23% and 50%, whereas the shrinkage strain of the PC was reduced by 19% and increased by 75% for the composite wall panels with a smooth interface in indoor and outdoor environments, respectively. It can be concluded that, compared with the indoor environment, the GRC-PC panel with a smooth interface has a wider variation range of shrinkage strain in the outdoor environment. It was proven that the shrinkage of composite wall panels is significantly affected by the temperature.

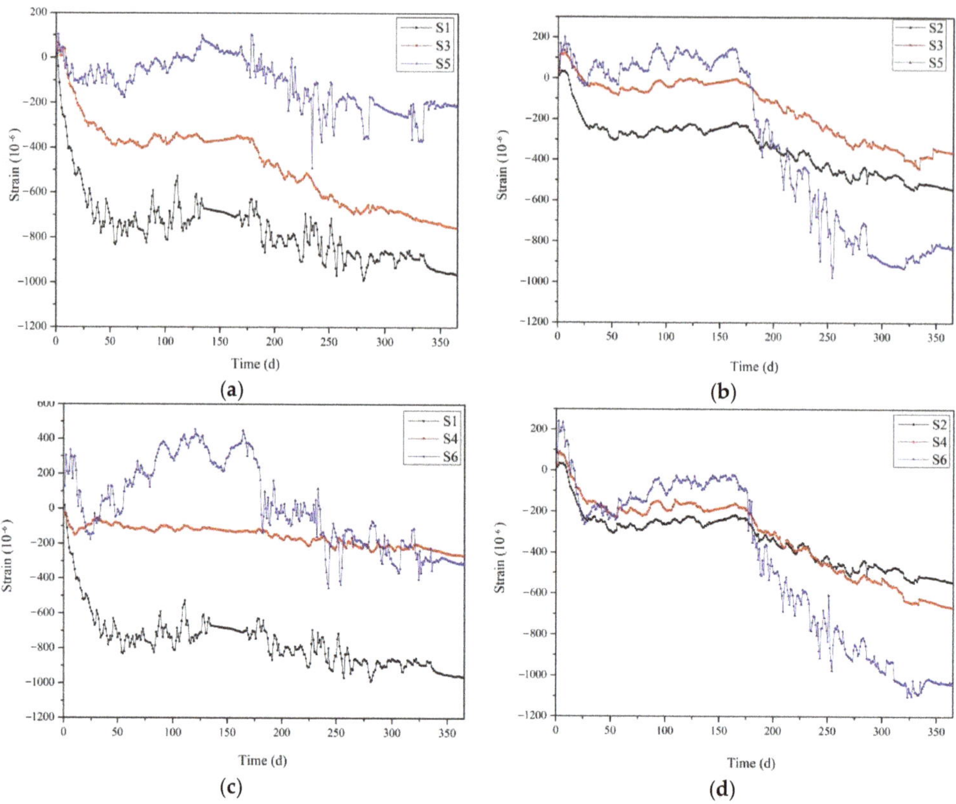

Figure 12. Strain curves for wall panels in different environments: (**a**) GRC strain curves of S1, S3, and S5; (**b**) PC strain curves of S2, S3, and S5; (**c**) GRC strain curves of S1, S4, and S6; (**d**) PC strain curves of S2, S4, and S6.

Figure 12c shows that the strain of S6 gradually increased, whereas that of S4 did not change significantly, and the shrinkage strains of both S4 and S6 were lower than that of S1. Figure 12d shows that S2, S4, and S6 had similar stress change trends in the early stages, and the stress change in S6 was greater than those in S2 and S4 in the later stages because of the significant change in the outdoor temperature. From the data listed in Table 6, we found that the shrinkage strain of the GRC was reduced by 72% and 54%, whereas the shrinkage strain of the PC was increased by 22% and 101% for the composite wall panels with a rough interface in indoor and outdoor environments, respectively. It can be concluded that, compared with the indoor environment, the GRC-PC with a rough interface has a wider variation range of shrinkage strain in the outdoor environment.

In summary, the shrinkage deformation degree of GRC-PC composite wall panels with two types of interfaces is greater in the outdoor environment than in the indoor environment. Although the GRC-PC has a greater shrinkage strain amplitude in a relatively harsh outdoor environment than in an indoor environment with suitable temperature and humidity, there was no sharp increase or decrease in the strain value due to component cracking, which indicates that the cracking resistance of the GRC-PC composite wall panel made of the new GRC material meets the requirements of the outdoor environment.

4. Conclusions

This research investigated the crack resistance and the facade effect of the GRC-PC integrated composite wall panels under different environments through an experimental research. The following conclusions can be drawn.

(1) According to the experimental results of S0, the cracks of wall panels are concentrated in the center position, where the shrinkage stress value is also the largest. In addition, fiber is an indispensable material to improve the crack resistance of GRC by comparing the cracks of S0 and S1.

(2) By studying the shrinkage performance of GRC-PC composite wall panels with different types of interfaces, we can conclude that the shrinkage deformation amplitude of the composite wall panel with a smooth interface is lower than that of the composite wall panel with a rough interface in both indoor and outdoor environments. The strain law of pure GRC and PC panels indicates that the processing method with the smooth interface is more beneficial to the crack resistance of composite wall panels in practice.

(3) The shrinkage deformation amplitude of GRC-PC composite wall panels with two types of interfaces was found to be greater outdoors than indoors. The shrinkage strain of the composite wall panels in the outdoor environment was in line with the free shrinkage law of the material, and no cracking occurred in any of the wall panels during the monitoring period, indicating that the crack resistance of the GRC-PC composite wall panels can be ensured in both indoor and outdoor environments.

GRC-PC insulation composite wall panel is a new type of prefabricated wall which can greatly reduce pollution, shorten the construction period, and improve the construction efficiency. The research in this paper provides an experimental basis for the large-scale application of the wall panel.

Due to the complexity of materials and the uncertainty of environmental changes, this paper was not able to find a reasonable and reliable finite element analysis model for the finite element analytical method, which is the research direction of future research.

Author Contributions: Conceptualization, D.C., P.L. and B.C.; methodology, D.C., P.L., H.C. and B.C.; formal analysis, D.C., P.L. and B.C.; investigation, D.C., H.C. and B.C.; resources, Q.W.; data curation, P.L. and B.C.; writing—original draft preparation, D.C., P.L., B.Z. and Q.W.; writing—review and editing, D.C., P.L. and B.C.; visualization, B.Z. and B.C.; supervision, B.Z.; project administration, D.C.; funding acquisition, D.C. All authors have read and agreed to the published version of the manuscript.

Funding: This work was supported by Natural Science Foundation of Anhui Province (19080885ME173), Research & Development project of China State Construction International Holdings Limited (CSCI-2020-Z-06-04), and Science and Technology Project of Anhui Province Housing and Urban-Rural Construction (2020-YF47).

Institutional Review Board Statement: Not applicable.

Informed Consent Statement: Not applicable.

Data Availability Statement: All data have been included in the manuscript.

Conflicts of Interest: The authors declare no conflict of interest.

References

1. Majumdar, A.J.; Laws, V. *Glass Fiber Reinforced Cement*; BSP Professional Books: Oxford, UK, 2001; pp. 46–53.
2. Enfedaque, A.; Gálvez, J.C.; Suárez, F. Analysis of fracture tests of glass fiber reinforced cement (GRC) using digital image correlation. *Constr. Build. Mater.* **2015**, *75*, 472–487. [CrossRef]
3. Cheng, C.; He, J.; Zhang, J.; Yang, Y. Study on the time-dependent mechanical properties of glass fiber reinforced cement (GRC) with fly ash or slag. *Constr. Build. Mater.* **2019**, *217*, 128–136. [CrossRef]
4. Baghdadi, A.; Heristchian, M.; Kloft, H. Design of prefabricated wall-floor building systems using meta-heuristic optimization algorithms. *Autom. Constr.* **2020**, *114*, 103156. [CrossRef]
5. Zhang, Z.; Chen, J.; Elbashiry, E.M.A.; Guo, Z.; Yu, X. Effects of changes in the structural parameters of bionic straw sandwich concrete beetle elytron plates on their mechanical and thermal insulation properties. *J. Mech. Behav. Biomed. Mater.* **2019**, *90*, 217–225. [CrossRef] [PubMed]

6. Noël, M.; Soudki, K. Estimation of the crack width and deformation of FRP-reinforced concrete flexural members with and without transverse shear reinforcement. *Eng. Struct.* **2014**, *59*, 393–398. [CrossRef]
7. João, R.C.; João, F.; Fernando, A.B. A rehabilitation study of sandwich GRC facade panels. *Constr. Build. Mater.* **2005**, *20*, 554–561. [CrossRef]
8. Brandt, M.A. Fiber reinforced cement-based (FRC) composites after over 40 years of development in building and civil engineering. *Compos. Struct.* **2008**, *1*, 86. [CrossRef]
9. Abaeian, R.; Behbahani, H.P.; Moslem, S.J. Effects of high temperatures on mechanical behavior of high strength concrete reinforced with high performance synthetic macro polypropylene (HPP) fibers. *Constr. Build. Mater.* **2018**, *165*, 631–638. [CrossRef]
10. Ferreira, J.; Branco, F. Structural application of GRC in telecommunication towers. *Constr. Build. Mater.* **2007**, *21*, 19–28. [CrossRef]
11. Liu, Y.F.; Wu, C.F. Experimental study on the elasticity modulus of glass fiber reinforced concrete. *Concrete* **2017**, *6*, 64–66+79. [CrossRef]
12. Zhao, W.H.; Su, Q.; Li, T.; Huang, J.J. Experimental research on mechanical properties of light-weight foamed concrete with glass fiber reinforcement. *Indust. Constr.* **2017**, *47*, 110–114+80. [CrossRef]
13. Shen, W.; Yang, D.Y.; Luo, J.J.; Huang, J.S.; Liu, X.; Zhu, Z.D. Research of the long-term mechanical properties of alkali resistant glass fiber concrete. *Concrete* **2017**, *6*, 102–106.
14. Qian, W.; He, W.X. The Study on Mechanical Properties of Glass Fiber and Fly Ash Composite Cement Soil. *J. Solid Waste Technol. Manag.* **2018**, *44*, 361–369. [CrossRef]
15. Lura, P.; Jensen, O.M.; Weiss, J. Cracking in cement paste induced by autogenous shrinkage. *Mater. Struct.* **2009**, *42*, 1089–1099. [CrossRef]
16. He, Z.; Li, Z.J. Influence of alkali on restrained shrinkage behavior of cement-based materials. *Cem. Concr. Res.* **2004**, *35*, 457–463. [CrossRef]
17. Sant, G. The influence of temperature on autogenous volume changes in cementitious materials containing shrinkage reducing admixtures. *Cem. Concr. Compos.* **2012**, *34*, 855–865. [CrossRef]
18. Yuan, Z.; Jia, Y. Mechanical properties and microstructure of glass fiber and polypropylene fiber reinforced concrete: An experimental study. *Constr. Build. Mater.* **2021**, *266*, 121048. [CrossRef]
19. Enfedaque, A.; Paradela, L.S.; Sánchez-Gálvez, V. An alternative methodology to predict aging effects on the mechanical properties of glass fiber reinforced cements (GRC). *Constr. Build. Mater.* **2012**, *27*, 425–431. [CrossRef]
20. Bian, G.R.; Li, B.Z.; Li, M.; Xu, W.; Tian, Q. Influence of Variable Temperature Condition on Crack Resistance of Concrete. *China Concr. Cem. Prod.* **2021**, *2*, 23–26. [CrossRef]
21. Ye, H.; Radlińska, A.; Neves, J. Drying and carbonation shrinkage of cement paste containing alkalis. *Mater. Struct.* **2017**, *50*, 132. [CrossRef]
22. Nguyen, T.-T.; Waldmann, D.; Bui, T.Q. Phase field simulation of early-age fracture in cement-based materials. *Int. J. Solids Struct.* **2020**, 191–192. [CrossRef]
23. Kumarappa, D.B.; Peethamparan, S.; Ngami, M. Autogenous shrinkage of alkali activated slag mortars: Basic mechanisms and mitigation methods. *Cem. Concr. Res.* **2018**, *109*, 1–9. [CrossRef]
24. Wu, Z.T.; Zhang, Y.S.; Liu, N.D.; Zhang, Y.T.; Yuan, D.F.; Wang, T.; Gu, J.S. Shrinkage properties of glass fiber reinforced cement-based materials. *Bull. Chin. Ceram. Soc.* **2019**, *38*, 2570–2577. [CrossRef]
25. Chylík, R.; Fládr, J.; Petr, B.; Trtík, T.; Vráblík, L. An analysis of the applicability of existing shrinkage prediction models to concretes containing steel fibers or crumb rubber. *J. Build. Eng.* **2019**, *24*. [CrossRef]
26. Guo, J.; Zhang, S.; Guo, T.; Zhang, P. Effects of UEA and MgO expansive agents on fracture properties of concrete. *Constr. Build. Mater.* **2020**, *263*, 120245. [CrossRef]
27. Moosa, M.; Hemin, K.; Mohammad, K. Fracture behavior of monotype and hybrid fiber reinforced self-compacting concrete at different temperatures. *Adv. Concr. Constr.* **2020**, *9*, 375–386.
28. He, J.H.; Cheng, C.M.; Zhang, Y.F.; Hong, S.L. Effect of glass fiber parameters on plastic shrinkage cracking of mortar. *Concrete* **2020**, *10*, 118–120.
29. Shen, D.; Wen, C.; Zhu, P.; Wu, Y.; Wu, Y. Influence of Barchip fiber on early-age autogenous shrinkage of high-strength concrete internally cured with super absorbent polymers. *Constr. Build. Mater.* **2020**, *264*, 119983. [CrossRef]
30. Kasagani, H.; Rao, C. Effect of graded fibers on stress strain behaviour of Glass Fiber Reinforced Concrete in tension. *Constr. Build. Mater.* **2018**, *183*, 592–604. [CrossRef]
31. Li, Q.R.; Xu, R. Experimental study on RAC performance enhancement and micro action mechanism of metakaolin. *Concrete* **2021**, *5*, 84–87. [CrossRef]
32. Almusallam, A.A. Effect of environmental conditions on the properties of fresh and hardened concrete. *Cem. Concr. Compos.* **2001**, *23*, 353–361. [CrossRef]

 crystals

MDPI

Article

Experimental Study on Durability of Hybrid Fiber-Reinforced Concrete in Deep Alluvium Frozen Shaft Lining

Zhishu Yao, Yu Fang, Ping Zhang and Xianwen Huang *

School of Civil Engineering and Architecture, Anhui University of Science and Technology, Huainan 232001, China; zsyao@aust.edu.cn (Z.Y.); fy18956925891@163.com (Y.F.); Zhangping940813@163.com (P.Z.)
* Correspondence: 162100003@stu.just.edu.cn; Tel.: +86-183-6289-1238

Abstract: This article proposes hybrid fiber-reinforced concrete (HFRC) mixed with polyvinyl alcohol fiber (PVA) and polypropylene steel fiber (FST) as a wall construction material to improve the bearing capacity and durability of frozen shaft lining structures in deep alluvium. According to the stress characteristics and engineering environment of the frozen shaft lining, the strength, impermeability, freeze–thaw damage, and corrosion resistance are taken as the evaluation and control indexes. The C60 concrete commonly used in freezing shaft lining is selected as the reference group. Compared to the reference group, the test results show that the compressive strength of HFRC is similar to that of the reference concrete, but its splitting tensile strength and flexural strength are higher; according to the strength test, the optimum mixed content of 1.092 kg/m^3 PVA and 5 kg/m^3 FST are obtained. According to the impermeability test results, the mixing of PVA and FST can improve the impermeability resistance of concrete. For the freeze–thaw cycle test results, the mixing of PVA and FST can improve the frost resistance of concrete; based on the 120 days sulfate corrosion test, the mixing of PVA and FST will improve the corrosion resistance of concrete.

Keywords: freezing shaft sinking; shaft lining structure; uneven pressure; hybrid fiber-reinforced concrete; durability

Citation: Yao, Z.; Fang, Y.; Zhang, P.; Huang, X. Experimental Study on Durability of Hybrid Fiber-Reinforced Concrete in Deep Alluvium Frozen Shaft Lining. *Crystals* **2021**, *11*, 725. https://doi.org/10.3390/cryst11070725

Academic Editors: Cesare Signorini, Antonella Sola, Sumit Chakraborty, Valentina Volpini and Ing. José L. García

Received: 11 May 2021
Accepted: 21 June 2021
Published: 23 June 2021

1. Introduction

In coal mining, the vertical shaft is an important channel for moving personnel up and down, ventilation, drainage, and coal lifting [1–3], and its supporting structure is called the shaft lining structure [4,5]. When a newly established shaft passes through deep alluvium, it needs to be constructed by a special drilling method. Freezing shaft sinking has the advantages of strong construction adaptability, flexible supporting structure, and fast completion speed; therefore, it has become the main method of construction in China [6–8]. In order to resist strong external loads, high-strength concrete is usually selected as the building material for walls in the structural design of frozen shaft lining in deep alluvium, and the outer shaft wall is required to have the characteristics of early strength [9,10]. With the increase in the depth of the well construction, the thickness of the inner and outer sidewalls of the deep alluvium shaft is greatly increased, the hydration heat generated during the pouring of the shaft is high, and the environmental temperature varies greatly during construction, so the high-strength concrete of the shaft lining easily produces temperature cracks [11]. Especially in the method of shaft sinking by freezing in deep alluvium, the freezing pressure on the outer wall of the shaft is obviously unevenly distributed, due to the uneven distribution of the saline flow at the freezing hole deviation, which is extremely adverse to the force on the shaft lining and can easily cause the tensile failure of the high-strength concrete [12,13]. Figure 1 shows the cracks of high-strength concrete in shaft lining.

Frozen shaft lining concrete in deep alluvium has been used under a harsh engineering environment for a long time [14–16]. In addition to bearing loads, such as high ground pressure and freezing pressure, the built shaft was also faced with problems, such as water seepage, freezing, and thawing, and the erosion of harmful ions in groundwater [17–20]. Traditional high strength concrete has obvious deficiencies in impermeability, frost resistance, and corrosion resistance, which is difficult to be applied to such formation conditions as deep alluvium. Therefore, scholars have carried out a series of studies on the durability of shaft lining concrete. In order to solve the problems of many cracks and serious water seepage caused by thermal stress, Yang, L. et al. [21] prepared excellent concrete as a building material for walls by adding hybrid fiber and an expansive agent, and impermeability, crack resistance, and microscopic tests were carried out. The results show that it can solve the problems of shaft lining cracking and water seepage. Zhao, X. M. et al. [22] carried out the study on the frost resistance of fiber reinforced concrete and found that the incorporation of hybrid fiber could improve the compressive strength of concrete after freeze–thaw cycles and reduce the mass loss rate. Yang, D. et al. [23] carried out dry–wet cycle corrosion tests on concrete cube specimens at different corrosion periods. The results show that the splitting tensile strength of concrete decreases continuously with the increase in corrosion depth. Hou, Y. F. et al. [24] analyzed the evolution of concrete corrosion depth by means of ultrasonic testing and microstructure observation, and discussed the factors causing shaft lining concrete corrosion. Liu, J. H. et al. [25] studied the mechanical properties of polypropylene steel fiber-reinforced shaft lining concrete under the coupling action of early stress and negative temperature through a self-made test device, and concluded that concrete's load-bearing capacity and Cl^{-1} corrosion resistance were improved.

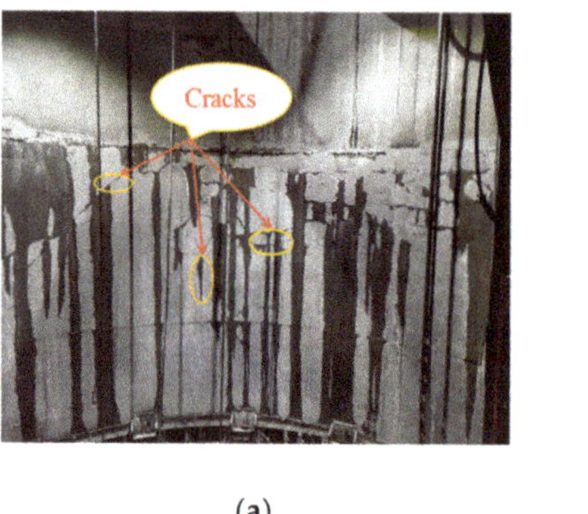

(a)

(b)

Figure 1. The cracks of high-strength concrete in shaft lining: (**a**) Temperature cracks caused by temperature stress; (**b**) Rupture cracks caused by uneven freezing pressure.

Based on the above analysis, in this research, experimental studies on the static properties and durability of hybrid fiber-reinforced concrete were carried out to improve the bearing capacity and durability of a frozen shaft lining structure in deep alluvium, based on the frozen wall, Polyvinyl alcohol fiber and polypropylene steel fiber were added to commonly used C60 high-strength concrete to create a high-performance hybrid fiber concrete as a building material for the wall. The research results can provide a scientific basis for engineering applications.

2. Preparation of Hybrid Fiber-Reinforced Concrete for Shaft Lining

2.1. Raw Materials

This study used P.O52.5R cement, which was produced by the Fengyang Conch Cement Plant. The main parameters are shown in Table 1. The coarse aggregate was continuously graded basalt gravel with a 5~20 millimeter particle size; the crushability index was 6.7%, and the mud content was detected before the test to meet the requirements of the specification. The fine aggregate was natural river sand with a mud content of less than 1.6%, which belongs to medium and coarse sand. The fineness modulus is shown in Table 2. Slag powder and silica fume were selected as the mineral admixtures; The slag powder (Anhui Qingya Building Material Co., Ltd., Huainan, China) had a specific surface area greater than 350 m^2/kg; The silica fume (Shanxi Dongyi Ferroalloy Factory, Shanxi, China) had a specific surface area greater than 18,000 m^2/kg; Table 3 summarizes the mineral composition of the admixture. Moreover, a high-efficiency NF water-reducing agent was used, and its main properties are presented in Table 4. The fiber materials used were polyvinyl alcohol fiber (PVA) and polypropylene steel fiber (FST). Their basic performance parameters are shown in Table 5, and the physical appearance is shown in Figure 2.

Table 1. Main performance parameters of P.O52.5R cement.

Cement Type	Stability	Setting Time (min)		Flexural Strength (MPa)		Compressive Strength (MPa)	
		Initial setting	Final setting	3d	28d	3d	28d
P.O 52.5R	Qualified	120	280	7.2	10.9	34.5	57.8

Table 2. Fineness modulus of sand.

Square Screen Size (mm)	Remaining Percentage of Screen (%)	Cumulative Sieve Residual Percentage (%)
4.75	1.3	1.3
2.36	11.0	12.3
1.18	13.1	25.4
0.6	26.5	51.9
0.3	41.8	93.7
0.15	3.8	97.5
Amount at the bottom of the screen	2.5	100.0
Total	100.0	100.0
Fineness modulus	2.934	

Table 3. Mineral composition of admixture.

Admixtures	Component (%)						
	SiO_2	Al_2O_3	Fe_2O_3	CaO	MgO	SiO_2	SO_3
Slag powder	32.41	9.99	1.50	40.32	6.86	32.41	2.51
Silicon powder	93.60	0.78	0.65	0.82	1.30	93.60	0.10

Table 4. Properties of water reducer.

Water Reducing Rate (%)	Color	Density (kg/m^3)	Chlorideion Content (%)	Solid Content (%)
21	Yellow	1410	$\leq$0.3	$\geq$90

Table 5. Basic performance parameters of fibers.

Fibers	Length (mm)	Elastic Modulus (GPa)	Elongation (%)	Density ($g{\cdot}cm^{-3}$)	Tensile Strength (MPa)
FSTF	50	5	24	0.91	570
PVAF	18	39	6.9	1.30	1830

(a) (b)

Figure 2. Appearance of fibers: (**a**) FSTF; (**b**) PVAF.

2.2. Test Mix Ratio

In order to better resist the effect of uneven freezing pressure, based on C60 high-strength concrete commonly used in freezing walls, a high-performance hybrid fiber concrete was prepared with the addition of PVA fiber and FST fiber. The PVA content and FST content were selected as the two influencing factors of the orthogonal test, and three levels were set for each factor. The PVA fiber content levels were set to 0.728 kg/m^3, 1.092 kg/m^3, and 1.456 kg/m^3; FST fiber content levels were set to 4.0 kg/m^3, 5.0 kg/m^3, and 6.0 kg/m^3. At the same time, another group of reference concrete was set as the control group, resulting in 10 groups of tests in total. The reference mix ratio of the C60 high-strength concrete is shown in Table 6.

Table 6. The reference mix ratio of the C60 high-strength concrete.

Strength Grade	Water/Binder Ratio	Cement (kg·m^{-3})	Admixtures (kg·m^{-3})	Stone (kg·m^{-3})	Sand (kg·m^{-3})	Water (kg·m^{-3})
C60	0.28	410	130	1121.5	630.8	151.2

Annotation: Admixtures = Slag powder (102.12 kg·m^{-3}), Silicon powder (20.5 kg·m^{-3}), NF (7.38 kg·m^{-3}).

2.3. Preparation and Curing of Specimens

Firstly, after weighing, the gravel and sand were poured into a mixer for dry mixing for 2 min; then, cement and admixture were added, and the dry mixing was continued for 2 min; Secondly, the fiber was evenly added in batches, and then stirred for 2 min; Finally, water was added to the mixture quickly followed by slow wet mixing for 2 min.

The slump of the mixture was measured during mixing to evaluate the influence of the hybrid fiber on the fluidity of the concrete. Then, the concrete mixture was poured into a test mold and transported to a shaking table for vibration. The mixture was added or removed according to the situation during the vibration process. It was placed on indoor flat ground for natural curing for 24 h, and the mold was then removed accordingly. After numbering, it was transported to a standard curing box (temperature 20 $\pm$ 2 °C; relative humidity 97%) to the age.

2.4. Test Results and Analysis

The specimens were taken out after standard curing for 28 days, and cube compression, splitting tensile, and flexural strength tests were carried out. A CSS-YAW3000 electro-hydraulic servo press was adopted to determine the loading method and rate according to the relevant provisions of CECS13-2009 [26]. Cube specimens of 100 mm $\times$ 100 mm $\times$ 100 mm were selected for the compression and splitting tensile strength tests, and the results were multiplied by size conversion coefficients of 0.95 and 0.85, respectively. The orthogonal test results are shown in Table 7.

Table 7. Orthogonal test results.

Specimen	Concrete Type	PVA (kg·m⁻³)	FST (kg·m⁻³)	Slump (mm)	CS (MPa)	Maximum Error Rate of CS (%)	TS (MPa)	Maximum Error Rate of TS (%)	FS (MPa)	Maximum Error Rate of FS (%)
J-1	Reference concrete	0	0	195	72.4	5.6	4.54	4.3	6.34	5.9
H-1	Hybrid fiber concrete	0.728	5	178	72.0	6.4	5.55	5.1	7.12	6.8
H-2		1.456	4	165	69.5	4.5	5.31	5.6	6.98	5.2
H-3		1.092	5	178	74.6	5.2	6.01	5.8	7.96	6.3
H-4		1.092	6	174	71.5	7.1	5.25	6.9	7.01	5.7
H-5		0.728	6	185	71.1	4.9	5.22	4.2	6.94	6.1
H-6		1.092	4	182	70.9	5.6	5.16	6.3	6.89	5.8
H-7		0.728	4	187	70.5	6.7	5.58	5.5	7.28	4.9
H-8		1.456	5	168	69.6	4.2	5.97	4.7	7.87	5.8
H-9		1.456	6	160	68.4	3.9	5.32	4.4	7.06	4.7

Annotation: CS = Compressive strength, TS = Splitting tensile strength, FS = Flexural strength.

As shown in Table 7, the fluidity of the hybrid fiber concrete is significantly lower than that of the reference group, and the greater the fiber content is, the greater the decrease is. At the maximum content level, the slump is only 160 mm, which is 35 mm lower than that of the reference group; therefore, the amount of hybrid fiber should not be too high, as otherwise, it is difficult to meet the working performance requirements of concrete. The test results of each group show that the compressive strength of the hybrid fiber group is the same as that of the reference group, with an increase of –5.5~3.0%, while the splitting tensile strength and flexural strength are significantly increased compared with those of the reference group, with a maximum increase of 32.4% and 25.6%, respectively. The failure patterns of each group of specimens are shown in Figures 3–5.

Figure 3. Comparison of compression failure modes of specimens: (a) Reference group; (b) hybrid fiber group.

The comparison of the compressive failure patterns of the specimens in Figure 3, shows that the base concrete was completely crushed, and the fragments fell off, showing the characteristics of brittle failure, while the hybrid fiber-reinforced concrete specimen remained intact and did not break down; only the outer surface was slightly raised and several vertical cracks appeared, similar to hoop failure in the circumferential direction. As can be seen from the comparison of the splitting and tensile failure modes of the specimens in Figure 4, the reference concrete was directly split into two along the center line, while the hybrid fiber-reinforced concrete had vertical cracks along the center line, and the bridging fiber could be seen at the crack section. As shown in Figure 5, flexural strength test, the reference group concrete was directly broken, while the hybrid fiber group showed failure cracks. Thus, hybrid fiber can improve the brittleness of concrete during failure, make it show ductility; and effectively improve the crack resistance of concrete. The analysis shows that PVA fiber has good crack resistance in the early stage, and FST fiber can inhibit crack expansion in the late stage. When the two fibers are mixed, they show a positive hybrid effect, improving toughness, crack resistance, and deformation constraint. Meanwhile, based on the range analysis of the test results in Table 7, obtained using the SPSS software, it was found that the optimal contents of the two kinds of fibers in this experiment were 1.092 kg/m^3 PVA fiber and 5 kg/m^3 FST fiber, which was determined to be the fiber content of the shaft lining hybrid fiber concrete to be prepared.

(a)

(b)

Figure 4. Comparison of splitting failure modes of specimens: (**a**) Reference group; (**b**) hybrid fiber group.

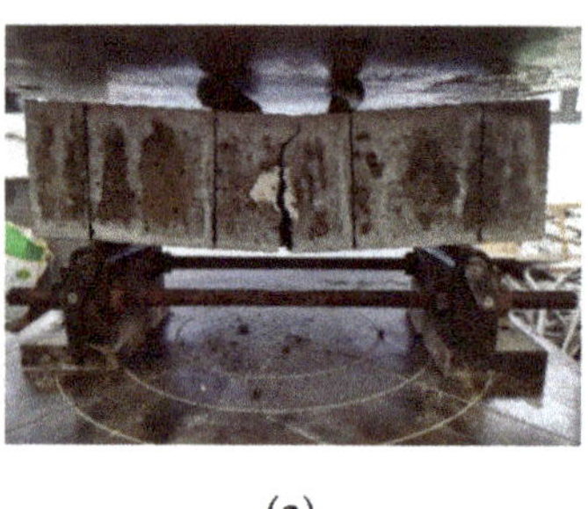

(a)

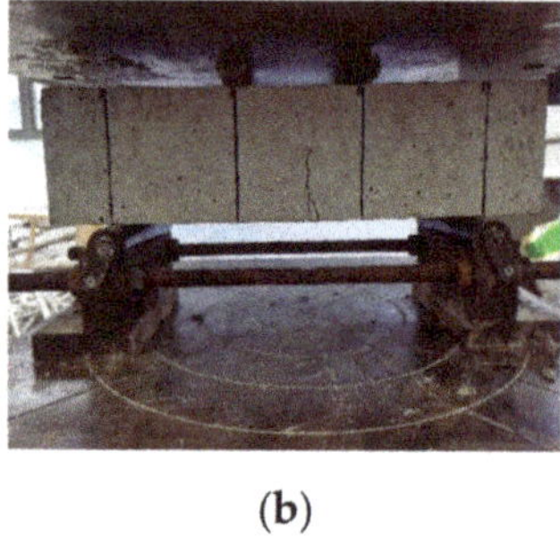

(b)

Figure 5. Comparison of flexural failure modes of specimens: (**a**) Reference group; (**b**) hybrid fiber group.

3. Durability Test of Shaft Lining Hybrid Fiber-Reinforced Concrete

As the supporting structure of the shaft, the shaft lining will be affected by high-pressure groundwater after the freezing wall is thawed [27,28]. Therefore, the impermeability of high-performance shaft lining concrete must meet certain requirements. The design life of the shaft is usually more than 50 years, and a large number of SO_4^{2-} and Cl^{1-} are gathered in soil and groundwater in many areas, and the shaft often suffered from freeze–thaw action, so the harsh working environment and long service life put forward higher requirements for the durability of wall material [29,30]. At present, there is a significant amount of research on the durability of fiber-reinforced concrete, but little work on the durability of hybrid fiber-reinforced shaft lining concrete. Therefore, to make the prepared hybrid fiber-reinforced concrete of shaft lining better applied in engineering practice, the durability experiment on shaft lining concrete of reference group and hybrid fiber group was carried out by taking C60 high-strength concrete as the object.

3.1. Impermeability Test

3.1.1. Experimental Design

The ability of concrete to resist seepage under the action of water pressure is called impermeability, which is an important index that reflects the durability of concrete. Based on the GBJ82-2009 [31] specification, a water penetration test was carried out using the penetration height method. A digital display automatic adjusting concrete impervious meter was selected as the measuring device. Additionally, Φ 175 × 185 × 150 cylindrical specimens were selected, with six pieces in each group. The test was carried out after 28 days of standard maintenance of the specimens, and the loading device is shown in Figure 6.

Figure 6. Impermeable loading device.

Considering that shaft lining structures are subjected to high underground water pressure in engineering practice, 3.6 MPa was selected to stabilize water pressure, and the following measures were taken during the impermeability test.

1. After the maintenance of specimens was completed, the surface of each specimen was polished to remove the influence of surface floating slurry.
2. Paraffin was heated and melted in a shallow dish, and then, a layer of paraffin with a thickness of about 1.5 mm was applied around the specimen to seal the water.
3. The cylindrical paraffin-wrapped specimen was placed into a preheated metal sleeve die; and left to sink slowly, and then pressed to level with the bottom of the sleeve die.
4. After cooling, the specimens were installed in turn, the fixing screws were tightened; and the parameters of the antipermeability meter were adjusted. The loading was set as follows: Upper limit of 4 MPa, lower limit of 3.6 MPa; and the pressure was continuously stabilized for 24 h.
5. After the test was completed, the sleeve mold was removed, and a press was used to demold and split the specimen to observe the section. Ten measuring points were taken from each specimen, and the water seepage height at each point was measured. A flow chart of the operation steps is shown in Figure 7.

Figure 7. Process chart of impermeability test. (**a**) Sample grinding; (**b**) Sample sealing wax; (**c**) Sample installation; (**d**) impermeability test; (**e**) Take out the sample; (**f**) Split specimen.

3.1.2. Test Results and Analysis

After measuring the seepage height of the concrete, the relative permeability coefficient of the concrete was calculated using Formula (1).

$$K_r = \frac{\alpha D_m^2}{2TH} \tag{1}$$

where K_r is the relative permeability coefficient of concrete, α is the water absorption of concrete (0.03), D_m is the average seepage height of concrete (mm), T is the constant pressure time (h), and H is the water column height corresponding to the water pressure (1 MPa corresponds to 100 m water column height).

Table 8 shows the average height of water seepage and the calculated relative permeability coefficient of the two groups of specimens after 24 h of pressure stabilization at 3.6 MPa water pressure. It can be seen from Table 8 and Figure 8 that the water seepage height of the two groups of specimens was not large, which indicates that impermeability of both specimen groups is good and meets the impermeability requirements of high-performance concrete. This is mainly because the water–binder ratio is relatively low; at the same time, in the admixture, the slag and silicon powder generated hydration products to improve the compactness of concrete and reduce the number of pores. In addition, the average water permeability height and relative permeability coefficient of the concrete in the hybrid fiber group were significantly lower than those in the reference group, and the average water permeability height was 31.7% lower than that in the reference group. Compared with the reference group, the relative permeability coefficient decreased by 53.3%. This shows that the incorporation of hybrid fiber can further improve the impermeability of concrete.

Table 8. Concrete impermeability test results.

Specimen Groups	Water Penetration Height (cm)						Average Value	Relative Permeability Coefficient (10^{-7}cm/h)
	1	2	3	4	5	6		
Reference group	2.02	1.81	1.53	1.87	1.82	1.75	1.80	0.563
Hybrid fiber group	1.15	0.96	1.34	1.25	1.42	1.27	1.23	0.263

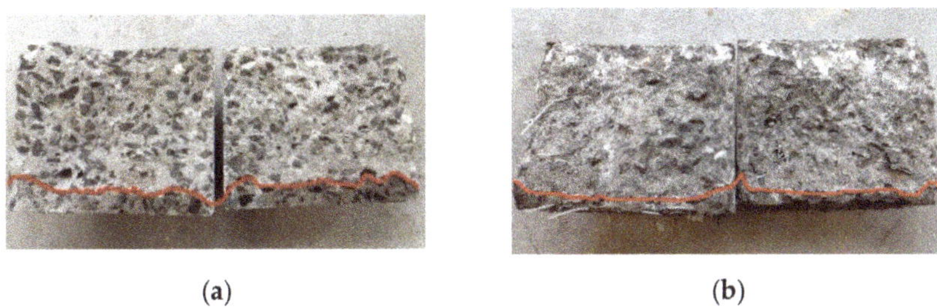

Figure 8. Concrete seepage height chart: (**a**) Reference group; (**b**) hybrid fiber group.

3.2. Freeze–Thaw Resistance Cycle Test

3.2.1. Experimental Design

Freezing resistance capacity is another important indicator reflecting the durability of concrete. Considering the particularity of the environment in which the structure of a frozen shaft is located, the concrete of the shaft lining should meet certain requirements of freezing resistance [32,33]. The TEST-1000 high and low temperature TEST chamber was selected, and the temperature control range can reach −60 °C~+150 °C. As shown in Figure 9, small cube specimens with a side length of 100 mm were selected for the test. The slow freezing method was adopted to first maintain the specimens in the curing box and then in the water tank. At this time, the water surface temperature no lower than that of the surface of the specimens, and then, they were placed in the test box for freeze–thaw cycle testing. The specimens were frozen at −15 °C for 4 h and then melted in a 20 °C water tank for 6 h. After freezing and thawing, it was regarded as a freeze–thaw cycle.

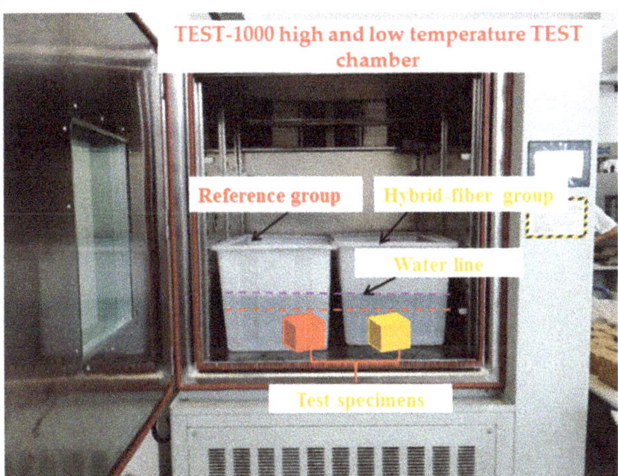

Figure 9. Freeze–thaw cycle test device.

In this test, the freezing resistance of the C60 benchmark group specimens and those in the hybrid fiber group was studied. The number of freeze–thaw cycles were set as 25, 50, 75, and 100, with three specimens in each freeze–thaw cycle.

3.2.2. Test Results and Analysis

The mass loss rate of concrete specimens was calculated using the following formulas.

$$\Delta m_{ni} = \frac{m_{0i} - m_{ni}}{m_{0i}} \times 100\% \quad (2)$$

$$\Delta m_n = \frac{\sum_i^3 \Delta m_{ni}}{3} \times 100\% \tag{3}$$

where Δm_{ni} stands for the mass loss rate of the ith specimen after $n(n = 0, 25, 50, 75$, and $100)$ freeze–thaw cycles, m_{0i} stands for the mass of the ith specimen before the freeze–thaw cycle m_{ni} stands for the mass of the ith specimen after n freeze–thaw cycles, and Δm_n stands for the average mass loss rate of each group of specimens after n freeze–thaw cycles (the mass decreases when $\Delta m_n > 0$; and increases when $\Delta m_n < 0$).

The results of mass loss are shown in Table 9.

Table 9. Concrete mass loss results after freeze–thaw cycles (%).

Specimen Groups	Number of Freeze–Thaw Cycles (Times)				
	0	**25**	**50**	**75**	**100**
Reference group	0	–0.10	–0.16	0.02	0.23
Hybrid fiber group	0	–0.04	–0.06	–0.02	0.05

Meanwhile, the relationship of concrete compressive strength and mass loss rate with the number of freeze–thaw cycles were drawn, as shown in Figure 10.

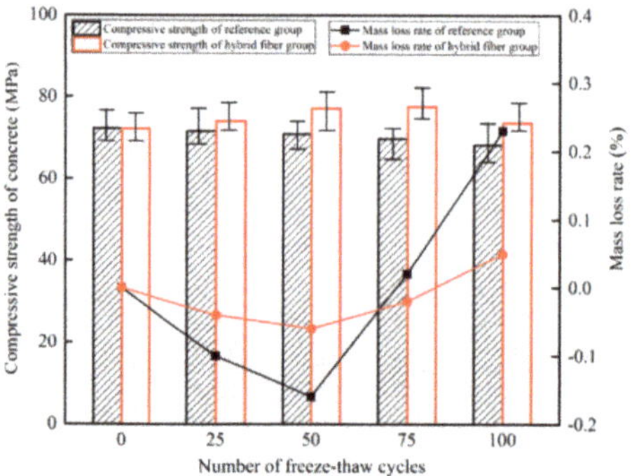

Figure 10. Relation chart of compressive strength and mass loss factors with the number of freeze–thaw cycles.

It can be seen from Figure 10 that the mass loss rate of concrete showed a negative value in the early stage and changed slowly, that is, the mass increased slowly. After 75 freeze–thaw cycles, the concrete surface began to shed mud, thus reducing the mass, and the change speed of the specimen mass accelerated. The mass loss in the hybrid fiber group was not significant during the whole freeze–thaw test period, and the mass loss rate after 100 freeze–thaw cycles was 78.3% lower than that of the reference group. With the increase in the number of freeze–thaw cycles, the compressive strength of the two groups of specimens continued to decrease slowly, and the specimens remained in a high–strength state after 100 freeze–thaw cycles. The strength loss rates of the reference group after 25, 50, 75, and 100 cycles were 1.56%, 3.13%, 4.98%, and 8.11%, respectively, and those of the hybrid fiber group were 0.83%, 1.80%, 3.46%, and 5.54%, respectively. The strength loss rate of the hybrid fiber group was 0.73%, 1.33%, 1.52%, and 2.57% lower than that of the reference group, respectively. This can be explained by the fact that the mixture of FST fiber and PVA fiber has a positive hybrid effect, which can effectively reduce the number of micropores in the matrix and improve the antispalling ability of the concrete. Therefore, it can be seen that the incorporation of hybrid fiber can effectively reduce the strength loss of concrete after freeze–thaw cycles and improve the frost resistance of concrete.

A morphological comparison of each group of specimens after 100 freeze–thaw cycles is shown in Figure 11. Many gullies appeared on the surface of the reference concrete group, especially in the corner zone, which is more serious. This is because the concrete freeze–thaw damage generally starts with the spalling of the surface cement mortar, when the deterioration of the reference concrete has already begun. On the contrary, the surface of the hybrid fiber concrete was relatively intact, which indicates that the PVA-FST fiber hybrid can effectively reduce the spalling of concrete subjected to freezing and thawing.

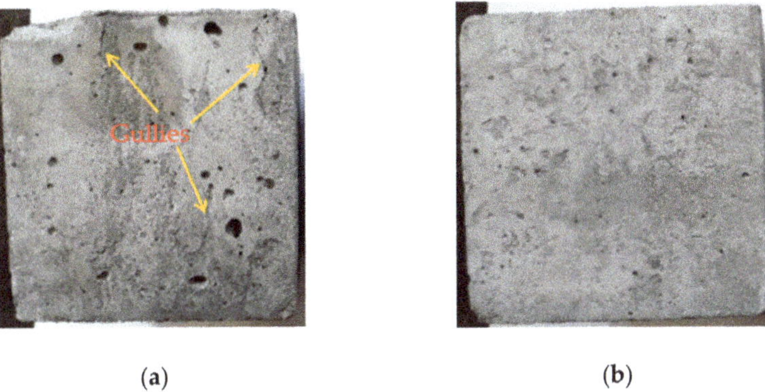

(a) (b)

Figure 11. Contrast of morphology of concrete after 100 freeze–thaw cycles: (**a**) Reference group; (**b**) hybrid fiber group.

3.3. Sulfate Corrosion Resistance Test

3.3.1. Experimental Design

In addition to bearing complex loads, the shaft lining structure in a long-term deep formation environment is also affected by various harmful corrosive substances gathered in soil and water, among which sulfate is a typical one [34,35]. A long-term immersion method was adopted in this sulfate corrosion resistance test. In order to accelerate corrosion, a 10% $NaSO_4$ solution with mass concentration was selected as the immersion solution, which was prepared from anhydrous sodium sulfate and tap water. And the specimens were soaked in $NaSO_4$ solution after curing, and the soaking time was set as 30d, 60d, 90d, and 120d. In order to ensure that the concentration of the solution was not reduced by crystal precipitation and water evaporation, the solution was replaced regularly (every 30 days in this test). After reaching the expected soaking time, the specimens were washed and wiped dry, and then, the mass loss and strength loss were measured.

3.3.2. Test Results and Analysis

In order to analyze the mass change of concrete after sulfate corrosion, the mass change factor S was defined, and the expression is as follows:

$$S = \frac{m_t - m_0}{m_0} \times 100\% \tag{4}$$

where S is the mass variation factor (the mass increases when $S > 0$, and decreases when $S < 0$), m_t is the mass of the specimen after tage corrosion, and m_0 is the mass of the uncorroded specimen.

The results of concrete mass change are shown in Table 10.

Table 10 shows that with the increase in corrosion exposure time, the change factor of concrete mass increases at first and then decreases as the concrete reacts with $NaSO_4$ solution and gradually generates ettringite, gypsum, and other corrosive substances in the initial stage of corrosion [36]. At the same time, some salt crystals invade the specimen and fill the micropores inside the concrete, thus improving the density and mass of the specimen. As the corrosion continues, the filling material inside the specimen continues to accumulate and expand, which destroys the pore structure, produces microcracks and gradually expands; and is accompanied by exfoliation of the epidermis, which finally leads to an increase in the number of cracks and easier invasion of harmful erosion materials; the corrosion is also more serious.

Table 10. Mass change factors of specimens at different corrosion exposure (%).

Specimen Groups	Corrosion Exposure Time (d)				
	0	30	60	90	120
Reference group	0	0.17	0.41	0.56	0.32
Hybrid fiber group	0	0.08	0.24	0.31	0.14

Additionally, compressive strength tests were carried out on the specimens soaked for various periods of time, and the test results at different exposure times are shown in Table 11.

Table 11. Results of compressive strength after sulfate corrosion (MPa).

Specimen Groups	Corrosion Exposure Time (d)				
	0	30	60	90	120
Reference group	70.3	72.3	76.1	73.4	68.5
Hybrid fiber group	72.2	74.1	77.2	77.6	73.7

As shown in Table 11, when the corrosion exposure time were 30d, 60d, 90d, and 120d, the strong growth rates of the specimens in the reference group and the hybrid fiber group were 2.8%, 8.3%, 4.4%, and –2.6%, and 2.6%, 6.9%, 7.5%, and 2.1%, respectively. After 120d of corrosion, the strength of the reference group specimens decreased by 2.6%, while the strength of the hybrid fiber group specimens was still in the growth stage, and the strength was 1.076 times that of the reference group, which indicated that the hybrid fiber could effectively slow down the strength loss of concrete after sulfate corrosion. In order to analyze the relationship between the change factors of the compressive strength and mass of concrete specimens and the change in corrosion exposure time more intuitively, the results in Tables 10 and 11 were drawn into a double-Y-axis, columnar-broken line diagram, as shown in Figure 12.

As can be seen from Figure 12, the mass factor of concrete increased rapidly after 30 days of corrosion exposure time, which involved the process of the filling and compaction of erosive materials in the first stage, but this process obviously slowed down after 60 days. This is mainly due to the increase in expansion stress in the specimen, resulting in the generation and expansion of microcracks, and the decline in mass along with the shedding of the outer skin and cement mortar, and the corrosion of concrete intensified, due to the increase in cracks. Meanwhile, the compaction process of erosive material filling continued; therefore, the mass change factor of concrete still shows an increasing trend, but the rate decreases obviously. At 90d of corrosion, the mass variation factor of concrete in each group reached the maximum value, and at this time, the mass variation factor of concrete in the hybrid fiber group was 44.6% lower than that in the reference group. However, after 90d of corrosion, the deterioration cracking and shedding process intensified, and the mass variation factor of concrete decreased sharply. When the corrosion exposure time was 120 days, the mass variation factors of the two groups of concrete were still positive. However, if the trend is developed, the mass of the two groups of concrete will decrease sharply when the corrosion exposure time is further increased. The compressive strength of the specimens in each group increased first and then decreased with the increase in corrosion exposure time. The initial decrease in compressive strength in the reference group occurred after corrosion for 60 days, while that in the hybrid fiber group occurred after 90 days. This is because the erosive material invades and fills the concrete, the number of pores in the matrix is reduced, the compactness is increased, and the strength is also improved to a certain extent. With continuous corrosion, the filling material in the matrix continues to accumulate and expand, and the expansion stress increases continuously, which eventually exceeds the tensile strength of the concrete, resulting in internal deterioration and cracking, and the pressure-bearing capacity is also reduced. The hybrid fiber can improve the tensile capacity of the matrix, hinder the expansion of cracks and increase the compactness of the concrete; therefore, it slows down the rate of corrosion deterioration to a certain extent.

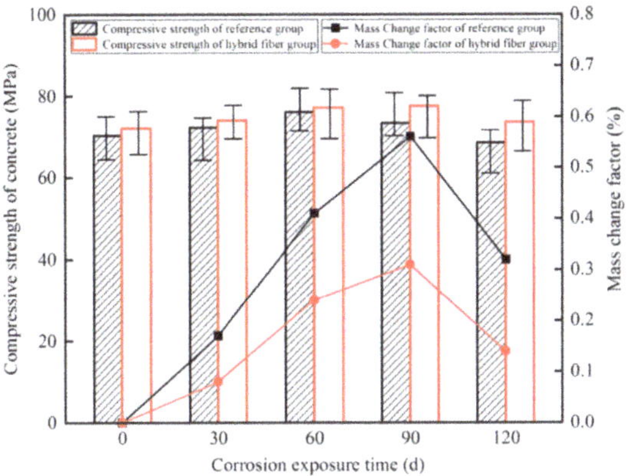

Figure 12. Relation chart of compressive strength and mass change factor with corrosion.

Overall, after 120 days of sulfate corrosion, there was no obvious mass loss in the reference group and hybrid fiber group, and the compressive strength of the hybrid fiber-reinforced concrete did not decrease but rather even slightly improved. Additionally, the apparent integrity of specimen morphology was relatively high, while the concrete of reference group has obvious surface spalling, as shown in Figure 13. Therefore, the hybrid fiber concrete has good corrosion resistance.

Figure 13. Appearance of specimens after 120 d immersion: (**a**) Reference group; (**b**) hybrid fiber group.

4. Discussion

In this research, according to the static mechanical test, we can see that the flexural strength and splitting tensile strength of the hybrid fiber concrete were significantly higher than those of the reference group, because the bridging effect of the hybrid fiber improves the bond strength between the matrix and the fiber, and hinders the expansion of microcracks and macro cracks in the concrete. This is consistent with the discussion of the strength mechanism of steel fiber reinforced concrete by Liu, J. H. et al. [25]. In the durability test, various properties of the hybrid fiber concrete group were better than those of the reference group. This can be explained by the fact that the mixture of FST fiber and PVA fiber has a positive hybrid effect, which can effectively reduce the number of micropores in the matrix and improves the compactness of concrete so as to resist water pressure penetration, improve the freeze–thaw resistance of concrete, and slow down the speed of sulfate corrosion cracking [21,22]. Therefore, the hybrid fiber concrete prepared is an excellent wall building material for the shaft, which can be applied in engineering practice. Unfortunately, we also acknowledge that there are some shortcomings in this paper. For example, we failed to consider the performance of hybrid fiber concrete under the combined action of freeze–thaw cycles and sulfate. Additionally, we can increase the number of freeze–thaw cycles and extend the exposure time of corrosion tests in the subsequent studies to achieve better test results.

5. Conclusions

In order to improve the bearing capacity and durability of frozen shaft lining structures in deep alluvium, experimental studies on the static properties and durability of hybrid fiber-reinforced concrete were carried out. The static mechanical properties, impermeability, frost resistance, and corrosion resistance of hybrid fiber concrete were studied, and the conclusions are as follows:

1. Under the design strength of C60, the compressive strength of the hybrid fiber group was the same as that of the reference group, with an increase of −5.5~3.0%, while the splitting tensile strength and flexural strength were significantly higher than those of the reference group, with a maximum increase of 32.4% and 25.6%, respectively. At the same time, the optimum content of hybrid fiber in this experiment was determined as 1.092 kg/m^3 PVA fiber and 5 kg/m^3 FST fiber.
2. According to the impermeability test results, the average impermeability height and relative permeability coefficient of the hybrid fiber concrete were reduced by 31.7% and 53.3%, respectively, compared with the reference group, which indicates that the hybrid fiber can significantly improve the impermeability of concrete; and enable it to meet the impermeability requirements for frozen shaft lining concrete, which is in accordance with the conclusion of the impermeability test of hybrid fiber conducted by Yang, L. et al. [21].
3. Through the freeze–thaw resistance cycle test, it found that after 100 freeze–thaw cycles, the mass loss rate and strength loss of the hybrid fiber concrete were reduced by 78.3% and 2.57%, respectively, compared with those of the reference concrete, which is similar to the results of the freeze–thaw test of fiber-reinforced concrete carried out by Zhao, X. M. et al. [22]. Additionally, the group of hybrid fiber concrete still maintained a higher bearing capacity and a better apparent morphology, which indicates that the hybrid fiber can improve the frost resistance of concrete.
4. According to the sulfate corrosion resistance test results, after soaking in sulfate solution for 120 days, the mass and strength of the hybrid fiber-reinforced concrete increased rather than decreasing; its strength still maintained a high level. It shows that hybrid fiber-reinforced concrete has good corrosion resistance and can meet the requirements for long-term use in harsh underground environments.

Author Contributions: Investigation, Y.F. and P.Z.; methodology, Z.Y., Y.F. and X.H.; software, P.Z. and X.H.; data curation, Y.F. and P.Z.; formal analysis, Z.Y, Y.F. and X.H.; writing—originaldraft, Z.Y. and Y.F.; visualization, P.Z. and X.H.; writing—review and editing, Y.F., P.Z. and X.H.; conceptualization, Z.Y.,Y.F. and X.H.; resources, Z.Y.; supervision, Z.Y. and P.Z.; project administration, Z.Y., Y.F. and P.Z.; funding acquisition, Z.Y. and X.H. All authors have read and agreed to the published version of the manuscript.

Funding: This research was supported by the Anhui University Discipline Professional Talented Person (No.gxbjZD09), Anhui Provincial Natural Science Foundation Youth Project (1908085QE185),

Anhui Provincial College of Natural Science Research Key Project (KJ2018A0098), Project Funded by China Postdoctoral Science Foundation (2018M642502), and the Science Research Foundation for Young Teachers in Anhui University of Science and Technology (QN2017211).

Institutional Review Board Statement: Not applicable.

Informed Consent Statement: Not applicable.

Data Availability Statement: Not applicable.

Conflicts of Interest: The authors declare no conflict of interest.

References

1. Xie, H.P. Research review of the state key research development program of China: Deep rock mechanics and mining theory. *J. China Coal Soc.* **2019**, *44*, 1283–1305.
2. He, M.C. Research progress of deep shaft construction mechanics. *J. China Coal Soc.* **2021**, *46*, 726–746.
3. Yang, H. Design of shaft freezing construction in thick surface soil. *Coal Sci. Technol.* **2019**, *40*, 62–64.
4. Wang, X.S.; Cheng, H.; Wu, T.L.; Yao, Z.S.; Huang, X.W. Numerical analysis of a novel shaft lining structure in coal mines consisting of hybrid-fiber-reinforced concrete. *Crystals* **2020**, *10*, 928. [CrossRef]
5. Jia, C.G. The design of shaft lining structure in extra thick alluvium of deep shaft by freeze sinking method. *Mine Constr. Technol.* **2019**, *40*, 1–4.
6. Wang, J.P.; Liu, W.M.; Wang, H. Comparisions on ground freezing constructions in 1000m depth mine shaft. *Mine Constr. Technol.* **2017**, *38*, 34–37.
7. Li, J.Z.; Gao, W.; Chen, Z.P.; Peng, F. Status and outlook of over C80 high strength and high performance concrete technology applied to shaft liner by freezing sinking method in China coal mine. *Mine Construct Techno.* **2018**, *39*, 45–48.
8. Li, J.Z.; Gao, W.; Li, F.Z. New progress of theory and technology in deep shaft sinking by artificial ground freezing method. *Mine Constr. Technol.* **2020**, *41*, 10–14.
9. Zhang, S.P.; He, P.L.; Niu, L.L. Mechanical properties and permeability of fiber-reinforced concrete with recycled aggregate made from waste clay brick. *J. Clean. Prod.* **2020**, *26*, 918–922. [CrossRef]
10. Yao, Z.S.; Chen, H.; Sun, W.R. Experimental study on high strength composite shaft lining in deep alluvium. *J. Rock. Soil Mech.* **2003**, *5*, 739–743.
11. Zhang, T.; Yang, W.H.; Chen, G.H.; Huang, J.H.; Hang, T.; Zhang, C. Monitoring and analysis of hydration heat temperature field for high performance mass concrete freezing shaft lining. *J. Min. Saf. Eng.* **2016**, *33*, 290–296.
12. Guo, L. Study on the Inhomogeneity of Horizontal Lateral pressure on Shaft Lining. Doctor's Thesis, University of Mining and Technology, Beijing, China, 2010.
13. Dai, C. Analysis of Deep Alluvium Freezing Shaft Wall Mechanical Properties. Master's Thesis, Anhui University of Science and Technology, Huainan, China, 2017.
14. Jiang, L.H.; Xu, H.D.; Chu, H.Q.; Wang, M.J.; Xu, L. Development and application of high strength and high performance concrete to mine freezing shaft. *Coal Sci. Technol.* **2010**, *38*, 38–41.
15. Peng, S.L.; Rong, C.X.; Cheng, H.; Wang, X.J.; Li, M.J.; Tang, B.; Li, X.M. Mechanical Properties of High-Strength High-Performance Reinforced Concrete Shaft Lining Structures in Deep Freezing Wells. *Adv. Civ. Eng.* **2019**, *21*, 69–78. [CrossRef]
16. Yang, Y. The Preparation and Performance Research of Freezing Shaft Lining C80 HPC in Extraordinary Depth Surface Soil. Master's Thesis, Anhui University of Science and Technology, Huainan, China, 2006.
17. Shan, F.D. Common problems and countermeasures in design and construction of mine shaft equipment. *Mine Constr. Technol.* **2016**, *37*, 38–40.
18. Li, Y.; Li, B.; Zhang, L.Y.; Ma, C.; Zhu, J.; Li, M.; Pu, H. Chloride Ion Corrosion Pattern and Mathematical Model for C60 High-Strength Concrete after Freeze-Thawing Cycles. *Adv. Civ. Eng.* **2021**, *6*, 32–40.
19. Xue, W.P. Research on Coupled Damage Evolution Mechanism and Strength Characteristics of Shaft Lining Concrete Under High Pressure Water. Doctor's Thesis, Anhui University of Science and Technology, Huainan, China, 2017.
20. Jiang, X.Q.; Cao, D.F.; Ge, W.J.; Zhang, Y. Effect of freeze-thaw cycle and chloride corrosion on bond properties between steel bar and concrete. *J. Yangzhou Univ. Nat. Sci. Ed.* **2017**, *20*, 69–74.
21. Yang, L.; Yao, Z.S.; Xue, W.P.; Wang, X.S.; Kong, W.H.; Wu, T.L. Preparation performance test and microanalysis of hybrid–fibers and microexpansive high-performance shaft lining concrete. *Constr. Build. Mater.* **2019**, *223*, 431–440. [CrossRef]
22. Zhao, X.M.; Li, A.Y.; Qiao, H.X.; Li, J.C.; Wang, X.K. Frost resistance and damage deterioration model of fiber-reinforced concrete. *Bull. Chin. Ceram. Soc.* **2020**, *39*, 3196–3202.
23. Yang, D.; Yan, C.; Liu, S. Spliting Tensile Strength of Concerte Corroded by Saline Soil. *ACI. Mater. J.* **2020**, *32*, 137–145.
24. Hou, Y.F.; You, S.; Ji, H.G. Experimental study on corrosion characteristics of concrete shaft lining by ultrasonic detection. In Proceedings of the 3rd ISRM Young Scholars' Symposium on Rock Mechanics, Xian, China, 8–10 November 2014; pp. 285–290.
25. Liu, J.H.; Chen, Z.M.; Ji, H.G. Study of the performance of shaft concrete mixed with imitation steel fiber under the coupling of early age load and negative temperature. *J. China Coal Soc.* **2013**, *38*, 2140–2145.

26. China Standards Publication. *Standard Test Methods for Fiber Reinforced Concrete*; CECS/13-2009; Standard Press of China: Beijing, China, 2009.
27. Han, J.H.; Zou, J.Q.; Yang, W.H.; Hu, C.C. Mechanism of Fracturing in Shaft Lining Caused by High-Pressure Pore Water in Stable Rock Strata. *Math. Probl. Eng.* **2019**, *12*, 132–140. [CrossRef]
28. Liu, T.Y.; Zhang, P.; Li, Q.F.; Hu, S.W.; Ling, Y.F. Durability Assessment of PVA Fiber-Reinforced Cementitious Composite Containing Nano-SiO_2 Using Adaptive Neuro-Fuzzy Inference System. *Crystals* **2020**, *10*, 347. [CrossRef]
29. Yu, X.F.; Dong, Y.W.; Wang, G.; Lu, H.L. Durability of fly ash concrete against sulfate erosion in complex environment. *Concrete* **2020**, *6*, 58–69.
30. Anwar, A.Y.; Moetaz, E.H.; Khallad, N.; Khan, P.B. Corrosion resistance of recycled aggregate concrete incorporating slag. *ACI. Mater. J.* **2020**, *117*, 111–122.
31. China Standards Publication. *Standard for Test Methods of Long–Term Performance and Durability of Ordinary Concrete*; GBT/50082-2009; Standard Press of China: Beijing, China, 2009.
32. Yang, M.F.; Ni, X.Q.; Wang, X. Complex multi function anti-freeze influenced to durability of minus temperature mine shaft liner concrete. *Coal Sci. Technol.* **2010**, *38*, 47–49.
33. Ni, X.Q.; Yang, M.F. Research and application of frost-resistant, impermeable, high-strength and high-flow shaft lining concrete. *J. Min. Saf. Eng.* **2004**, *1*, 27–32.
34. Jiang, J.H.; Wen, W.X.; Qiu, J.Q. Corrosion behaviors of concrete exposed to sulfate attack with simulated groundwater pressure. *J. Hebei Univ. Eng. Nat. Sci. Ed.* **2019**, *3*, 11–15.
35. Limeira, J.; Etxeberria, M.; Agulló, L.; Molina, D. Mechanical and durability properties of concrete made with dredged marine sand. *Constr. Build. Mater.* **2011**, *25*, 4165–4174. [CrossRef]
36. Zhao, L.; Liu, J.H.; Zhou, W.J.; Ji, H.G. Damage evolution and mechanism of concrete erosion at sulfate environment in undergroundmine. *J. China Coal Soc.* **2016**, *41*, 1422–1428.

MDPI

Article

Experimental Research on Interfacial Bonding Strength between Vertical Cast-In-Situ Joint and Precast Concrete Walls

Changyong Li, Yabin Yang, Jiuzhou Su, Huidi Meng, Liyun Pan * and Shunbo Zhao *

International Joint Research Laboratory for Eco-Building Materials and Engineering of Henan, School of Civil Engineering and Communications, North China University of Water Resources and Electric Power, Zhengzhou 450045, China; lichang@ncwu.edu.cn (C.L.); yangyabin@ncwu.edu.cn (Y.Y.); z201710313228@stu.ncwu.edu.cn (J.S.); Z201910311259@stu.ncwu.edu.cn (H.M.)

* Correspondence: ply67@ncwu.edu.cn (L.P.); sbzhao@ncwu.edu.cn (S.Z.); Tel.: +86-371-65665160 (S.Z.)

Abstract: In the monolithic precast concrete shear-wall structure, the bonding property of cast-in-situ joints to precast concrete walls is important to ensure the entire structural performance. Aiming to the vertical joint of precast concrete walls, an experimental study was carried out considering the factors including the strength of precast and joint concretes, as well as the interface processing and casting age of precast concrete. The micro-expansion self-compacting concrete was used for the cast-in-situ joints. The interfacial bonding strength between joint and precast concrete was measured by splitting tensile test. Results show that the interfacial bonding strength was benefited from the increasing strength of joint concrete and the spraying binder paste on the interface of precast concrete, and unbenefited from the overtime storage of precast concrete. The washed rough surface with exposed aggregates improved the interfacial bonding strength, which increased with the increasing roughness. Based on the test results, the limits of the strength grade of joint concrete and the roughness of washed rough surface are proposed to get the interfacial bonding strength equivalent to the tensile strength of precast concrete. Meanwhile, the spraying of binder paste on precast concrete is a good choice, the storage time of precast components is a better limit within 28 days.

Keywords: precast concrete wall; interfacial bonding strength; joint concrete; interface processing; washed rough surface; roughness; storage time

Citation: Li, C.; Yang, Y.; Su, J.; Meng, H.; Pan, L.; Zhao, S. Experimental Research on Interfacial Bonding Strength between Vertical Cast-In-Situ Joint and Precast Concrete Walls. *Crystals* **2021**, *11*, 494. https://doi.org/10.3390/cryst11050494

Academic Editors: Antonella Sola and Amir H. Mosavi

Received: 4 March 2021
Accepted: 25 April 2021
Published: 28 April 2021

1. Introduction

In recent years in China, the assembling of buildings with precast concrete structure has become an advanced construction technology with features of green, environmental protection and energy conservation for the building industry [1,2]. This produces an importance to the construction process of the cast-in-situ joint which relates to the entire performance and quality of monolithic precast concrete structure. In view of the monolithic precast shear-wall structures, great concerns have been made on how to safely anchorage of rebars in precast shear-walls. Methods for rebar splicing by grout-filled coupling sleeve, slurry anchor lap joints, closed-loop anchoring and their composites have been applied [3–6]. However, acting as the linking of the precast concrete walls together to subject the loads and seismic actions, the bonding performance of cast-in-situ concrete joints to precast concrete walls have not been studied sufficiently. This may affect the deformation and energy dissipation capacity of the structure. In practice, cracks along interfaces appear due to the weak bonding of cast-in-situ concrete to precast concrete. This forms a weak section of monolithic precast concrete structure. Meanwhile, the map cracking presents due to large drying shrinkage of the cast-in-situ concrete, and the cast quality problems of spongy surface and internal voids exist due to difficult compaction in narrow joint space.

To achieve the design criteria of equivalent cast-in-place for precast concrete shear-wall structures, the joint connection between precast concrete walls has been specified in China code JGJ 1 [3]. The strength grade of precast concrete should not be less than C30, and that of cast-in-situ concrete is better, one grade higher than precast concrete. Meanwhile, the interface of the precast concrete wall should be roughened or treated with groove keys. If a rough surface washed by pressure water is used, the rough area of the interface should be larger than 80%, and the roughness should not be less than 6 mm. However, by looking up the published literature, only a few studies were performed on the interface between cast-in-situ joint and precast concrete component [6–9].

Coming from the same bonding mechanism of concrete to concrete, studies on the bond of new to old concrete for the strengthening of existing concrete structures can be referenced to find the main influencing factors [10–13]. Firstly, the factors relate to the quality of the concrete interface. The interfacial bonding strength increases with the increasing strength of concrete, especially new concrete [10,14–16], and benefits from the spraying of cement paste on the original surface of precast concrete [10–13,17]. Secondly, the factors relate to the condition of the concrete interface. The roughening of old concrete surface is necessary to further improve the bonding strength of new to old concrete [18–20]. During the research process, several kinds of roughening methods have been applied, including indentation with steel bars, scraping with iron combs of different-shaped saw-teeth, artificial chipping, mechanical napping, sand blasting, washing to expose aggregates with pressure water, groove keys and rough formwork. This makes the interface zigzag with concrete protuberances or turns into a zone with certain thickness composited by the cohesive layer of binder paste and the permeable layer of interaction [13,17,21]. The bonding strength increases with the increase in the roughness of the old concrete. Comparatively, the best effect can be obtained with an exposed aggregate surface and mechanical napping surface [7–9,22–24]. Meanwhile, the interfacial bonding strength is also affected by the degradation of surface condition depended on the environmental actions such as carbonation, freezing and thawing, and chemical erosion. Even in a short time after casting (within 90 days), the bonding strength of new to old concretes decreases whatever the surface of old concrete is processed with different methods [24–26].

Therefore, the three kinds of factors mentioned above should be considered for the experimental study on interfacial bonding strength of cast-in-situ concrete to precast concrete. Differing from the artificial post-roughening of existing concrete surface for strengthening purpose, the interface roughening of precast concrete components should be industrialized in the precast factory. Therefore, the joint surface of precast concrete is always roughened by using the methods of mechanical napping, rough formwork, key groove or washing to expose aggregates by pressure water [7,8]. Meanwhile, a storage time after casting of precast concrete components is always created due to the out-of-sync of production and installation.

To make up for the lack of systematical evaluation of the bonding rationale for cast-in-situ joint to precast concrete walls, an experimental study was carried out in this paper. The precast concrete specimens were prepared in strength grade of C30 and C40, the micro-expansion self-compacting concrete was used for cast-in-situ joints in strength grade of C30, C35, C40 and C45. Three kinds of interfaces of precast concrete were made: the original, the closing net formed and the washed rough to expose aggregates. The interface of washed rough to expose aggregates was made with four levels of roughness. The storage time of precast concrete after demolding was considered at 14, 28, 56 and 90 days. The interfacial bonding strength of joint to precast concretes was experimentally studied for 32 groups of specimens by using the splitting tensile test. Results are analyzed, and the measures to satisfy the interfacial bonding strength equivalent to tensile strength of precast concrete are suggested.

2. Experimental Work

2.1. Preparation of Concretes

The cement (PC) was ordinary silicate cement of grade 42.5 produced by Henan Xinxiang Mengdian Cement Co. Ltd., Xinxiang, China. As presented in Table 1, the properties of cement met the specification of China code GB 175 [27]. Class-II fly ash (FA) and ground limestone (GL) were used as mineral admixtures to improve the workability of fresh concrete, the properties presented in Table 2 met the specification of China codes GB/T 1596 and JGJ/T 318 [28,29]. The HEM-V expansive agent (EA) produced by Jiangsu Subote New Materials Co. Ltd., Nanjing, China, was used for cast-in-situ joint concrete, the properties are presented in Table 3. The chemical compositions of cement, fly ash, ground limestone and expansive agent are presented in Table 4. LOI is the loss on ignition.

Table 1. Physical and mechanical properties of cement.

Density (g/cm^3)	Water for Standard Consistency (%)	Specific Surface Area (m^2/kg)	Setting Time (min)		Compressive Strength (MPa)		Flexural Strength (MPa)	
			Initial	Final	3d	28d	3d	28d
3.09	27	360	170	215	27.8	58.4	5.2	8.3

Table 2. Physical properties of fly ash and ground limestone.

Material	Apparent Density (kg/m^3)	Specific Surface Area (m^2/kg)	Activity Index (%)	Water Demand Ratio (%)	Mobility Ratio (%)	Fineness: Residual on Sieve (%)	
						80 μm	45 μm
FA	2350	406	73.3	84	-	5.48	21.75
GL	2780	428	61.6	-	103	1.2	25

Table 3. Physical and mechanical properties of the expansion agent (HEM-V).

Fineness		Water of Standard Consistency (%)	Setting Time (min)		Restrained Expansion Rate (%)		Compressive Strength (MPa)	
Specific Surface Area (m^2/kg)	Residual on 1.18 mm Sieve (%)		Initial	Final	In Water 7d	In Air 21d	7d	28d
375	0.155	30	260	351	0.042	0.075	29.5	44.6

Table 4. Chemical compositions of cementitious materials (unit: %).

Material	SiO_2	Al_2O_3	Fe_2O_3	CaO	MgO	SO_3	f-Cao	Na_2O	K_2O	LOI	Others
PC	20.81	5.99	3.28	60.12	2.13	2.23	0.67	0.11	0.55	3.52	0.59
FA	55.92	17.31	5.91	6.95	3.82	1.93	0.26	0.48	1.96	2.63	2.83
GL	0.89	0.51	0.29	47.56	1.15	0.06	0.02	0.67	0.27	40.71	4.57
EA	3.48	9.27	1.44	42.78	0.48	27.38	6.65	0.62	0.47	5.51	1.92

The crushed limestones in continuous grading with particle size of 5–20 mm and 5–16 mm were used for precast concrete and cast-in-situ joint concrete, respectively. The fine aggregate was manufactured sand with fineness modulus of 2.85, stone powder of 8.8% and Methylene Blue value of 1.3. The properties are presented in Table 5.

Table 5. Physical properties of crushed limestone and manufactured sand.

Particle Size (mm)	Apparent Density (kg/m^3)	Bulk Density (kg/m^3)	Closed-Compact Density (kg/m^3)	Moisture Content (%)	Water Absorption (%)	Porosity (%)
5~20	2730	1548	1613	0.30	1.17	42
5~16	2760	1554	1678	0.23	1.05	41
Sand	2689	1583	1726	0.6	2.0	

The water reducer was PCA-I high-performance polycarboxylic acid type with a water reduction of 30%, which was produced by Jiangsu Subote New Materials Co. Ltd., Nanjing, China. The mix water was tap water of Zhengzhou city.

The mix proportions of concrete were designed by the absolute volume method [30,31], and results are presented in Table 6. The conventional concrete was used for precast concrete with two strength grades. The slump of fresh mixture was kept at 80–100 mm. Due to operating in a limited narrow space for the cast-in-situ concrete of joints with disturbing of reinforcements, the self-compacting concrete was used to ensure the compactness without vibration. Four strength grades of self-compacting concrete were prepared with the slump extension of fresh mixture kept at 650–750 mm [32,33]. Based on previous study, the EA content was 10% of total weight of cementitious materials [34,35].

Table 6. Mix proportions of concrete for precast components and cast-in-situ joints.

Concrete	Water to Binder Ratio	Dosage of Raw Materials (kg/m^3)							
		Water	Cement	FA	GL	Crushed Limestone	Sand	Water Reducer	EA
Precast components	0.47	175	335	37	-	1086	786	3.7	-
	0.57	185	292	32	-	1060	831	3.2	-
Cast-in-situ joints	0.37	190	308	51	103	885	816	5.6	51.4
	0.34	185	326	54	109	873	806	5.4	54.4
	0.31	185	358	60	119	851	786	7.2	59. 7
	0.28	185	396	66	132	816	754	7.3	66.1

2.2. Mechanical Properties of Concretes

The mechanical properties of conventional concrete and self-compacting concrete were measured by using the test methods specified in China code GB/T50081 [36]. Six cubes with a dimension of 150 mm, with three of them as a group, were used for each concrete to measure the cubic compressive strength and the splitting tensile strength. Six cylinders, with a diameter of 150 mm and height of 300 mm, with three of them as a group, were used for each concrete to measure the axial compressive strength and the modulus of elasticity. Tests were carried out for concretes at a curing age of 28 days. The loading speed was controlled at 0.5 MPa/s for testing of cubic and axial compressive strengths, while that was 0.05 MPa/s for testing of splitting tensile strength.

Test results are presented in Table 7. The test values met the requirement of a corresponded strength grade, and tended the common regularity increasing with the decrease in the water to binder ratio [30,31].

Table 7. Mechanical properties of precast concrete and cast-in-situ concrete.

Concrete	Water to Binder Ratio	Strength Grade	Cubic Compressive Strength (MPa)	Axial Compressive Strength (MPa)	Splitting Tensile Strength (MPa)	Modulus of Elasticity (GPa)
Precast	0.57	C30	33.1	28.7	2.38	30.5
	0.47	C40	48.4	35.5	3.18	31.7
Cast-in-situ	0.37	C30	37.5	28.7	2.08	28.9
	0.34	C35	43.6	30.1	2.77	30.5
	0.31	C40	47.0	32.4	2.94	31.3
	0.28	C45	54.7	35.7	3.71	33.3

2.3. Formation of Interface

Three kinds of interface of precast concrete were made in this study. The first was the original surface. As exhibited in Figure 1a, the original surface was flat with some small pores after demolding.

The second was the washed rough surface with exposed aggregates by pressure water washing, as exhibited in Figure 1b. An agent was brushed on the interface formwork. After demolding, the interface of concrete was washed by pressure water to expose aggregates.

The agent produced by Henan Meilitong New Materials Co. Ltd., Zhengzhou, China, was a water-soluble homogeneous viscous substance composited by non-toxic organic matter; the density was 1.10 g/cm^3.

Figure 1. Two kinds of interface in sectional dimension of 150 mm × 150 mm: (**a**) original surface; (**b**) washed rough surface.

The third was a rough surface formed with a fast-ribbed closing net, which was used as an interface formwork of precast concrete. As presented in Figure 2, the fast-ribbed closing net is a sheet steel plate rolled from thin galvanized steel with depth of 0.2 mm and swelling of 5 mm, which could be cut into the size used. The rough surface was formed with concrete protuberances after demolding of the fast ribbed closing net.

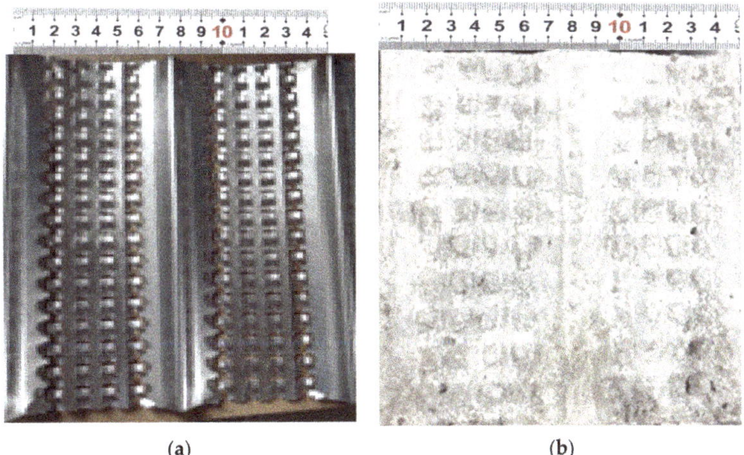

Figure 2. The third interface: (**a**) fast-ribbed closing net cut to be 150 mm × 150 mm when used for interface; (**b**) rough surface in sectional dimension of 150 mm × 150 mm.

The surface roughness of precast concrete was determined by using a sand patch test [13,23,37]. As presented in Figure 3, the precast specimen with dry surface was placed in a salver, and enveloped with transparent plastic sheets. The top surface of plastic sheets was taken at the highest point of protuberances. The sand weighted as m_1 was filled on the rough surface and finished flat by a ruler along the surface of the plastic sheet. The residual sand scraped into the salver was weighted as m_2, where the mass of patched sand is $M = m_1 - m_2$. The roughness expressed by the average sand depth can be computed as,

$$\overline{y} = \frac{M}{\rho A} \quad (1)$$

where, $\overline{y}$ is the average sand depth, mm; M is the mass of patched sand, g; ρ is the density of sand, g/cm^3; A is the sectional area of rough surface, mm^2.

Figure 3. Roughness measuring of surface in sectional dimension of 150 mm × 150 mm: (**a**) specimen preparation; (**b**) fit to plastic board; (**c**) sand replacement.

2.4. Preparation of Bond Specimens and Test Method

Based on previous studies, splitting tensile test is always applied for the bonding performance of new to old concrete [12–14,22–26,38]. In this study, the splitting tensile test was in accordance with the specification of China code GB/T50081 [36]. The composite cubic specimens with a single interface between precast concrete and cast-in-situ concrete was made in a dimension of 150 mm. Along sides of the interface was precast concrete and cast-in-situ concrete, respectively. The interface was 150 mm × 150 mm. Three specimens were made as a group.

The precast concrete in dimension of 75 mm × 150 mm × 150 mm was first cast and cured in standard curing room at a temperature of (20 ± 2) °C and relative humidity of 65% for the required curing age. The interface of precast concrete was pretreated according to the requirement. Except the six groups of precast concrete used for the research of the effect of casting age at 14, 56 and 90 days, others were cured for 28 days. As presented in Figure 4, the precast concrete was first placed into the cube mold, then the cast-in-situ concrete was poured into the mold, covered by plastic after finishing smooth of surface, and cured in a standard curing room for 28 days. Before testing, the load surface needed to be polished for the uniform loading.

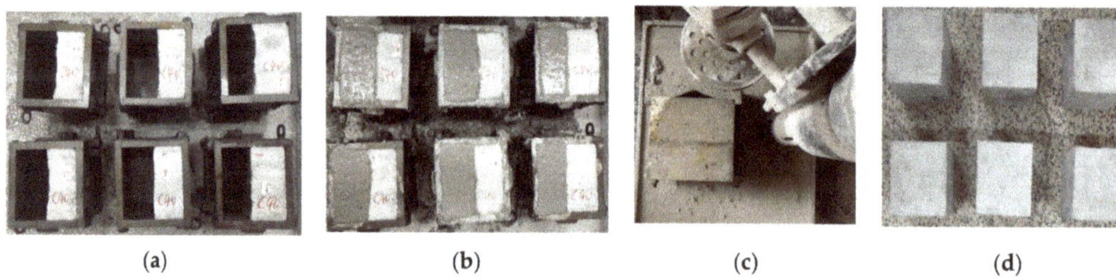

Figure 4. Main process of specimen formation: (**a**) precast concrete placed into module; (**b**) cast-in-situ concrete poured into mold; (**c**) polishing load surface; (**d**) prepared specimens in dimension of 150 mm.

As shown in Figure 5, the load was directly exerted along the interface section with steel strips at bottom and top surfaces on the universal testing machine produced by SNS Testing Machine Co. Ltd., Shanghai, China. The capacity of the testing machine is 600 kN, and the loading speed was 0.05 MPa/s. The interfacial bonding strength is computed as [36],

$$f_{\mathrm{b}} = 0.637\frac{P}{A} \tag{2}$$

where, f_{b} is the interfacial bonding strength, MPa; P is the peak load at failure, N; A is the area of splitting section, mm^2.

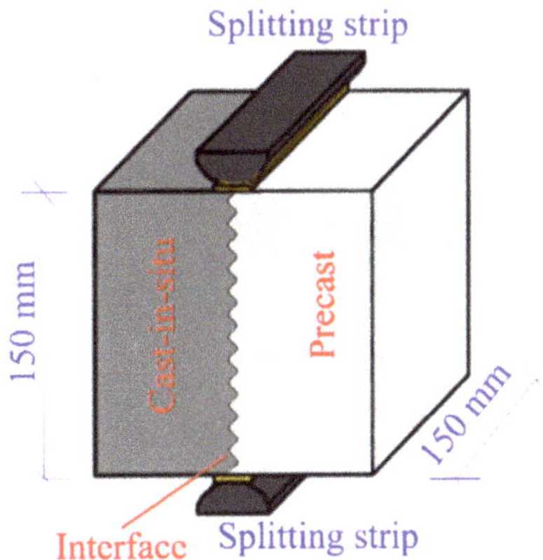

Figure 5. Loading diagram for bond strength of interface in sectional dimension of 150 mm × 150 mm.

According to the specification of China code GB/T50081 [36], test data of bonding strength for three specimens are dealt with the following criteria: (1) The arithmetic mean value of three test data is taken as the group strength; (2) If the difference between one of the maximum or minimum values and the median value exceeds 15% of the median value, the maximum and minimum values are discarded together, and the median value is taken as the group strength; (3) If the difference between the maximum and minimum values and the median value is over 15% of the median value, the test results of this group are invalid.

To evaluate the bond efficiency of interface, the interfacial bonding strength divided by the tensile strength of precast concrete was defined as the equivalent coefficient of bonding strength, that is,

$$\beta_{\mathrm{e}} = \frac{f_{\mathrm{b}}}{f_{\mathrm{t,p}}} \tag{3}$$

where, β_{e} is the equivalent coefficient of bonding strength; $f_{\mathrm{t,p}}$ is the tensile strength of precast concrete, MPa.

3. Analyses of Test Results

3.1. *Effect of Cast-In-Situ Concrete Strength*

The strength matching of cast-in-situ concrete to precast concrete was explored. The precast concrete was fixed at strength grade of C40, the cast-in-situ concrete was changed with strength grade of C30, C35, C40 and C45, the interface was the original surface of precast concrete with roughness of 1.55 mm.

All specimens broke at the interface section with smooth splitting, a peeling of precast concrete took place on the splitting section of some specimens with the original surface of precast concrete. The entrance of binder paste into the pores of precast concrete was observed on some splitting interface. As presented in Figure 6, the interfacial bonding strength increased with the increasing strength of cast-in-situ concrete. This is the macroscopic response of the meshing forces due to the interlaced crystals formed by the hydration of cast-in-situ concrete and precast concrete. As in previous studies [10–13,17], the hydration products $Ca(OH)_2$, AFt and C-H-S of new concrete grow in the holes or defects of old concrete, the skin needling of C-H-S and thinner needle-like AFt enter into the pores of old concrete, and the unhydrated and incomplete hydration composites of old concrete continuously hydrate in the new concrete. Due to the domination of mix proportion of concrete to the hydration process, the microscopic effect is directly represented by the strength of concrete in macroscopic. With the increasing strength of cast-in-situ concrete, the binder paste was higher of strength with fewer pores adhered to the interface to improve the bonding behavior of cast-in-situ concrete with precast concrete. However, the equivalent coefficient of bonding strength β_e was only 0.32~0.45. In this condition, the bonding strength of the joint interface has a large gap to the tensile strength of precast concrete. This could not meet the requirement of equivalent monolithic concrete [3,5]. Therefore, other measures should be adopted to improve the bonding strength.

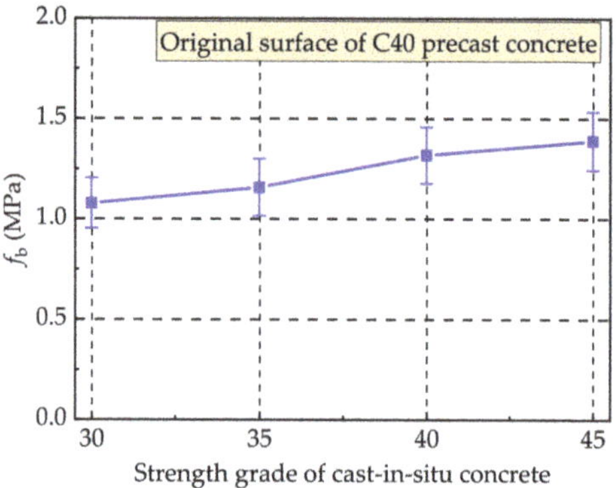

Figure 6. Bonding strength of interface changed with different strength of cast-in-situ concrete.

3.2. *Effect of Interface of Precast Concrete*

In this trial of testing, the effect of interfaces of the original, the closing net formed and the washed rough of precast concrete was examined. The roughness of the interfaces was 1.55 mm, 5.25 mm and 6.60 mm respectively. All specimens failed in splitting at the interface section. For specimens with closing net formed interface, some of the concrete protuberances were scraped to expose aggregates due to the binder paste peeled off. In this condition, except the meshing forces of interlaced crystals formed by the hydration of cast-in-situ concrete and precast concrete, the built-in effect of precast concrete protuberances to the cast-in-situ concrete takes part in the bond of interface [18–20]. This further promotes the interfacial bonding strength of cast-in-situ to precast concrete.

For specimens with washed rough interface, the interlocked coarse aggregates of cast-in-situ and precast concretes broke on the splitting section, the failure mode of interface was similar to the splitting of monolithic concrete. In this condition, the aggregates of cast-in-situ concrete interlocked with the aggregates exposed on the interface of precast concrete and bonded by the binder paste into entirety. The meshing force and the interlock force work together on the interface, which participates in the main function of enhancing the interfacial bonding strength [13,17,21,22].

With the above interfacial bonding mechanisms, the interfacial bonding strength increased in the order of the original interface, closing net formed interface and washed rough interface. As presented in Table 8, compared to the specimens with original interface, the bonding strength of specimens with closing net formed interface increased by 19.4% and 26.5%, respectively, accompanied by a strength grade of C30 and C40 for cast-in-situ concrete, while that of specimens with washed rough interface increased by 87.0% and 125.8%.

Table 8. Bonding strength of interface with different surface of precast concrete.

Surface of Precast Concrete	Strength Grade of Concrete		f_b (MPa)	β_e
	Precast	Cast-In-Situ		
Original surface	C40	C30	1.08	0.34
		C40	1.32	0.41
Closing net formed surface	C40	C30	1.29	0.40
		C40	1.67	0.53
Washed rough surface	C40	C30	2.02	0.64
		C40	2.98	0.94

Meanwhile, a higher interfacial bonding strength was provided with the higher strength of cast-in-situ concrete, whatever the interfaces of precast concrete. With the strength grade of cast-in-situ concrete increased from C30 to C40, the increments of interfacial bonding strength are 22.2%, 29.4% and 47.5% respectively corresponded to the original, closing net formed and washed rough surfaces of precast concrete. This once again indicates the effect of cast-in-situ concrete strength on the interfacial bonding strength [10,14–16].

In this study, an equivalent coefficient β_e was 0.94 only for the specimens with washed rough surface of precast concrete and C40 cast-in-situ concrete. This means that the washed rough surface is optimum to enhance the interfacial bonding strength, other interfaces were difficult to have an equivalent tensile strength of precast concrete.

3.3. Effect of Interface Adhesion Agent

Accompanied by the trials of test for specimens in Section 3.2, a parallel trial of test was carried out on specimens with spraying adhesion agent on the surface of precast concrete. For the specimens of this trial, the adhesion agent was sprayed on the surface of precast concrete before cast-in-situ concrete was poured into mold. The adhesion agent was the cement paste with the same water to binder ratio of cast-in-situ concrete, the binder was composite of 60% cement, 20% GL, 10% FA and 10% EA. Test results of interfacial bond strength are comparatively presented in Figure 7.

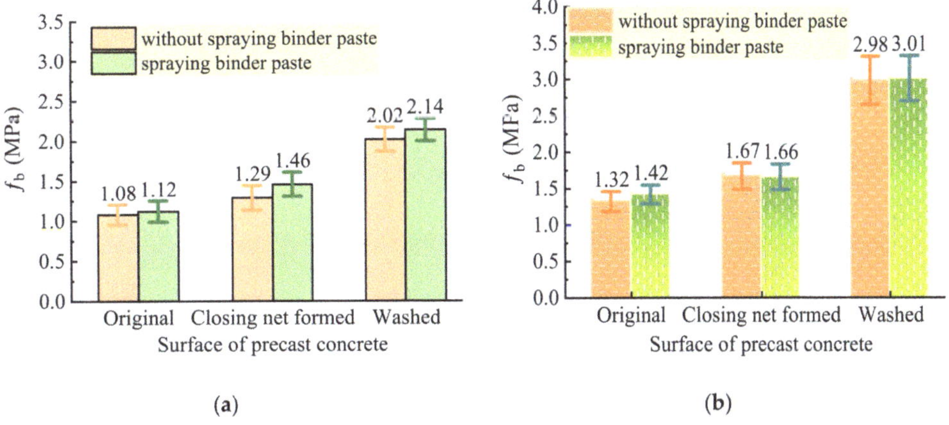

Figure 7. Comparison of bond strength of interface with or without adhesion agent: (**a**) C30 cast-in-situ concrete; (**b**) C40 cast-in-situ concrete.

Due to only three specimens as a group for each trial, the statistical result of test data has limitation to clarify the changes of bonding strength for interfaces with and without spraying adhesion agent. This leads the difference of bonding strength for specimens with and without spraying adhesion agent, which may be less than the error bar, which indicates the dispersion of test data of a group of specimens. However, due to almost equal error bars for specimens with and without spraying adhesion agent, the test results are comparable between these two kinds of specimens by using the statistical results. Compared with the specimens without spraying adhesion agent, the specimens sprayed adhesion agent had a higher bond strength, and the more beneficial effect appeared on the interface with C30 than C40 cast-in-situ concrete. With C30 cast-in-situ concrete, the interfacial bond strength with original, closing net formed and washed rough surfaces of precast concrete increased by 3.7%, 13.2% and 5.9%, while that with C40 cast-in-situ concrete increased by 7.6%, 0% and 1.0%. This indicates a favorable effect of the sprayed interface agent on the formation of interlaced crystals in hydration of cast-in-situ concrete with adequate humidity on the surface of precast concrete [17,21,23]. At the same time, spraying cement paste containing fly ash can improve the chemical force due to the rehydration of active SiO_2 of fly ash with $Ca(OH)_2$ of old concrete [10,17,39].

3.4. *Effect of Roughness of Washing Exposed Aggregates*

The formation of washed rough surface depends on the amount of agent that was brushed on the interface formwork, the curing age of precast concrete, the pressure of washing water and the washing time. Based on practice, the roughness measured by sand patch test is better to limit within 8 mm [6,7]. Therefore, a research was pointed on the roughness from 4 mm to 8 mm. The amount of water washing rough agent was 0.2~0.4 kg per square-meter of surface. The washing began at 24 h after demolding at a room temperature of 20 ± 5 °C. The working pressure of jetting machine was 8~10 MPa, the washing time was 12~15 min per square-meter of surface. Table 9 presents the test results of roughness of washed rough surface for precast concrete specimens. The roughness can be controlled by the washing technique. Four zones of roughness were divided into 4~5 mm, 5~6 mm, 6~7 mm and 7~8 mm. The photos are exhibited in Figure 8.

Table 9. Test results of roughness of washed rough surface of precast concrete.

Roughness Range	Average Depth of Filled Sand (mm)					
	C40 Precast Concrete			C30 Precast Concrete		
	1	2	3	1	2	3
4~5 mm	4.57	4.73	4.81	4.20	4.44	4.73
5~6 mm	5.04	5.36	5.73	5.14	5.33	5.95
6~7 mm	6.30	6.73	6.94	6.13	6.36	6.92
7~8 mm	7.02	7.24	7.46	7.31	7.33	7.96

Figure 8. Washed rough surfaces of specimens in sectional dimension of 150 mm × 150 mm with different roughness: (**a**) 4~5 mm; (**b**) 5~6 mm; (**c**) 6~7 mm; (**d**) 7~8 mm.

Figure 9 presents the failure mode of eight group specimens with different roughness of washed rough surface of precast concrete. Four groups were made with C40 precast concrete and C45 cast-in-situ concrete, others were made with C30 precast concrete and C35 cast-in-situ concrete. Most specimens failed along the interface with the interlaced mortar among exposed aggregates and part fractured aggregates. Some specimens appeared two cracks on the compression zone of the loading surface, one crack was along the interface, another was near the interface. This indicates that with the increase in interface roughness, the possibility increased that the bond failure of interface transfer to the weak side of precast or cast-in-situ concrete. In this study, the test result of larger splitting load without along the interface was discarded.

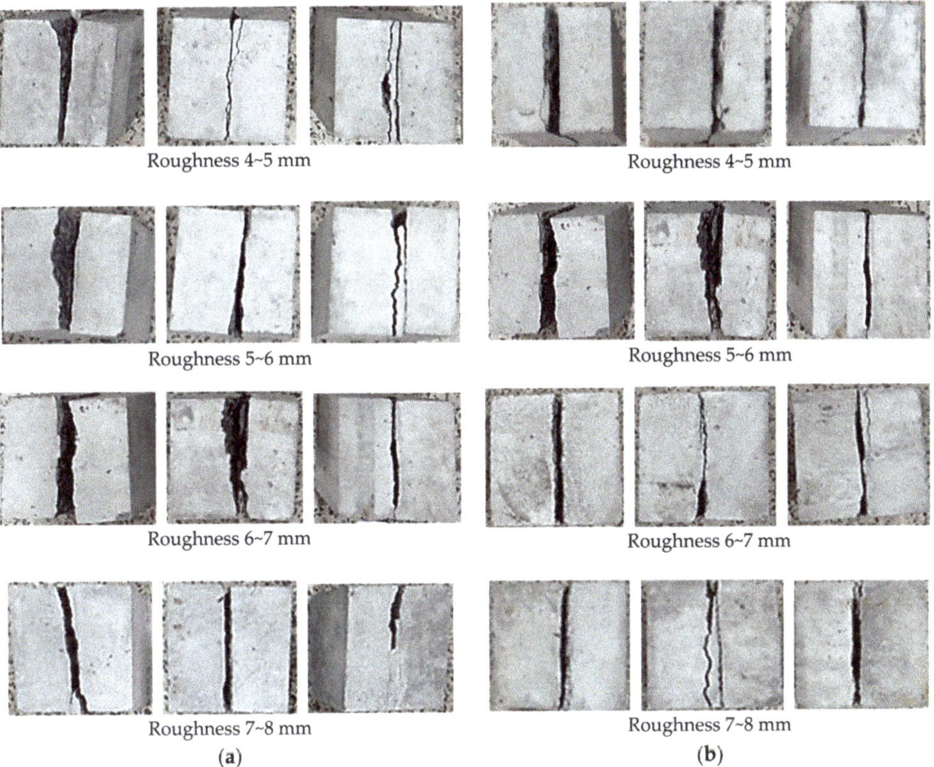

Figure 9. Failure mode of specimens with different roughness of precast concrete in dimension of 150 mm × 150 mm: (**a**) C45 cast-in-situ concrete to C40 precast concrete; (**b**) C35 cast-in-situ concrete to C30 precast concrete.

Test results of bonding strength of interface are presented in Table 10. The interfacial bonding strength increased obviously with the increasing roughness of washed rough surface of precast concrete. This is easy to be understood that the interlock effect of aggregates on interface became stronger with the increase in interface roughness [21–23]. When the roughness was over 6 mm, the equivalent coefficient of bonding strength was close to 1.00. In this condition, the exposed size of aggregate was about one-third to four-fifths of the maximum particle size of 20 mm for precast concrete. Therefore, the interfacial bonding strength equivalent to tensile strength of precast concrete can be provided by the washed rough interface of precast concrete at the roughness of 6~8 mm, accompanied by the strength grade of cast-in-situ concrete over precast concrete.

Table 10. Bond strength of interface with different roughness of precast concrete.

Strength Grade of Concrete	C40 Precast, C45 Cast-In-Situ		C30 Precast, C35 Cast-In-Situ	
Roughness of Precast Concrete	f_b **(MPa)**	β_e	f_b **(MPa)**	β_e
4~5 mm	2.69	0.85	1.99	0.84
5~6 mm	2.72	0.86	2.08	0.87
6~7 mm	3.09	0.97	2.32	0.97
7~8 mm	3.24	1.02	2.46	1.03

3.5. *Effect of Storage Time of Precast Concrete*

Table 11 presents the bonding strength of the interface at different storage time after casting of precast concrete. The washed rough surface of precast concrete was made with roughness of 7~8 mm. The specimens failed in splitting at interface with paste peeled off and aggregates fractured, except one failed in splitting inclined into the precast concrete. This indicates the interface was still weak without enough strength resisting the splitting tensile force. Due to the higher strength of cast-in-situ concrete than precast concrete, the possible failure may be happened in precast concrete.

Table 11. Bond strength of interface with different storage time of precast concrete (MPa).

Strength Grade of Concrete	C40 Precast, C45 Cast-In-Situ		C30 Precast, C35 Cast-In-Situ	
Age of Precast Concrete (d)	f_b **(MPa)**	β_e	f_b **(MPa)**	β_e
14	3.24	1.02	2.46	1.03
28	3.16	0.99	2.40	1.01
56	2.88	0.91	2.08	0.87
90	2.68	0.84	1.93	0.81

The interfacial bonding strength decreased with the increasing age of precast concrete. After the age of 28 days, the equivalent coefficient of bonding strength was lower than 1.00. Similar to those studies on bonding strength of new to old concrete [24–26], the unhydrated and incomplete hydrated binders on the surface of precast concrete has more activity to continuously hydrate with the binders of cast-in-situ concrete before the casting age of 28 days. However, the early-age carbonation of precast concrete consumes the hydrate product $Ca(OH)_2$ and filled the interfacial pores and defects [40,41]. This is unbeneficial to the interlaced crystals formed by the hydration of cast-in-situ concrete and precast concrete [10]. Meanwhile, the hydration of cement and mineral admixtures is continuous to keep the time-dependent strength development of concrete [42–44]. This is also unbeneficial to the interaction of cast-in-situ concrete to precast concrete, due to the decrease in the amount of unhydrated and incomplete hydrated binders on the surface of precast concrete. Therefore, a largest storage time no more than 28 days should be considered for the production and installation cycle of precast concrete components.

4. Conclusions

The strength grades of precast and cast-in-situ concretes, the interface conditions and storage time of precast concrete were considered as the experimental factors in this paper. The interface roughness of precast concrete was measured by sand patch test. The interfacial bonding strength of cast-in-situ to precast concrete was measured by the splitting tensile test. The equivalent coefficient of bonding strength was computed with the interfacial bonding strength divided by the tensile strength of precast concrete. Based on the experimental research, conclusions can be drawn as follows:

(1) With the premise of higher strength grade of cast-in-situ concrete than precast concrete, the interfacial bonding strength increased with the increasing strength of cast-in-situ concrete. In practice, the cast-in-situ concrete is better one strength grade higher than precast concrete.
(2) The interfacial bonding strength increased with sequence of original interface, closing net formed interface and washed rough interface of precast concrete. The washed rough interface of precast concrete is a good choice in practice to ensure the interfacial bonding strength. When the roughness was over 6 mm that is about one-third of exposed aggregates with maximum particle size of 20 mm, the interfacial bonding strength can reach the tensile strength of precast concrete.
(3) Spraying binder paste on surface of precast concrete has beneficial effect on the bond of interface. This can be considered as a choice for the quality promotion of joints in practice.
(4) The interfacial bonding strength decreased with the increasing storage time of precast concrete. The bonding strength could be equivalent to the tensile strength of precast concrete when the casting age of precast concrete was not over than 28 days. In practice, the production and installation cycle of precast concrete components should limit within 28 days.
(5) The research of this paper has limitations only in macroscopic phenomena and index of bonding strength. Due to the complexity of interfacial bonding performance influenced by multi factors, further systematical researches should be carried out combined the macroscopic with the microscopic indices to revel the truth and accumulate research results.

Author Contributions: Methodology, L.P. and S.Z.; tests, data interpretation and writing—original draft, C.L., Y.Y., J.S. and H.M.; writing—review and funding acquisition, L.P. and S.Z. All authors have read and agreed to the published version of the manuscript.

Funding: This research was funded by [State Key Research and Development Plan, China] grant number [2017YFC0703904], Key Scientific Research Project in Universities of Henan, China (grant number 19A560011), and Innovative Sci-Tech Team of Eco-building Material and Structural Engineering of Henan Province, China (grant number YKRZ-6-066).

Institutional Review Board Statement: Not applicable.

Informed Consent Statement: Not applicable.

Data Availability Statement: Data are contained within the article.

Conflicts of Interest: The authors declare no conflict of interest.

References

1. Ministry of Housing and Urban-Rural Development of the People's Republic of China. *Technical Standard for Assembled Buildings with Concrete Structure*; GB/T 51231-2016; China Building Industry Press: Beijing, China, 2016.
2. Ministry of Housing and Urban-Rural Development of the People's Republic of China. *Standard for Design of Assembled Housing*; JGJ/T398-2017; China Building Industry Press: Beijing, China, 2017.
3. Ministry of Housing and Urban-Rural Development of the People's Republic of China. *Technical Specification for Precast Concrete Structures*; JGJ 1-2014; China Building Industry Press: Beijing, China, 2014.
4. Ministry of Housing and Urban-Rural Development of the People's Republic of China. *Technical Specification for Grout Sleeve Splicing of Rebars*; JGJ 355-2015; China Building Industry Press: Beijing, China, 2015.
5. Ministry of Housing and Urban-Rural Development of the People's Republic of China. *Technical Standard for Precast Reinforced Concrete Shear Wall Structure Assembled by Anchoring Closed Loop Reinforcement*; JGJ/T430-2018; China Building Industry Press: Beijing, China, 2018.
6. Kang, P.; Xu, D.W.; Li, N.J. Key construction techniques of monolithic precast concrete shear-wall structure. *Shanghai Build. Technol.* **2018**, *3*, 62–64.
7. Yin, B.Q.; Wu, Y. Research and application of construction technology for prefabricated component rough surface. *J. New Ind.* **2018**, *8*, 109–112.
8. Huang, X.M.; Zhang, X.J.; Liu, H.; Zhang, N. Split tensile test of prefabricated component laminated surface. *Build. Sci.* **2019**, *35*, 70–76. [CrossRef]

9. Cao, H.; Ma, Q.Y.; Zhang, R.R.; Wang, Y. Experimental study on adhesive splitting tensile performance of post pouring concrete on precast concrete. *Bull. China Ceram. Soc.* **2016**, *35*, 2926–2929.
10. Xie, H.C.; Li, G.Y.; Xiong, G.J. The mechanism formed the bonding force between new and old concrete. *Bull. China Ceram. Soc.* **2003**, *24*, 7–10. [CrossRef]
11. Dong, S.S.; Feng, K.C.; Shi, W.Z. New agent based on different interface tensile strength of old concrete. *Concrete* **2011**, *2*, 14–16. [CrossRef]
12. Hu, L.M.; Ma, C.M.; Xu, C.G.; Cao, H.L. Effect of interface agent on splitting tensile strength of old and new concrete in second age. *Water Power* **2017**, *43*, 53–55.
13. Gao, D.Y.; Cheng, H.Q.; Zhu, H.T. Splitting tensile bonding strength of steel fiber reinforced concrete to old concrete. *J. Build. Mater.* **2007**, *10*, 505–509.
14. Santos, P.M.D.; Julio, E.N.B.S. Factors affecting bond between new and old concrete. *ACI Mater. J.* **2011**, *108*, 449–456.
15. Diab, A.M.; Elmoaty, A.E.M.A.; Eldin, M.R.T. Slant shear bond strength between self compacting concrete and old concrete. *Constr. Build. Mater.* **2017**, *130*, 73–82. [CrossRef]
16. Gadri, K.; Guettala, A. Evaluation of bond strength between sand concrete as new repair material and ordinary concrete substrate (The surface roughness effect). *Constr. Build. Mater.* **2017**, *157*, 1133–1144. [CrossRef]
17. He, Y.; Zhang, X.; Hooton, R.D.; Zhang, X.W. Effects of interface roughness and interface adhesion on new-to-old concrete bonding. *Constr. Build. Mater.* **2017**, *151*, 582–590. [CrossRef]
18. British Standards Institution. *Design of Concrete Structures: Part 1-1: General Rules and Rules for Buildings*; Eurocode 2; British Standards Institution: London, UK, 2004.
19. MC2010. *Fib Model Code for Concrete Structures*; Ernst & Sohn Publishing House: Hoboken, NJ, USA, 2013.
20. ACI Committee 318. *Building Code Requirements for Structural Concrete*; ACI 318-14; ACI: Farmington Hill, MI, USA, 2014.
21. Bentz, D.P.; De la Varga, I.; Munoz, J.F.; Spragg, R.P.; Graybeal, B.A.; Hussey, D.S.; Jacobson, D.L.; Jones, S.Z.; LaManna, J.M. Influence of substrate moisture state and roughness on interface microstructure and bond strength: Slant shear vs. pull-off testing. *Cem. Concr. Compos.* **2018**, *87*, 63–72. [CrossRef]
22. Zhao, Z.F.; Zhao, G.F. Experimental research on treating interface of young on old concrete with high-pressure water-jet method. *J. Dalian Univ. Tech.* **1999**, *39*, 559–561.
23. Zhao, Z.F.; Zhao, G.F.; Liu, J.; Yu, Y.H. Experimental study on adhesive tensile performance of young on old concrete. *J. Build. Struct.* **2001**, *22*, 51–55.
24. Fan, J.; Wu, L.; Zhang, B. Influence of old concrete age, interface roughness and freeze-thawing attack on new-to-old concrete structure. *Materials* **2021**, *14*, 1057. [CrossRef]
25. Zhang, M.J.; Chu, L.S.; Zhao, J.F.; Liu, X.H. Experimental study on splitting strength of new-to-old concrete in short age. *Sichuan Build. Sci.* **2018**, *44*, 106–108.
26. Shi, C.C.; Wang, D.H.; Yuan, Q.; Cao, H.L.; Ma, Y. Study of splitting tensile strength for bond of new to old concrete in short age. *China Concr. Cem. Prod.* **2015**, *5*, 15–18. [CrossRef]
27. General Administration of Quality Supervision, Inspection and Quarantine of the People's Republic of China. *Common Portland Cement*; GB 175-2007; China Standard Press: Beijing, China, 2007.
28. General Administration of Quality Supervision, Inspection and Quarantine of the People's Republic of China. *Fly Ash Used for Cement and Concrete*; GB/T 1596-2017; China Standard Press: Beijing, China, 2017.
29. Ministry of Housing and Urban-Rural Development of the People's Republic of China. *Technical Specification for Application of Ground Limestone in Concrete*; JGJ/T 318-2014; China Building Industry Press: Beijing, China, 2014.
30. Ministry of housing and urban rural development of the people's Republic of China. *Specification for Mix Proportion Design of Ordinary Concrete*; JGJ 55-2011; China Building Industry Press: Beijing, China, 2011.
31. Ministry of Housing and Urban-Rural Construction of the People's Republic of China. *Technical Specification for Application of Self-Compacting Concrete*; JGJ/T 283-2012; China Building Industry Press: Beijing, China, 2002.
32. Zhao, M.L.; Ding, X.X.; Li, J.; Law, D. Numerical analysis of mix proportion of self-compacting concrete compared to ordinary concrete. *Key Eng. Mater.* **2018**, *789*, 69–75. [CrossRef]
33. Ding, X.X.; Zhao, M.L.; Zhou, S.Y.; Fu, Y.; Li, C.Y. Statistical analysis and preliminary study on the mix proportion design of self-compacting steel fiber reinforced concrete. *Materials* **2019**, *12*, 637. [CrossRef]
34. Li, C.Y.; Shang, P.R.; Li, F.L.; Feng, M.; Zhao, S.B. Shrinkage and mechanical properties of self-compacting SFRC with calcium sulfoaluminate expansive agent. *Materials* **2020**, *13*, 588. [CrossRef] [PubMed]
35. Liu, S.M.; Zhu, M.M.; Ding, X.X.; Ren, Z.G.; Zhao, S.B.; Zhao, M.S.; Dang, J.T. High-durability concrete with supplementary cementitious admixtures used at severe environment. *Crystals* **2021**, *11*, 196. [CrossRef]
36. Ministry of Construction of the People's Republic of China. *Standard for Test Methods on Mechanical Properties of Ordinary Concrete*; GB/T 50081-2002; China Building Industry Press: Beijing, China, 2002.
37. Santos, P.M.D.; Julio, E.N.B.S. Comparison of methods for texture assessment of concrete surfaces. *ACI Mater. J.* **2010**, *107*, 433–440.
38. Espeche, A.D.; Leon, J. Estimation of bond strength envelopes for old-to-new concrete interfaces based on a cylinder splitting test. *Constr. Build. Mater.* **2011**, *25*, 1222–1235. [CrossRef]

39. Ma, J.T.; Wang, D.G.; Zhao, S.B.; Duan, P.; Yang, S.T. Influence of particle morphology of ground fly ash on the fluidity and strength of cement paste. *Materials* **2021**, *14*, 283. [CrossRef]
40. Zhang, Y.; Jiang, L.; Zhang, W.; Qu, W. *Durability of Concrete Structures*; Shanghai Science and Technology Press: Shanghai, China, 2003.
41. Zhao, S.B.; Liang, N.; Ma, X.L.; Yang, S. Experiment of carbonization for wet-sieving fine aggregate concrete made from ordinary concrete. *Adv. Mater. Res.* **2011**, *201–203*, 2287. [CrossRef]
42. Ding, X.X.; Li, C.Y.; Xu, Y.Y.; Li, F.L.; Zhao, S.B. Experimental study on long-term compressive strength of concrete with manufactured sand. *Constr. Build. Mater.* **2016**, *108*, 67–73. [CrossRef]
43. Li, C.Y.; Wang, F.; Deng, X.S.; Li, Y.Z.; Zhao, S.B. Testing and prediction of the strength development of recycled-aggregate concrete with large particle natural aggregate. *Materials* **2019**, *12*, 1891. [CrossRef]
44. Zhao, S.B.; Ding, X.X.; Zhao, M.S.; Li, C.Y.; Pei, S.W. Experimental study on tensile strength development of concrete with manufactured sand. *Constr. Build. Mater.* **2017**, *138*, 247–253. [CrossRef]

MDPI
St. Alban-Anlage 66
4052 Basel
Switzerland
Tel. +41 61 683 77 34
Fax +41 61 302 89 18
www.mdpi.com

Crystals Editorial Office
E-mail: crystals@mdpi.com
www.mdpi.com/journal/crystals

www.ingramcontent.com/pod-product-compliance
Lightning Source LLC
LaVergne TN
LVHW070136170726
843515LV00007B/1808